Workshop on Biomedical Natural Language Processing 2011

(BioNLP 2011)

Held at ACL HLT 2011

Portland, Oregon, USA
23-24 June 2011

ISBN: 978-1-61839-233-6

Printed from e-media with permission by:

Curran Associates, Inc.
57 Morehouse Lane
Red Hook, NY 12571

Some format issues inherent in the e-media version may also appear in this print version.

Printed by Curran Associates, Inc. (2011)

Association for Computational Linguistics
209 N. Eighth Street
Stroudsburg, Pennsylvania 18360

Phone: 1-570-476-8006
Fax: 1-570-476-0860

acl@aclweb.org

Additional copies of this publication are available from:

Curran Associates, Inc.
57 Morehouse Lane
Red Hook, NY 12571 USA
Phone: 845-758-0400
Fax: 845-758-2634
Email: curran@proceedings.com
Web: www.proceedings.com

Table of Contents

Introduction

BioNLP 2011 received 31 submissions that with very few exceptions maintain the tradition of excellence established by the BioNLP authors over the past 10 years. Eleven submissions were accepted as full papers and 14 as poster presentations.

The themes in this year's papers and posters cover complex NLP problems in biological and clinical language processing.

Acknowledgments

We are profoundly grateful to the authors who chose BioNLP as venue for presenting their innovative research.

The authors' willingness to share their work through BioNLP consistently makes the workshop noteworthy and stimulating.

We are equally indebted to the program committee members (listed elsewhere in this volume) who produced at least two thorough reviews per paper on a tight review schedule and with an admirable level of insight.

We are particularly grateful to reviewers who reviewed late submissions from Japan in even shorter period of time. And we admire our Japanese colleagues who continued focusing on their research in the middle of an unfathomable natural disaster.

Organizers:

Kevin Bretonnel Cohen, University of Colorado School of Medicine
Dina Demner-Fushman, US National Library of Medicine
Sophia Ananiadou, University of Manchester and National Centre for Text Mining, UK
John Pestian, Computational Medical Center, University of Cincinnati,
Cincinnati Children's Hospital Medical Center
Jun'ichi Tsujii, University of Tokyo
and University of Manchester and National Centre for Text Mining, UK
Bonnie Webber,University of Edinburgh, UK

Program Committee:

Alan Aronson
Emilia Apostolova
Olivier Bodenreider
Wendy Chapman
Aaron Cohen
Nigel Collier
Noemie Elhadad
Marcelo Fiszman
Filip Ginter
Su Jian
Halil Kilicoglu
Jin-Dong Kim
Marc Light
Zhiyong Lu
Aurelie Neveol
Sampo Pyysalo
Thomas Rindflesch
Andrey Rzhetsky
Daniel Rubin
Hagit Shatkay
Matthew Simpson
Larry Smith
Yuka Tateisi
Yoshimasa Tsuruoka
Karin Verspoor
W. John Wilbur
Limsoon Wong
Pierre Zweigenbaum

Not all links are equal: Exploiting Dependency Types for the Extraction of Protein-Protein Interactions from Text

Philippe Thomas [1], **Stefan Pietschmann** [1], **Illés Solt** [2], **Domonkos Tikk** [1,2] **and Ulf Leser** [1]
[1]Knowledge Management in Bioinformatics, Institute for Computer Science,
Humboldt-University of Berlin,
Unter den Linden 6, 10099 Berlin, Germany
[2]Department of Telecommunications and Media Informatics,
Budapest University of Technology and Economics,
Magyar tudósok körútja 2, 1117 Budapest, Hungary
{thomas,pietschm,solt,tikk,leser}@informatik.hu-berlin.de

Abstract

The extraction of protein-protein interactions (PPIs) reported in scientific publications is one of the most studied topics in Text Mining in the Life Sciences, as such algorithms can substantially decrease the effort for databases curators. The currently best methods for this task are based on analyzing the dependency tree (DT) representation of sentences. Many approaches exploit only topological features and thus do not yet fully exploit the information contained in DTs. We show that incorporating the grammatical information encoded in the types of the dependencies in DTs noticeably improves extraction performance by using a pattern matching approach. We automatically infer a large set of linguistic patterns using only information about interacting proteins. Patterns are then refined based on shallow linguistic features and the semantics of dependency types. Together, these lead to a total improvement of 17.2 percent points in F_1, as evaluated on five publicly available PPI corpora. More than half of that improvement is gained by properly handling dependency types. Our method provides a general framework for building task-specific relationship extraction methods that do not require annotated training data. Furthermore, our observations offer methods to improve upon relation extraction approaches.

1 Introduction

Insights about protein-protein interactions (PPIs) are vital to understand the biological processes within organisms. Accordingly, several databases, such as IntAct, DIP, or MINT, contain detailed information about PPIs. This information is often manually harvested from peer reviewed publications (Ceol et al., 2010). However, it is assumed that a high amount of PPIs is still hidden in publications. Therefore, the automated extraction of PPIs from text has attracted considerable attention from biology research.

A number of different techniques have been proposed to solve the problem of extracting PPIs from natural language text. These can be roughly organized into one of three classes: co-occurrence, machine learning, and pattern matching (for a recent survey, see (Zhou and He, 2008)). The co-occurrence based approaches use only information on the co-existence of protein mentions in a given scope. They are easy to implement and allow for efficient processing of huge amounts of texts, but they are also prone to generate many false positives because they cannot distinguish positive from negative pairs. The second class is based on machine learning. Here, a statistical model is learned from a set of positive and negative examples and then applied to unseen texts. In general, machine learning-based methods to relation extraction perform very well for any task where sufficient, representative and high quality training data is available (Kazama et al., 2002). This need for training data is their major drawback, as annotated texts are, especially in the Life Sciences, rather costly to produce. Furthermore, they are prone to over-fit to the training corpus, which renders evaluation results less inferable to real applications. A third class of methods is based on pattern matching. Such methods work with patterns constructed from linguistically anno-

Proceedings of the 2011 Workshop on Biomedical Natural Language Processing, ACL-HLT 2011, pages 1–9,
Portland, Oregon, USA, June 23-24, 2011.

tated text, which are matched against unseen text to detect relationships. Patterns can either be inferred from examples (Hakenberg et al., 2010; Liu et al., 2010) or can be defined manually (Fundel et al., 2007). Systems based on manually defined patterns typically use few patterns, leading to high precision but low recall (Blaschke et al., 2002). In contrast, systems that learn patterns automatically often produce more patterns and exhibit a better recall, at the cost of a decrease in precision. To circumvent this penalty, several works have tried to improve patterns. E.g., SPIES (Hao et al., 2005) filters patterns using the minimum description length (MDL) method which improves its F_1 by 6.72%.

Another classification of PPI extraction methods is based on the sentence representation that is applied. The simplest such representation is the bag of words (BoW) that occur in the sentence; more complex representations are constituent trees, capturing the syntactic structure of the sentence, and dependency trees (DTs), which represent the main grammatical entities and their relationships to each other. PPI extraction methods use various sentence representation, e.g., are based only on BoW (Bunescu and Mooney, 2006; Giuliano et al., 2006), use only DTs (Erkan et al., 2007), or combine representations (Airola et al., 2008; Miwa et al., 2008).

In the last years, dependency trees have become the most popular representation for relation extraction. DTs characterize, via their dependency links, grammatical relationships among words. They are particularly favored by kernel-based learning approaches, see e.g. (Culotta and Sorensen, 2004; Erkan et al., 2007; Airola et al., 2008; Miwa et al., 2008; Kim et al., 2010) but also graph matching approaches using DTs have been proposed (Liu et al., 2010). However, these methods do not further utilize the grammatical information encoded in the dependency types (edge labels). Recently proposed methods like (Buyko et al., 2009; Rinaldi et al., 2010) modify the DTs by e.g. trimming irrelevant dependencies. In contrast to these approaches, our method *exploits* the dependency types of DTs and performs basic transformations on DTs; we use Stanford dependencies, which are presumably the most often used DT representation in PPI extraction.

The rest of this paper is organized as follows. We describe our novel method for extracting PPIs from text that is based on pattern matching in dependency graphs. We evaluate our method against benchmark PPI corpora, and discuss results with a focus on dependency type information based methods.

2 Methods

Our approach consists of a series of steps: First, we extract sentences from Medline and PMC open access that contain pairs of genes/proteins known to interact. Second, we convert each of those sentences into DTs and derive putative tree patterns for each pair. Having a set of such patterns, we apply a number of generalization methods to improve recall and filtering methods to improve precision. We discern between methods that are purely heuristic (termed shallow) and steps that incorporate dependency type information (termed grammatical). To predict PPIs in unseen text, the resulting patterns are matched against the corresponding DTs.

2.1 Extraction of PPI sentences

We apply the method described in (Hakenberg et al., 2006) to extract a set of sentences from Medline and PMC potentially describing protein interactions. Essentially, this method takes a database of PPIs (here IntAct; (Aranda et al., 2010)) and searches all sentences in Medline and PMC containing any of those pairs. Proteins were tagged and normalized using GNAT (Hakenberg et al., 2008). To avoid a possible bias, articles contained in any of the five evaluation corpora are excluded. This resulted in 763,027 interacting protein pairs.

2.2 Pattern generation and matching

For each protein pair we generate a new sentence and apply entity blinding, meaning that named entities are replaced by placeholders to avoid systemic bias. Specifically, the mentions of the two proteins known to interact are replaced by the placeholder *ENTITY_A* and any additional proteins in the same sentence are replaced by *ENTITY_B*. Tokens are tagged with their part-of-speech (POS) using MedPost (Smith et al., 2004), which is specifically optimized for biomedical articles. Constituent parse trees are generated using the Bikel parser (Bikel, 2002) and converted to DTs by the Stanford converter (De Marneffe et al., 2006). In a DT, the shortest path between two tokens is often assumed to contain the most valuable information about their mutual relationship. Therefore, we generate a pattern from each DT by extracting the shortest, undirected path between the two occurrences of *ENTITY_A*. The set of initial patterns is denoted by S_{IP}.

We employ several methods to improve the quality of this initial set of patterns. We systematically evaluated possible constellations and identified those that help in improving performance of PPI extraction. The modifications are of two kinds. Pattern generalizers are intended to elevate recall, whereas pattern filters should raise precision. We present two types of methods: Shallow methods are simple heuristics whereas grammatical methods are rules that exploit the information in dependency types.

We use a strict graph matching approach for pattern matching. We consider a pattern to match a subgraph of a DT iff all their nodes and edges match exactly, including edge labels and edge directions.

2.3 Pattern generalization

It is a common practice in NLP to apply some preprocessing on patterns to reduce corpus-specificity. In particular, we perform stemming (G_{ST}), removal of common protein name prefixes and suffixes (G_{PN}), and replacement of interaction phrases by single words (G_{IW}). We summarize these steps as *shallow generalization* steps. We only describe the latter two (G_{PN}, G_{IW}) in more detail.

Protein names are often modified by suffixes like *-kinase*, *-receptor* or *-HeLa* or by prefixes like *phospho-* or *anti-*. These affixes are usually not covered by entity blinding as the entity recognition method does not consider them as part of the protein name. As such affixes are not relevant for relation extraction but do interfere with our exact graph matching approach, we apply the G_{PN} heuristic to remove them.

Interactions between proteins can be expressed very diversely in natural language. In almost all cases there is at least one word that specifies the interaction semantically, called the *interaction word*; this is often a verb, such as "binds" or "phosphorylates", but can as well be a noun, such as "[induced] phosphorylation", or an adjective, such as "binding". The G_{IW} heuristic generalizes patterns by substituting all contained interaction words with generic placeholders. We assembled a list of 851 interaction words (including inflection variants) based on (Temkin and Gilder, 2003; Hakenberg et al., 2006) that was further enriched manually. Based on their POS-tags, interaction words are assigned to one of the three placeholders *IVERB*, *INOUN*, *IADJECTIVE*. We also experimented with a single interaction word placeholder, *IWORD*, handling the case of incorrect POS-tags.

Unifying dependency types (G_{UD}): The Stanford typed dependency format contains 55 grammatical relations organized in a generalization hierarchy. Therefore, it is a natural idea to treat similar (e.g., sibling) dependency types equally by replacing them with their common parent type. We manually evaluated all dependency types to assess whether such replacements are viable. The final list of replacements is listed in Table 1. Note that we used the so-called collapsed representation of dependency types of the Stanford scheme. This means that prepositional and conjunctive dependencies are collapsed to form a single direct dependency between content words, and the type of this dependency is suffixed with the removed word. In the G_{UD} generalizer, these dependency subtypes are substituted by their ancestors (e.g., `prep`, `conj`).

Dependency types	Common type
`subj, nsubj*, csubj*`	`subj`
`obj, dobj, iobj, pobj`	`obj`
`prep_*, agent, prepc`	`prep`
`nn, appos`	`nn`

Table 1: Unification of specific dependency types to a single common type by the generalizer G_{UD}. Note that `agent` is merged with dependency type `prep` as it is inferred for the preposition "by".

Collapsing dependency links (G_{CD}): In addition to the collapsing performed by Stanford converter, we remove edges that likely are irrelevant for PPIs. We focused on removing the dependencies `nn` (noun compound modifier) and `appos` (appositional modifier). These grammatical constructions have the same syntactic role but they carry somewhat different meaning. They function as noun phrase modifiers and often specify the subtype of an entity, which is irrelevant for our task. As these two dependency types convey no information about the interaction itself, the dependency itself and the corresponding noun can be removed, as long as the noun is not an entity itself. As an example, this generalizer is applied on the dependency parse tree of the sentence "*ENTITY_A* protein recognized antibody (*ENTITY_A*)" shown on Figure 1a. The modified parse tree is depicted on Figure 1b.

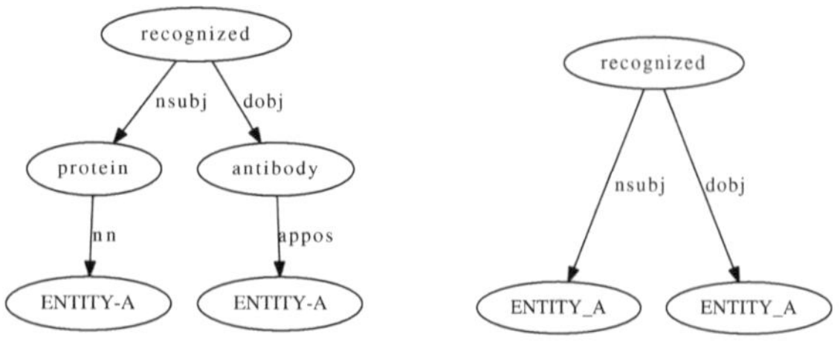

(a) Original pattern (b) Generalized pattern

Figure 1: Dependency pattern before and after collapsing `nn` and `appos` dependency links using the generalizer G_{CD}.

2.4 Pattern constraints

Due to the automatic construction method, our set of patterns also contains samples derived from sentences that do not actually describe an interaction between proteins, although it does contain a pair of interacting proteins. Such patterns lead to wrong matches. As a countermeasure, we define constraints an extracted pattern has to fulfill. Patterns not adhering to these constraints are removed from the pattern set, thus increasing precision. Standard (shallow) heuristics for doing so are the exclusion of negation words (C_{NW}) and the restriction to patterns containing interaction-related words from a predefined set (C_{IW}). Patterns containing negations potentially match two negative protein pairs. Such pattern can be removed to prevent wrong extractions. For negation words, the list described in (Fundel et al., 2007) was used. Additionally, patterns containing the dependency type *conj_no**, *conj_or*, or *prep_without* are also removed. On top of those previously known approaches, we developed two new filter to leverage the semantic richness of the DTs.

Dependency combination (C_{DC}): Interaction words are organized into the following categories: *verb*, *adjective* and *noun*. Based on linguistic considerations we define "dependency patterns" for the different word types. For example we assume that interaction verbs describe an action that originates in one protein and affects the other protein. Obviously, the dependency combination `subj` with `obj` fulfills this consideration (for an example see Figure 1b). We manually evaluated a few DTs containing PPI for each interaction word category (verb, noun, adjective) and determined all combinations of dependency types that are valid for the given category. The resulting combinations are listed in Table 2.

Part of speech	Dependency type combination	
Noun	prep	prep
	prep	nn
	prep	amod
	nn	nn
	nn	amod
Verb	prep	subj
	prep	infmod
	prep	partmod
	obj	subj
	obj	infmod
	obj	partmod
Adjective	amod	

Table 2: Allowed dependency type combinations based on classes of POS classes (constraint C_{DC}). `subj` = {`nsubj`, `nsubjpass`, `xsubj`, `csubj`, `csubjpass`}, `obj` = {`dobj`, `pobj`, `iobj`} and `prep` = {`prep_*`, `agent`}

Syntax Filter (C_{SF}): A particular case in PPI extraction are sentences with enumerations, as shown in Figure 2. Such (possibly quite long; the longest enumeration we found contains not less than 9 proteins) enumerations greatly increase the number of protein pairs.

We observed that sentences in which the common dependency type is `prep_between` or `nn` often do describe an association between the connected proteins. Accordingly, such patterns are retained.

The remaining pairs inside enumerations often do not describe an interaction between each other. Therefore, we developed a special handling of enumerations, based on dependency types. If two proteins have a common ancestor node connected by the same dependency type, we assume that those proteins do not interact with each other. Accordingly, we remove all such patterns.

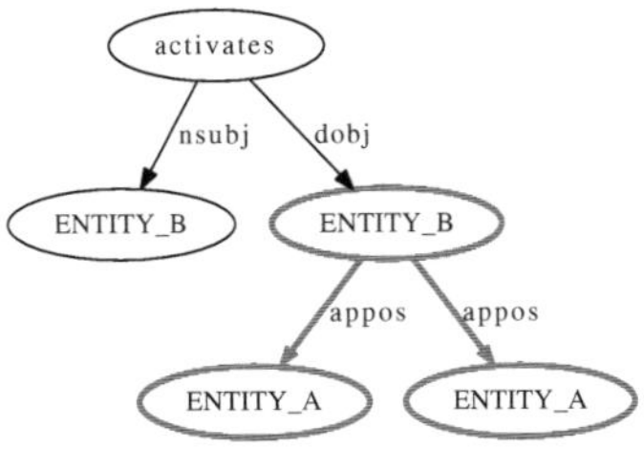

Figure 2: Dependency tree (DT) for the entity blinded sentence "*ENTITY_B* activates *ENTITY_B*, *ENTITY_A*, *ENTITY_A*." with the initial pattern highlighted in bold red. Application of C_{SF} removes this pattern.

3 Results

For evaluation we use five manually annotated benchmark corpora: AIMed, BioInfer, HPRD50, IEPA, and LLL. Those have been unified to the "greatest common factors" by Pyysalo et al. (2008). This means that protein names in the corpora are tagged and that interactions are undirected and binary. A basic overview of the corpora can be found in Table 1 of (Airola et al., 2008).

A sentence with n entities contains $\binom{n}{2}$ different undirected entity pairs. For each entity pair in a sentence, we generate a separate evaluation example, apply entity blinding and generate the DT in the same manner as previously described for generating the pattern set. All patterns are then matched against the DTs of the sentences from the evaluation corpora. If at least one pattern matches, the pair is counted as *positive* otherwise as *negative*. From this information we calculate precision, recall, and F_1 scores.

Table 3 shows results using the initial pattern set and the different configurations of generalized / filtered pattern sets. We evaluate the impact of shallow and grammar-based methods separately. Recall that $S_{shallow}$ encompasses stemming (G_{ST}), substitution of interaction words (G_{IW}), suffix/prefix removal at entity names (G_{PN}), and interaction (C_{IW}) and negation word filtering (C_{NW}), while $S_{grammar\text{-}based}$ encompasses unification of dependency types (G_{UD}), collapsing dependency links (G_{CD}), the dependency combination constraint (C_{DC}) and the syntax filter (C_{SF}). In addition, results after application of all generalizers $S_{generalizers}$, all constraints $S_{constraints}$ and the combination of both S_{all} are also included. Corpus-specific results for the best setting in terms of F_1 (S_{all}) are shown in Figure 3.

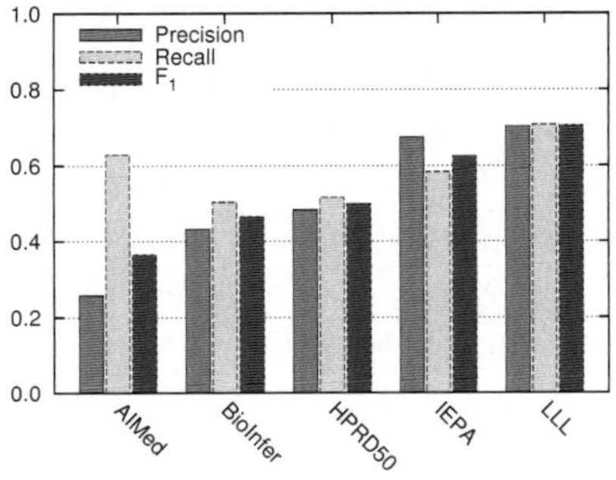

Figure 3: Results for the five corpora using the setting achieving highest overall F_1 (S_{all}).

Setting		P	R	F_1	#
Baseline	$S_{initial}$	23.2	34.8	27.8	478 k
Generalizers	G_{PN}	23.4	37.0	28.7	423 k
	G_{IW}	26.2	45.3	33.2	453 k
	G_{ST}	24.1	37.4	29.3	471 k
	G_{UD}	24.0	38.3	29.5	467 k
	G_{CD}	26.3	48.9	34.2	418 k
Constraints	C_{NW}	23.4	34.8	28.0	455 k
	C_{IW}	42.5	17.2	24.5	322 k
	C_{DC}	39.5	15.9	22.7	318 k
	C_{SF}	28.2	32.6	30.3	419 k
Combinations	$S_{generalizers}$	28.2	69.0	39.9	290 k
	$S_{constraints}$	68.3	12.7	21.4	224 k
	$S_{shallow}$	40.9	31.4	35.5	253 k
	$S_{grammar\text{-}based}$	33.2	42.1	37.2	264 k
	S_{all}	38.2	54.8	45.0	152 k

Table 3: Performance of pattern sets on the ensemble of all five corpora. # denotes the pattern set size.

4 Discussion

We presented a pattern-based approach to extract protein-protein interactions from text. Our main contribution in this paper was a systematic study on the usage of dependency types within this approach. We showed that using such knowledge, F_1 on average improves by 9.4 percentage points (pp) (27.8 % to 37.2 %) as measured on the five benchmark PPI corpora.

Apart from this result, we note that our method

also has a number advantageous features: First, patterns are learned from co-mentions of pairs of proteins known to interact, and hence no annotated corpus is necessary. This does not only make an application of the method for new tasks easier and cheaper, but also prevents over-fitting to a training corpus. Note, that as shown recently by (Airola et al., 2008; Tikk et al., 2010), essentially all state-of-the-art machine learning methods show large performance differences depending on whether or not the evaluation and training examples are drawn from the same corpus. In particular, cross-validation results of those are consistently more optimistic than the more realistic cross-learning results. In contrast, a pattern-based approach like ours is not prone to over-fitting. Furthermore, debugging our method is rather simple. Unlike when using a black-box machine learning method, whenever a false positive match is found, one can pinpoint the specific pattern producing it and take action.

The work most closely related to ours is RelEx (Fundel et al., 2007). RelEx uses a small set of fixed rules to extract directed PPIs from dependency trees. Some of these rules also take advantage of dependency types, for instance, to properly treat enumerations. A reimplementation of RelEx (Pyysalo et al., 2008) was recently evaluated on the same corpora we used (see Table 7) and was found to be on par with other systems, though some of its measures were considerably worse than those reported in the original publication. Compared to our approach, RelEx is similar in that it performs pattern matching on DTs using information encoded in dependency types, however, there are some notable differences: First, RelEx rules were defined manually and are highly specific to protein-protein interactions. It is not clear how these could be adapted to other applications; in contrast, we described a general method that performs pattern learning from automatically generated examples. Second, it is not clear how RelEx results could be further improved except by trial-and-error with more rules. In contrast, our pattern learning method offers a natural way of improvement by simply increasing the number of examples and hence the number of patterns. We compared the results of our approach using an increasingly larger pattern set and observed a continuous increase in F_1, due to a continuous improvement in recall. Consequently, using more PPI databases would likely produce better results. Third, our generalization methods can be seen as graph rewriting rules. The result of applying them to a DT is, again, a DT; thus, they can easily be used as pre-processing coupled with other PPI extraction methods (a direction we are currently exploring). In contrast, RelEx matches patterns against DTs, but cannot be used to transform DTs.

In the following, we discuss the impact of the refinement methods individually and provide a brief error analysis based on a random sample of false negative pairs and a random sample of low precision patterns. We also compare our best results with those of several state-of-the-art methods.

4.1 Generalizers and constraints

As can be seen in Table 3, all of the generalizers increased recall and even provide minor improvement in precision. For the combination of all five generalizers ($S_{generalizers}$), an overall increase of 34.2 pp in recall and 5 pp in precision was observed. From the shallow generalizers, merging interaction phrases (G_{IW}) was proven to be the most effective, accounting for an increase of 10.5 pp in recall and 3 pp in precision. As shown in Table 4, the general variant, which merges all interaction phrases to a common word, is slightly superior to the variant in which interaction words are merged class-wise.

G_{IW} variant	P	R	F_1
Specific	26.1	44.7	33.0
General	26.2	45.4	33.2

Table 4: Results for collapsing interaction word variants (G_{IW}).

For the grammar-based generalizer unifying dependency types (G_{UD}), each of the different variants was evaluated separately (see Table 5). The combination of the all different variants leads to an increase of 3.5 pp in recall. As shown in Table 6, collapsing the dependency types `nn` and `appos` (G_{CD}) also provides the highest improvement when applied in combination.

In contrast to generalizers that alter patterns, constraints remove patterns from the pattern set. As shown in Table 3, application of all constraints

G_{UD} variant	P	R	F_1
subj	23.4	35.1	28.1
obj	23.3	34.9	27.9
prep	24.0	37.0	29.1
nn	23.1	35.6	28.1
sopn	24.0	38.3	29.5

Table 5: Dependency type aggregations used in generalizer G_{UD}. sopn combines the dependency aggregations for subj, obj, prep, and nn.

G_{CD} variant	P	R	F_1
appos	23.6	38.1	29.2
nn	25.8	45.3	32.9
appos+nn	26.3	48.9	34.2

Table 6: Impact of collapsing the dependency types appos and nn using generalizer G_{CD}.

($S_{constraints}$) leads to an increase in precision of 45.1 pp at the cost of 22.1 pp reduced recall.

The shallow constraint that disallows patterns with negation words (C_{NW}) has comparably little impact and removes only 5 % of the patterns. In contrast, the interaction word constraint (C_{IW}) is less conservative and removes more than 32.6 % of the patterns, trading off an increase of 19.3 pp in precision to a decrease of 17.6 pp in recall.

Among the grammar-based constraints, the dependency combination constraint C_{DC} is superseded by the syntax filter constraint (C_{SF}) that removes 12 % of the patterns and increases precision about 5 pp while recall drops by only 2.2 pp. Note that C_{SF} is the only constraint substantially increasing F_1.

As seen in Table 3, combinations of generalizers and constraints yield almost fully additive improvements. The combination of all shallow refinements only ($S_{shallow}$) leads to an increase of 7.7 pp in F_1, whereas with the addition of our grammar-based refinements (S_{all}) a total increase of 17.2 pp in F_1 is achieved. We justify that the inclusion of dependency type information adds a source of knowledge that further improves on conventional methods such as stemming or negation filtering.

4.2 Comparison with other methods

We compare the results of our best setting (S_{all}) with the results of the recently published analysis of nine state-of-the-art machine learning methods for PPI extraction (Tikk et al., 2010). For these methods, a cross-learning evaluation by training each kernel on the ensemble of four corpora and evaluating it on the fifth has been performed. Detailed results are given in Table 7. In terms of F_1 we achieve on BioInfer, the corpus with most protein pairs, the second-best result. On IEPA and LLL we achieve mid-range results and on AIMed and HPRD50 we yield results below average. Overfitting remains a severe problem in ML based methods as these results are inferior to those measured in cross-validation (Tikk et al., 2010), though there are suggestions to overcome this issue even in a ML setting (Miwa et al., 2009).

4.3 Error analysis

We randomly picked 30 gold standard sentences (all corpora) containing false negatives pairs and investigated all 72 false negative pairs included therein. For 29 positive pairs, possibly matching pattern were removed by C_{DC}, as the corresponding dependency combination was not covered in our rule set. Further 16 graphs passed the filtering, but our set of sentences contained no adequate pattern. The third largest fraction of errors (13 cases) are pairs which, by our understanding, hardly describe an interaction. In 11 cases, the dependency parse trees are incorrect and therefore they do not provide the correct syntactic information. For 7 pairs, the shortest path covers insufficient syntactic information to decide whether two proteins interact. For instance Figure 4 provides not enough information on the shortest path, whereas the second shortest path would provide sufficient information. Finally, three pairs were filtered by the C_{IW} filter, as their interaction words were missing from our list.

We conclude that some constraints (especially C_{DC} and C_{IW}) are too aggressive. Relaxation of these syntactic rules should lead to higher recall.

We also analyzed the 30 patterns producing the most false positives matches. 20 of them contained an interaction verb, the remaining 10 an interaction noun. The 10 noun patterns produced more than twice as many false positives as the 20 verb patterns while matching about 50 % less true positives. The single noun pattern producing the most false positives (356) can be seen on Figure 5a. Among the 10, four additional patterns can be seen as an extension

Method	AIMed			BioInfer			HPRD50			IEPA			LLL		
	P	R	F_1	P	R	F_1	P	R	F_1	P	R	F_1	P	R	F_1
Shallow linguistic (Giuliano et al., 2006)	28.3	**86.6**	42.6	**62.8**	36.5	46.2	56.9	68.7	62.2	71.0	52.5	60.4	79.0	57.3	66.4
Spectrum tree (Kuboyama et al., 2007)	20.3	48.4	28.6	38.9	48.0	43.0	44.7	**77.3**	56.6	41.6	9.6	15.5	48.2	**83.5**	61.2
k-band shortest path (Tikk et al., 2010)	28.6	68.0	40.3	62.2	38.5	**47.6**	61.7	74.2	67.4	72.8	**68.7**	**70.7**	83.7	75.0	**79.1**
Cosine distance (Erkan et al., 2007)	27.5	59.1	37.6	42.1	32.2	36.5	63.0	56.4	59.6	46.3	31.6	37.6	80.3	37.2	50.8
Edit distance (Erkan et al., 2007)	26.8	59.7	37.0	53.0	22.7	31.7	58.1	55.2	56.6	58.1	45.1	50.8	68.1	48.2	56.4
All-paths graph (Airola et al., 2008)	30.5	77.5	43.8	58.1	29.4	39.1	64.2	76.1	**69.7**	**78.5**	48.1	59.6	**86.4**	62.2	72.3
RelEx reimpl. (Pyysalo et al., 2008)	**40.0**	50.0	**44.0**	39.0	45.0	41.0	**76.0**	64.0	69.0	74.0	61.0	67.0	82.0	72.0	77.0
Our method (S_{all})	25.8	62.9	36.6	43.4	**50.3**	46.6	48.3	51.5	49.9	67.5	58.2	62.5	70.3	70.7	70.5

Table 7: Cross-learning results. Classifiers are trained on the ensemble of four corpora and tested on the fifth one (except for the rule-based RelEx). Best results are typeset in bold.

of this pattern leading to a total amount of 732 false positives while only 172 true positives. This phenomenon is caused by the way in which generalizers and constraints are currently applied. The unification of different `prep_*` dependency types to the general `prep` (G_{UD}) makes some dependency type combinations indistinguishable, e.g. (`prep_to`, `prep_to`) and (`prep_to`, `prep_of`). The dependency type combination constraint (C_{DC}) would disallow a pattern containing the first combination, but as it is not applied in the matching phase, its benefits cannot be realized. A lesson learned from this example is that constraints should also be applied in the matching step as follows. After a successful match, the constraints should be applied to the original ungeneralized counterparts of the matching subgraphs. Similar conclusions can be drawn from examining the verb pattern producing the most false positives shown in Figure 5b.

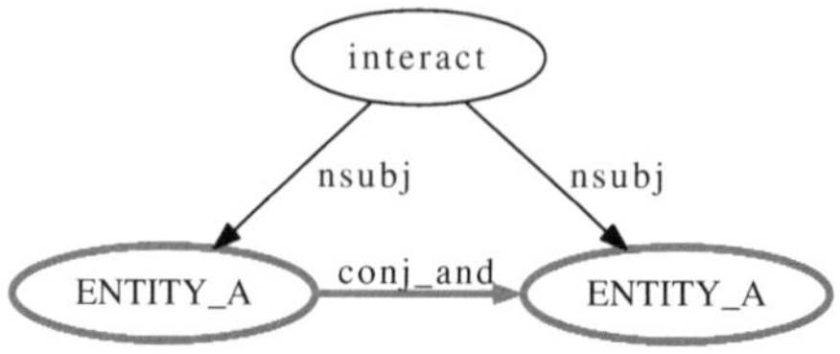

Figure 4: Example dependency parse where the information extracted by the shortest path (highlighted in bold red) is insufficient.

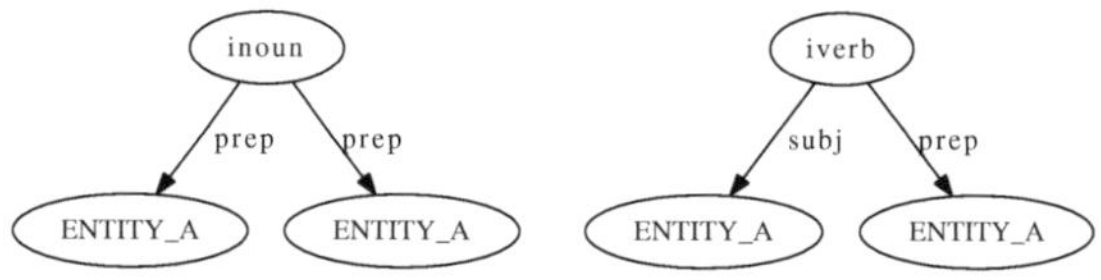

(a) Noun pattern (b) Verb pattern

Figure 5: Patterns producing the most false positives. Depicted dependency types are generalized according to G_{UD} and G_{IW}.

5 Conclusion

We presented a pattern-based approach to extract PPIs from literature. Its unique features are the capability of learning patterns from "cheap" training data, i.e., co-mentions of proteins known to interact, and the use of linguistically motivated refinements on the dependency structures of the DT it operates on. We utilized the fact that not all dependency types are of equal importance for relation extraction; for instance, collapsing dependency links (G_{CD}) or unifying dependencies (G_{UD}) considerably improved extraction performance. However, as our error analysis shows, especially our filtering methods deserve further study. Even without manually annotated training data, our approach performs on par with state-of-the-art machine learning methods when evaluated in a cross-learning setting. In particular, it reaches the second best performance (in terms of F-measure) of all approaches on the largest PPI corpus.

Apart from presenting a volatile and elegant new method for relationship extraction, we believe that our study on using dependency type information also will be helpful for advancing the performance of other methods. For instance, we are currently experimenting with using our DT-rewrite rules as a preprocessor for a kernel-based extraction method.

Acknowledgments

PT was supported by BMBF, grant No 0315417B; IS was supported by DAAD; DT was supported by Alexander von Humboldt-Foundation.

References

A. Airola, S. Pyysalo, F Ginter J. Björne, and T. Pahikkala. 2008. All-paths graph kernel for protein-protein interaction extraction with evaluation of cross-corpus learning. *BMC Bioinformatics*, 9:S2.

B. Aranda, P. Achuthan, Y. Alam-Faruque, I. Armean, A. Bridge, C. Derow, M. Feuermann, et al. 2010. The IntAct molecular interaction database in 2010. *Nucleic Acids Res*, 38:D525–D531, Jan.

DM. Bikel. 2002. Design of a Multi-lingual, Parallel-processing Statistical Parsing Engine. In *In Human Language Technology Conference*, pages 24–27.

C. Blaschke, L. Hirschman, and A. Valencia. 2002. Information extraction in molecular biology. *Briefings in Bioinformatics*, 3(2):154–165.

R. Bunescu and R. Mooney. 2006. Subsequence Kernels for Relation Extraction. In Y. Weiss, B. Schölkopf, and J. Platt, editors, *Advances in Neural Information Processing Systems 18*, pages 171–178. MIT Press, Cambridge, MA.

E. Buyko, E. Faessler, J. Wermter, and U. Hahn. 2009. Event extraction from trimmed dependency graphs. In *BioNLP'09*, pages 19–27.

A. Ceol, Chatr AA., L. Licata, D. Peluso, L. Briganti, L. Perfetto, L. Castagnoli, and G. Cesareni. 2010. MINT, the molecular interaction database: 2009 update. *Nucl. Acids Res.*, 38(suppl1):D532–539.

A. Culotta and JS. Sorensen. 2004. Dependency Tree Kernels for Relation Extraction. In *ACL'04*, pages 423–429.

MC. De Marneffe, B.Maccartney, and CD. Manning. 2006. Generating typed dependency parses from phrase structure parses. In *In LREC 2006*.

G. Erkan, A. Özgür, and DR. Radev. 2007. Semi-Supervised Classification for Extracting Protein Interaction Sentences using Dependency Parsing. In *Proc. of EMNLP–CoNLL'07*, pages 228–237.

K. Fundel, R. Küffner, and R. Zimmer. 2007. RelEx – relation extraction using dependency parse trees. *Bioinformatics*, 23(3):365–371, February.

A. Giuliano, A. Lavelli, and L. Romano. 2006. Exploiting Shallow Linguistic Information for Relation Extraction from Biomedical Literature. In *Proc. of EACL'06*, pages 401–408, Trento, Italy. The Association for Computer Linguistics.

J. Hakenberg, U. Leser, H. Kirsch, and D. Rebholz-Schuhmann. 2006. Collecting a large corpus from all of Medline. In *SMBM'06*, pages 89–92, April.

J. Hakenberg, C. Plake, R. Leaman, M. Schroeder, and G. Gonzalez. 2008. Inter-species normalization of gene mentions with GNAT. *Bioinformatics*, 24(16):i126–132.

J. Hakenberg, R. Leaman, NH. Vo, S.Jonnalagadda, R. Sullivan, C. Miller, L. Tari, C. Baral, and G. Gonzalez. 2010. Efficient extraction of protein-protein interactions from full-text articles. *IEEE/ACM Trans Comput Biol Bioinform*, 7(3):481–494.

Y. Hao, X. Zhu, M. Huang, and M. Li. 2005. Discovering patterns to extract protein-protein interactions from the literature: Part II. *Bioinformatics*, 21:3294–3300.

J. Kazama, T. Makino, Y. Ohta, and J. Tsujii. 2002. Tuning support vector machines for biomedical named entity recognition. In *Proc. of BioNLP at ACL'02*, page 8.

S. Kim, J. Yoon, J Yang, and S. Park. 2010. Walk-weighted subsequence kernels for protein-protein interaction extraction. *BMC Bioinformatics*, 11(1):107.

R. Kuboyama, K. Hirata, H. Kashima, KF. Aoki-Kinoshita, and H. Yasuda. 2007. A spectrum tree kernel. *Information and Media Technologies*, 2(1):292–299.

H. Liu, V. Keselj, and C. Blouin. 2010. Biological Event Extraction using Subgraph Matching. In *SMBM'10*, October.

M. Miwa, R. Sætre, Y. Miyao, T. Ohta, and J. Tsujii. 2008. Combining multiple layers of syntactic information for protein-protein interaction extraction. In *SMBM'08*, pages 101–108.

M. Miwa, R. Sætre, Y. Miyao, and J. Tsujii. 2009. A Rich Feature Vector for Protein-Protein Interaction Extraction from Multiple Corpora. In *EMNLP'09*, pages 121–130.

S. Pyysalo, A. Airola, J. Heimonen, J. Bjrne, F. Ginter, and T. Salakoski. 2008. Comparative analysis of five protein-protein interaction corpora. *BMC Bioinformatics*, 9 Suppl 3:S6.

F. Rinaldi, G. Schneider, K. Kaljurand, S. Clematide, T. Vachon, and M. Romacker. 2010. Ontogene in biocreative ii.5. *IEEE/ACM Trans Comput Biol Bioinform*, 7(3):472–480.

L. Smith, T. Rindflesch, and W. J. Wilbur. 2004. MedPost: a part-of-speech tagger for bioMedical text. *Bioinformatics*, 20(14):2320–2321, Sep.

JM. Temkin and MR. Gilder. 2003. Extraction of protein interaction information from unstructured text using a context-free grammar. *Bioinformatics*, 19(16):2046–2053, Nov.

D. Tikk, P. Thomas, P. Palaga, J. Hakenberg, and U. Leser. 2010. A comprehensive benchmark of kernel methods to extract protein-protein interactions from literature. *PLoS Comput Biol*, 6:e1000837.

D. Zhou and Y. He. 2008. Extracting interactions between proteins from the literature. *J Biomed Inform*, 41(2):393–407, April.

Unsupervised Entailment Detection between Dependency Graph Fragments

Marek Rei
Computer Laboratory
University of Cambridge
United Kingdom
Marek.Rei@cl.cam.ac.uk

Ted Briscoe
Computer Laboratory
University of Cambridge
United Kingdom
Ted.Briscoe@cl.cam.ac.uk

Abstract

Entailment detection systems are generally designed to work either on single words, relations or full sentences. We propose a new task – detecting entailment between dependency graph fragments of any type – which relaxes these restrictions and leads to much wider entailment discovery. An unsupervised framework is described that uses intrinsic similarity, multi-level extrinsic similarity and the detection of negation and hedged language to assign a confidence score to entailment relations between two fragments. The final system achieves 84.1% average precision on a data set of entailment examples from the biomedical domain.

1 Introduction

Understanding that two different texts are semantically similar has benefits for nearly all NLP tasks, including IR, IE, QA and Summarisation. Similarity detection is usually performed either on single words (synonymy) or full sentences and paragraphs (paraphrasing). A symmetric similarity relation implies that both elements can be interchanged (synonymy and paraphrasing), while directional similarity suggests that one fragment can be substituted for the other but not the opposite (hyponymy and entailment).

All of these language phenomena can be expressed using a single entailment relation. For paraphrases and synonyms the relation holds in both directions (*observe* $\leftrightarrow$ *notice*), whereas entailment and hyponymy are modelled as a unidirectional relation (*overexpress* $\rightarrow$ *express*). Such relations, however, can be defined between text fragments of any size and shape, ranging from a single word to a complete text segment. For example (*argues against* $\rightarrow$ *does not support*; *the protein has been implicated* $\leftrightarrow$ *the protein has been shown to be involved*).

We propose a new task – detecting entailment relations between any kinds of text fragments. A unified approach is not expected to perform better when compared to systems optimised only for a specific task (e.g. recognising entailment between sentences), but constructing a common theory to cover all text fragments has important benefits. A broader approach will allow for entailment discovery among a much wider range of fragment types for which no specialised systems exist. In addition, entailment relations can be found between different types of fragments (e.g. a predicate entailing an adjunct). Finally, a common system is much easier to develop and integrate with potential applications compared to having a separate system for each type of fragment.

In this paper we present a unified framework that can be used to detect entailment relations between fragments of various types and sizes. The system is designed to work with anything that can be represented as a dependency graph, including single words, constituents of various sizes, text adjuncts, predicates, relations and full sentences. The approach is completely unsupervised and requires only a large plain text corpus to gather information for calculating distributional similarity. This makes it ideal for the biomedical domain where the availability of annotated training data is limited. We apply these methods by using a background corpus

Proceedings of the 2011 Workshop on Biomedical Natural Language Processing, ACL-HLT 2011, pages 10–18,
Portland, Oregon, USA, June 23-24, 2011.

of biomedical papers and evaluate on a manually constructed dataset of entailing fragment pairs, extracted from biomedical texts.

2 Applications

Entailment detection between fragments is a vital step towards entailment generation – given text T, the system will have to generate all texts that either entail T or are entailed by T. This is motivated by applications in Relation Extraction, Information Retrieval and Information Extraction. For example, if we wish to find all genes that are synthesised in the larval tissue, the following IE query can be constructed (with x and y marking unknown variables):

(1) x is synthesised in the larval tissue

Known entailment relations can be used to modify the query: (*overexpression* $\rightarrow$ *synthesis*), (*larval fat body* $\rightarrow$ *larval tissue*) and (*the synthesis of x in y* $\leftrightarrow$ *x is synthesised in y*). Pattern (2) entails pattern (1) and would also return results matching the information need.

(2) the overexpression of x in the larval fat body

A system for entailment detection can automatically extract a database of entailing fragments from a large corpus and use them to modify any query given by the user. Recent studies have also investigated how complex sentence-level entailment relations can be broken down into smaller consecutive steps involving fragment-level entailment (Sammons et al., 2010; Bentivogli et al., 2010). For example:

(3) **Text**: The mitogenic effects of the B beta chain of fibrinogen are mediated through cell surface calreticulin.

Hypothesis: Fibrinogen beta chain interacts with CRP55.

To recognise that the hypothesis is entailed by the text, it can be decomposed into five separate steps involving text fragments:

1. *B beta chain of fibrinogen* $\rightarrow$ *Fibrinogen beta chain*
2. *calreticulin* $\rightarrow$ *CRP55*
3. *the mitogenic effects of x are mediated through y* $\rightarrow$ *y mediates the mitogenic effects of x*
4. *y mediates the mitogenic effects of x* $\rightarrow$ *y interacts with x*
5. *y interacts with x* $\rightarrow$ *x interacts with y*

This illustrates how entailment detection between various smaller fragments can be used to construct an entailment decision between more complicated sentences. However, only the presence of these constructions has been investigated and, to the best of our knowledge, no models currently exist for automated detection and composition of such entailment relations.

3 Modelling entailment between graph fragments

In order to cover a wide variety of language phenomena, a fragment is defined in the following way:

Definition 1. *A fragment is any connected subgraph of a directed dependency graph containing one or more words and the grammatical relations between them.*

This definition is intended to allow extraction of a wide variety of fragments from a dependency tree or graph representation of sentences found using any appropriate parser capable of returning such output (e.g. Kübler et al., 2009). The definition covers single- or multi-word constituents functioning as dependents (e.g. *sites*, *putative binding sites*), text adjuncts (*in the cell wall*), single- or multi-word predicates (* *binds to receptors in the airways*) and relations (* *binds and activates* *) including ones with 'internal' dependent slots (* *inhibits* * *at* *), some of which may be fixed in the fragment (* *induces autophosphorylation of* * *in* * *cells*), and also full sentences[1]. An example dependency graph and some selected fragments can be seen in Figure 1.

Our aim is to detect semantically similar fragments which can be substituted for each other in text, resulting in more general or more specific versions of the same proposition. This kind of similarity can be thought of as an entailment relation and we define entailment between two fragments as follows:

[1]The asterisks (*) are used to indicate missing dependents in order to increase the readability of the fragment when represented textually. The actual fragments are kept in graph form and have no need for them.

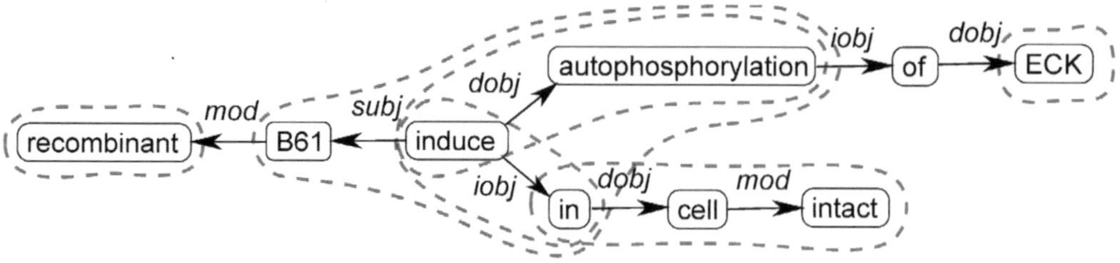

Figure 1: Dependency graph for the sentence: *Recombinant B61 induces autophosphorylation of ECK in intact cells.* Some interesting fragments are marked by dotted lines.

Definition 2. *Fragment A entails fragment B (A → B) if A can be replaced by B in sentence S and the resulting sentence S' can be entailed from the original one (S → S').*

This also requires estimating entailment relations between sentences, for which we use the definition established by Bar-Haim et al. (2006):

Definition 3. *Text T entails hypothesis H (T → H) if, typically, a human reading T would infer that H is most likely true.*

We model the semantic similarity of fragments as a combination of two separate directional similarity scores:

1. **Intrinsic similarity**: how similar are the components of the fragments.

2. **Extrinsic similarity**: how similar are the contexts of the fragments.

To find the overall score, these two similarity measures are combined linearly using a weighting parameter α:

$$Score(A \rightarrow B) = \alpha \times IntSim(A \rightarrow B) + (1-\alpha) \times ExtSim(A \rightarrow B)$$

In this paper $f(A \rightarrow B)$ designates an asymmetric function between A and B. When referring only to single words, lowercase letters (a,b) are used; when referring to fragments of any size, including single words, then uppercase letters are used (A,B).

$Score(A \rightarrow B)$ is the confidence score that fragment A entails fragment B – higher score indicates higher confidence and 0 means no entailment. $IntSim(A \rightarrow B)$ is the intrinsic similarity between two fragments. It can be any function that compares them, for example by matching the structure of one fragment to another, and outputs a similarity score in the range of $[0, 1]$. $ExtSim(A \rightarrow B)$ is a measure of extrinsic similarity that compares the contexts of the two fragments. α is set to 0.5 for an unsupervised approach but the effects of tuning this parameter are further analysed in Section 5.

The directional similarity score is first found between words in each fragment, which are then used to calculate the score between the two fragments.

3.1 Intrinsic similarity

$IntSim(A \rightarrow B)$ is the intrinsic similarity between the two words or fragments. In order to best capture entailment, the measure should be non-symmetrical. We use the following simple formula for word-level score calculation:

$$IntSim(a \rightarrow b) = \frac{length(c)}{length(b)}$$

where c is the longest common substring for a and b. This measure will show the ratio of b that is also contained in a. For example:

$$IntSim(overexpress \rightarrow expression) = 0.70$$

$$IntSim(expression \rightarrow overexpress) = 0.64$$

The intrinsic similarity function for fragments is defined using an injective function between components of A and components of B:

$$IntSim(A \rightarrow B) = \frac{Mapping(A \rightarrow B)}{|B|}$$

where $Mapping(A \rightarrow B)$ is a function that goes through all the possible word pairs $\{(a, b)|a \in A, b \in B\}$ and at each iteration connects the one with the highest entailment score, returning the sum of those scores. Figure 2 contains pseudocode for the mapping process. Dividing the value of $Mapping(A \rightarrow B)$ by the number of components in B gives an asymmetric score that indicates how well B is covered by A. It returns a lower score if B contains more elements than A as some words cannot be matched to anything. While there are exceptions, it is common that if B is larger than A, then it cannot be entailed by A as it contains more information.

```
while unused elements in A and B do
  bestScore = 0
  for a ∈ A, b ∈ B, a and b are unused do
    if Score(a → b) > bestScore then
      bestScore = Score(a → b)
    end if
  end for
  total+ = bestScore
end while
return total
```

Figure 2: Pseudocode for mapping between two fragments

The word-level entailment score $Score(a \rightarrow b)$ is directly used to estimate the entailment score between fragments, $Score(A \rightarrow B)$. In this case we are working with two levels – fragments which in turn consist of words. However, this can be extended to a truly recursive method where fragments consist of smaller fragments.

3.2 Extrinsic similarity

The extrinsic similarity between two fragments or words is modelled using measures of directional distributional similarity. We define a context relation as a tuple (a, d, r, a') where a is the main word, a' is a word connected to it through a dependency relation, r is the label of that relation and d shows the direction of the relation. The tuple $f : (d, r, a')$ is referred to as a feature of a.

To calculate the distributional similarity between two fragments, we adopt an approach similar to Weeds et al. (2005). Using the previous notation, (d, r, a') is a feature of fragment A if (d, r, a') is a feature for a word which is contained in A. The general algorithm for feature collection is as follows:

1. Find the next instance of a fragment in the background corpus.

2. For each word in the fragment, find dependency relations which connect to words not contained in the fragment.

3. Count these dependency relations as distributional features for the fragment.

For example, in Figure 1 the fragment *(* induces * in *)* has three features: *(1, subj, B61)*, *(1, dobj, autophosphorylation)* and *(1, dobj, cell)*.

The BioMed Central[2] corpus of full papers was used to collect distributional similarity features for each fragment. 1000 papers were randomly selected and separated for constructing the test set, leaving 70821 biomedical full papers. These were tokenised and parsed using the RASP system (Briscoe et al., 2006) in order to extract dependency relations.

We experimented with various schemes for feature weighting and found the best one to be a variation of Dice's coefficient (Dice, 1945), described by Curran (2003):

$$w_A(f) = \frac{2P(A, f)}{P(A, *) + P(*, f)}$$

where $w_A(f)$ is the weight of feature f for fragment A, $P(*, f)$ is the probability of the feature appearing in the corpus with any fragment, $P(A, *)$ is the probability of the fragment appearing with any feature, and $P(A, f)$ is the probability of the fragment and the feature appearing together.

Different measures of distributional similarity, both symmetrical and directonal, were also tested and $ClarkeDE$ (Clarke, 2009) was used for the final system as it achieved the highest performance on graph fragments:

$$ClarkeDE(A \rightarrow B) = \frac{\sum_{f \in F_A \cap F_B} min(w_A(f), w_B(f))}{\sum_{f \in F_A} w_A(f)}$$

where F_A is the set of features for fragment A and $w_A(f)$ is the weight of feature f for fragment A. It quantifies the weighted coverage of the features of A by the features of B by finding the sum of minimum weights.

[2] http://www.biomedcentral.com/info/about/datamining/

The $ClarkeDE$ similarity measure is designed to detect whether the features of A are a proper subset of the features of B. This works well for finding more general versions of fragments, but not when comparing fragments which are roughly equal paraphrases. As a solution we constructed a new measure based on the symmetrical Lin measure (Lin, 1998).

$$LinD(A \rightarrow B) = \frac{\sum_{f \in F_A \cap F_B} [w_A(f) + w_B(f)]}{\sum_{f \in F_A} w_A(f) + \sum_{f \in F_A \cap F_B} w_B(f)}$$

Compared to the original, the features of B which are not found in A are excluded from the score calculation, making the score non-symmetrical but more balanced compared to $ClarkeDE$. We applied this for word-level distributional similarity and achieved better results than with other common similarity measures.

The LinD similarity is also calculated between fragment levels to help detect possible paraphrases. If the similarity is very high (greater than 85%), then a symmetric relationship between the fragments is assumed and the value of $LinD$ is used as $ExtSim$. Otherwise, the system reverts to the $ClarkeDE$ measure for handling unidirectional relations.

3.3 Hedging and negation

Language constructions such as hedging and negation typically invert the normal direction of an entailment relation. For example, (*biological discovery* → *discovery*) becomes (*no biological discovery* ← *no discovery*) and (*is repressed by* → *is affected by*) becomes (*may be repressed by* ← *is affected by*).

Such cases are handled by inverting the direction of the score calculation if a fragment is found to contain a special cue word that commonly indicates hedged language or negation. In order to find the list of indicative hedge cues, we analysed the training corpus of CoNLL 2010 Shared Task (Farkas et al., 2010) which is annotated for speculation cues and scopes. Any cues that occurred less than 5 times or occurred more often as normal text than as cue words were filtered out, resulting in the following list:

(4) *suggest, may, might, indicate that, appear, likely, could, possible, whether, would, think, seem, probably, assume, putative, unclear, propose, imply, possibly*

For negation cues we used the list collected by Morante (2009):

(5) *absence, absent, cannot, could not, either, except, exclude, fail, failure, favor over, impossible, instead of, lack, loss, miss, negative, neither, nor, never, no, no longer, none, not, rather than, rule out, unable, with the exception of, without*

This is a fast and basic method for estimating the presence of hedging and negation in a fragment. When dealing with longer texts, the exact scope of the cue word should be detected, but for relatively short fragments the presence of a keyword acts as a good indication of hedging and negation.

4 Evaluation

A "pilot" dataset was created to evaluate different entailment detection methods between fragments[3]. In order to look for valid entailment examples, 1000 biomedical papers from the BioMed Central full-text corpus were randomly chosen and analysed. We hypothesised that two very similar sentences originating from the same paper are likely to be more and less general versions of the same proposition. First, the similarities between all sentences in a single paper were calculated using a bag-of-words approach. Then, ten of the most similar but non-identical sentence pairs from each paper were presented for manual review and 150 fragment pairs were created based on these sentences, 100 of which were selected for the final set.

When applied to sentence-level entailment extraction, similar methods can suffer from high lexical overlap as sentences need to contain many matching words to pass the initial filter. However, for the extraction of entailing fragments most of the matching words are discarded and only the non-identical fragments are stored, greatly reducing the overlap problem. Experiments in Section 5 demonstrate that a simple bag-of-words approach performs rather poorly on the task, confirming that the extraction method produces a diverse selection of fragments.

[3] http://www.cl.cam.ac.uk/~mr472/entailment/

Two annotators assigned a relation type to candidate pairs based on how well one fragment can be substituted for the other in text while preserving meaning ($A \leftrightarrow B$, $A \rightarrow B$, $A \leftarrow B$ or $A \neq B$). Cohen's Kappa between the annotators was 0.88, indicating very high agreement. Instances with disagreement were then reviewed and replaced for the final dataset.

Each fragment pair has two binary entailment decisions (one in either direction) and the set is evenly balanced, containing 100 entailment and 100 non-entailment relations. An example sentence with the first fragment is also included. Fragment sizes range from 1 to 20 words, with the average of 2.86 words.

The system assigns a score to each entailment relation, with higher values indicating higher confidence in entailment. All the relations are ranked based on their score, and average precision (AP) is used to evaluate the performance:

$$\text{AP} = \frac{1}{R}\sum_{i=1}^{N}\frac{E(i) \times CorrectUpTo(i)}{i}$$

where R is the number of correct entailment relations, N is the number of possible relations in the test set, $E(i)$ is 1 if the i-th relation is entailment in the gold standard and 0 otherwise, and $CorrectUpTo(i)$ is the number of correctly returned entailment relations up to rank i. Average precision assigns a higher score to systems which rank correct entailment examples higher in the list.

As a secondary measure we also report the Break-Even Point (BEP) which is defined as precision at the rank where precision is equal to recall. Using the previous annotation, this can also be calculated as precision at rank R:

$$BEP = \frac{CorrectUpTo(R)}{R}$$

BEP is a much more strict measure, treating the task as binary classification and ignoring changes to the ranks within the classes.

5 Results

The test set is balanced, therefore random guessing would be expected to achieve an AP and BEP of 0.5 which can be regarded as the simplest (random) baseline. Table 1 contains results for two more basic approaches to the task. For the bag-of-words (BOW) system, the score of A entailing B is the proportion of words in B that are also contained in A.

$$Score_{bow}(A \rightarrow B) = \frac{|\{b|b \in A, B\}|}{|\{b|b \in B\}|}$$

We also tested entailment detection when using only the directional distributional similarity between fragments as it is commonly done for words. While both of the systems perform better than random, the results are much lower than those for more informed methods. This indicates that even though there is some lexical overlap between the fragments, it is not enough to make good decisions about the entailment relations.

System type	**AP**	**BEP**
Random baseline	0.500	0.500
BOW	0.657	0.610
Distributional similarity	0.645	0.480

Table 1: Results for basic approaches.

Table 2 contains the results for the system described in Section 3. We start with the most basic version and gradually add components. Using only the intrinsic similarity, the system performs better than any of the basic approaches, delivering 0.71 AP.

System type	**AP**	**BEP**
Intrinsic similarity only	0.710	0.680
+ Word ExtSim	0.754	0.710
+ Fragment ExtSim	0.801	0.710
+ Negation & hedging	0.831	0.720
+ Paraphrase check	0.841	0.720

Table 2: Results for the system described in Section 3. Components are added incrementally.

Next, the extrinsic similarity between words is included, raising the AP to 0.754. When the string-level similarity fails, the added directional distributional similarity helps in mapping semantically equivalent words to each other.

The inclusion of extrinsic similarity between fragments gives a further increase, with AP of 0.801. The 4.5% increase shows that while fragments are larger and occur less often in a corpus, their distributional similarity can still be used as a valuable component to detect semantic similarity and entailment.

Checking for negation and hedge cues raises the AP to 0.831. The performance is already high and a 3% improvement shows that hedging and negation affect fragment-level entailment and other components do not manage to successfully capture this information.

Finally, applying the fragment-level check for paraphrases with a more appropriate distributional similarity measure, as described in Section 3.2, returns an AP of 0.841. The results of this final configuration are significantly different compared to the initial system using only intrinsic similarity, according to the Wilcoxon signed rank test at the level of 0.05.

The formula in Section 3 contains parameter α which can be tuned to adjust the balance of intrinsic and extrinsic similarity. This can be done heuristically or through machine learning methods and different values can be used for fragments and words. In order to investigate the effects of tuning on the system, we tried all possible combinations of α_{word} and $\alpha_{fragment}$ with values between 0 and 1 at increments of 0.05. Table 3 contains results for some of these experiments.

α_{word}	$\alpha_{fragment}$	AP	BEP
0.5	0.5	0.841	0.720
*	0.0	0.656	0.480
0.0	1.0	0.813	0.720
1.0	1.0	0.765	0.690
0.45	0.65	0.847	0.740

Table 3: Results of tuning the weights for intrinsic and distributional similarity.

The best results were obtained with $\alpha_{word} = 0.45$ and $\alpha_{fragment} = 0.65$, resulting in 0.847 AP and 0.74 BEP. This shows that parameter tuning can improve the results, but the 0.6% increase is modest and a completely unsupervised approach can give competitive results. In addition, the optimal values of α are close to 0.5, indicating that all four components (intrinsic and distributional similarities between words and fragments) are all contributing to the performance of the final system.

6 Previous work

Most work on entailment has focused on comparing sentences or paragraphs. For example, Haghighi et al. (2005) represent sentences as directed dependency graphs and use graph matching to measure semantic overlap. This method also compares the dependencies when calculating similarity, which supports incorporation of extra syntactic information. Hickl et al. (2006) combine lexico-syntactic features and automatically acquired paraphrases to classify entailing sentences. Lintean and Rus (2009) make use of weighted dependencies and word semantics to detect paraphrases. In addition to similarity they look at dissimilarity between two sentences and use their ratio as the confidence score for paraphrasing.

Lin and Pantel (2001) were one of the first to extend the distributional hypothesis to dependency paths to detect entailment between relations. Szpektor et al. (2004) describe the TEASE method for extracting entailing relation templates from the Web. Szpektor and Dagan (2008) use the distributional similarity of arguments to detect unary template entailment, whilst Berant et al. (2010) apply it to binary relations in focused entailment graphs.

Snow et al. (2005) described a basic method of syntactic pattern matching to automatically discover word-level hypernym relations from text. The use of directional distributional similarity measures to find inference relations between single words is explored by Kotlerman et al. (2010). They propose new measures based on feature ranks and compare them with existing ones for the tasks of lexical expansion and text categorisation.

In contrast to current work, each of the approaches described above only focuses on detecting entailment between specific subtypes of fragments (either sentences, relations or words) and optimising the system for a single scenario. This means only limited types of entailment relations are found and they cannot be used for entailment generation or compositional entailment detection as described in Section 2.

MacCartney and Manning (2008) approach sentence-level entailment detection by breaking the problem into a sequence of atomic edits linking the premise to the hypothesis. Entailment relations are then predicted for each edit, propagated up through a syntax tree and then used to compose the resulting entailment decision. However, their system focuses more on natural logic and uses a predefined set of compositional rules to capture a subset of valid inferences with high precision but low recall. It also relies on a supervised classifier and information from WordNet to reach the final entailment decision.

Androutsopoulos and Malakasiotis (2010) have assembled a survey of different tasks and approaches related to paraphrasing and entailment. They describe three different goals (paraphrase recognition, generation and extraction) and analyse various methods for solving them.

7 Conclusion

Entailment detection systems are generally developed to work on specific text units – either single words, relations, or full sentences. While this reduces the complexity of the problem, it can also lead to important information being disregarded. In this paper we proposed a new task – detecting entailment relations between any kind of dependency graph fragments. The definition of a fragment covers the language structures mentioned above and also extends to others that have not been fully investigated in the context of entailment recognition (such as multi-word constituents, predicates and adjuncts).

To perform entailment detection between various types of dependency graph fragments, a new system was built that combines the directional intrinsic and extrinsic similarities of two fragments to reach a final score. Fragments which contain hedging or negation are identified and their score calculation is inverted to better model the effect on entailment. The extrinsic similarity is found with two different distributional similarity measures, first checking for symmetric similarity and then for directional containment of distributional features. The system was evaluated on a manually constructed dataset of fragment-level entailment relations from the biomedical domain and each of the added methods improved the results.

Traditionally, the method for entailment recognition is chosen based on what appears optimal for the task – either structure matching or distributional similarity. Our experiments show that the combination of both gives the best performance for all fragment types. It is to be expected that single words will benefit more from distributional measures while full sentences get matched by their components. However, this separation is not strict and evidence from both methods can be used to strengthen the decision.

The experiments confirmed that entailment between dependency graph fragments of various types can be detected in a completely unsupervised setting, without the need for specific resources or annotated training data. As our method can be applied equally to any domain and requires only a large plain text corpus, we believe it is a promising direction for research in entailment detection. This can lead to useful applications in biomedical information extraction where manually annotated datasets are in short supply.

We have shown that a unified approach can be used to detect entailment relations between dependency graph fragments. This allows for entailment discovery among a wide range of fragment types, including ones for which no specialised systems currently exist. The framework for fragment-level entailment detection can be integrated into various applications that require identifying and rewriting semantically equivalent phrases - for example, query expansion in IE and IR, text mining, sentence-level entailment composition, relation extraction and protein-protein interaction extraction.

References

Ion Androutsopoulos and Prodromos Malakasiotis. 2010. A survey of paraphrasing and textual entailment methods. *Journal of Artificial Intelligence Research*, 38(7):135–187.

Roy Bar-Haim, Ido Dagan, Bill Dolan, Lisa Ferro, Danilo Giampiccolo, Bernardo Magnini, and Idan Szpektor. 2006. The second pascal recognising textual entailment challenge. In *Proceedings of the Second PASCAL Challenges Workshop on Recognising Textual Entailment*, pages 1–9. Citeseer.

Luisa Bentivogli, Elena Cabrio, Ido Dagan, Danilo Giampiccolo, Medea Lo Leggio, and Bernardo Magnini. 2010. Building Textual Entailment Specialized Data Sets: a Methodology for Isolating Linguistic Phenomena Relevant to Inference. In *Proceedings of the Seventh conference on International Language Resources and Evaluation (LREC'10)*.

Jonathan Berant, Ido Dagan, and Jacob Goldberger. 2010. Global learning of focused entailment graphs. In *Proceedings of the 48th Annual Meeting of the Association for Computational Linguistics*, number Section 6, pages 1220–1229. Association for Computational Linguistics.

Ted Briscoe, John Carroll, and Rebecca Watson. 2006. The second release of the RASP system. In *Proceedings of the COLING/ACL 2006 Interactive Presentation Sessions*, number July, pages 77–80, Sydney, Australia. Association for Computational Linguistics.

Daoud Clarke. 2009. Context-theoretic semantics for natural language: an overview. In *Proceedings of the Workshop on Geometrical Models of Natural Language Semantics*, number March, pages 112–119. Association for Computational Linguistics.

James Richard Curran. 2003. *From distributional to semantic similarity*. Ph.D. thesis, University of Edinburgh.

Lee R. Dice. 1945. Measures of the amount of ecologic association between species. *Ecology*, 26(3):297–302.

Richárd Farkas, Veronika Vincze, György Móra, Jànos Csirik, and György Szarvas. 2010. The CoNLL-2010 Shared Task: Learning to Detect Hedges and their Scope in Natural Language Text. In *Proceedings of the Fourteenth Conference on Computational Natural Language Learning — Shared Task*, pages 1–12. Association for Computational Linguistics.

Aria D. Haghighi, Andrew Y. Ng, and Christopher D. Manning. 2005. Robust textual inference via graph matching. In *Proceedings of the conference on Human Language Technology and Empirical Methods in Natural Language Processing*, Morristown, NJ, USA. Association for Computational Linguistics.

Andrew Hickl, Jeremy Bensley, John Williams, Kirk Roberts, Bryan Rink, and Ying Shi. 2006. Recognizing textual entailment with LCC's GROUNDHOG system. In *Proceedings of the Second PASCAL Challenges Workshop*.

Lili Kotlerman, Ido Dagan, Idan Szpektor, and Maayan Zhitomirsky-Geffet. 2010. Directional distributional similarity for lexical inference. *Natural Language Engineering*, 16(04):359–389.

Sandra Kübler, Ryan McDonald, and Joakim Nivre. 2009. Dependency Parsing. *Synthesis Lectures on Human Language Technologies*, 2:1–127.

Dekang Lin and Patrick Pantel. 2001. Discovery of inference rules for question-answering. *Natural Language Engineering*, 7(04):343–360.

Dekang Lin. 1998. Automatic retrieval and clustering of similar words. In *Proceedings of the 17th international conference on Computational linguistics-Volume 2*, pages 768–774. Association for Computational Linguistics.

Mihain C. Lintean and Vasile Rus. 2009. Paraphrase Identification Using Weighted Dependencies and Word Semantics. In *Proceedings of the FLAIRS-22*, volume 22, pages 19–28.

Bill MacCartney and Christopher D. Manning. 2008. Modeling semantic containment and exclusion in natural language inference. In *Proceedings of the 22nd International Conference on Computational Linguistics-Volume 1*, pages 521–528. Association for Computational Linguistics.

Roser Morante. 2009. Descriptive analysis of negation cues in biomedical texts. In *Proceedings of the Seventh International Language Resources and Evaluation (LREC10)*, pages 1429–1436.

Mark Sammons, V.G. Vinod Vydiswaran, and Dan Roth. 2010. Ask not what textual entailment can do for you... In *Proceedings of the Annual Meeting of the Association of Computational Linguistics (ACL)*, pages 1199–1208. Association for Computational Linguistics.

Rion Snow, Daniel Jurafsky, and Andrew Y. Ng. 2005. Learning syntactic patterns for automatic hypernym discovery. In *Advances in Neural Information Processing Systems*.

Idan Szpektor and Ido Dagan. 2008. Learning entailment rules for unary templates. In *Proceedings of the 22nd International Conference on Computational Linguistics - COLING '08*, pages 849–856, Morristown, NJ, USA. Association for Computational Linguistics.

Idan Szpektor, Hristo Tanev, Ido Dagan, and Bonaventura Coppola. 2004. Scaling web-based acquisition of entailment relations. In *Proceedings of EMNLP*, volume 4, pages 41–48.

Julie Weeds, David Weir, and Bill Keller. 2005. The distributional similarity of sub-parses. In *Proceedings of the ACL Workshop on Empirical Modeling of Semantic Equivalence and Entailment*, pages 7–12, Morristown, NJ, USA. Association for Computational Linguistics.

Learning Phenotype Mapping for Integrating Large Genetic Data

Chun-Nan Hsu[1,2,*], **Cheng-Ju Kuo**[2], **Congxing Cai**[1]
Sarah A. Pendergrass[3], **Marylyn D. Ritchie**[3,4] and **Jose Luis Ambite**[1]
[1]USC Information Sciences Institute, Marina del Rey, CA, USA
[2]Institute of Information Sciences, Academia Sinica, Taipei, Taiwan
[3]Center for Human Genetics Research, [4]Dept. of Molecular Physiology and Biophysics, Vanderbilt University, Nashville, TN, USA
*chunnan@isi.edu

Abstract

Accurate phenotype mapping will play an important role in facilitating Phenome-Wide Association Studies (PheWAS), and potentially in other phenomics based studies. The PheWAS approach investigates the association between genetic variation and an extensive range of phenotypes in a high-throughput manner to better understand the impact of genetic variations on multiple phenotypes. Herein we define the phenotype mapping problem posed by PheWAS analyses, discuss the challenges, and present a machine-learning solution. Our key ideas include the use of weighted Jaccard features and term augmentation by dictionary lookup. When compared to string similarity metric-based features, our approach improves the F-score from 0.59 to 0.73. With augmentation we show further improvement in F-score to 0.89. For terms not covered by the dictionary, we use transitive closure inference and reach an F-score of 0.91, close to a level sufficient for practical use. We also show that our model generalizes well to phenotypes not used in our training dataset.

1 Introduction

There is a wealth of biomedical data available in public and private repositories (*e.g.* the database issue of *Nucleic Acids Research* (**?**).) Along with this explosion of information comes the need to integrate data from multiple sources to achieve sufficient statistical power for analyses and/or to characterize phenomena more precisely. This trend manifests itself in two primary ways: the formation of large multi-institution multi-study consortia and public repositories. Although this situation occurs across many areas of biomedicine and our techniques are general, in this paper we will illustrate the ideas with examples from genetic studies in which we are participating.

Consider the National Center for Biotechnology Information (NCBI) database of Genotypes and Phenotypes (dbGaP) (`www.ncbi.nlm.nih.gov/gap`), that was developed to archive and distribute the results of studies that have investigated the interaction of genotype and phenotype. This is a large repository that includes genome-wide association studies (GWAS), medical sequencing, molecular diagnostic assays, as well as association between genotype and non-clinical traits. Genetic studies funded by the National Institutes of Health (NIH) over a certain size are required to submit the genetic and phenotypic data to dbGaP. There are over 130 top-level studies, 1900 datasets, 5600 analyses, comprising about 125000 phenotypic variables. Unfortunately, each study uses its own set of variables, thus far dbGaP does not attempt to reconcile, match or harmonize any of these variables. For example, a variable called `'BMI'` in one study and `'Body Mass Index'` in another study are recorded as different variables. The task of matching or harmonizing these variables falls on each researcher that obtains dbGaP data from multiple studies.

Similarly, consider a large consortium, such as the Population Architecture Using Genomics and Epidemiology (PAGE) network. PAGE (`www.pagestudy.org`) is a consortium of four major studies with the goal of understanding the

Proceedings of the 2011 Workshop on Biomedical Natural Language Processing, ACL-HLT 2011, pages 19–27,
Portland, Oregon, USA, June 23-24, 2011.

association of genetic variants with complex diseases and traits across a variety of populations. The studies that comprise PAGE include: the Women's Health Initiative (WHI, www.whiscience.org/); the Multiethnic Cohort (MEC, www.crch.org/multiethniccohort/, www.uscnorris.com/mecgenetics/); the CALiCo Consortium, comprised in turn of the Atherosclerosis Risk In Communities (ARIC) study (www.cscc.unc.edu/aric/), the Coronary Artery Risk In Young Adults (CARDIA) study (www.cardia.dopm.uab.edu), the Cardiovascular Heart Study (www.chs-nhlbi.org/), the Hispanic Community Health Study (www.cscc.unc.edu/hchs/), the Strong Heart Cohort Study, and the Strong Heart Family Study (strongheart.ouhsc.edu/); and the Epidemiologic Architecture of Genes Linked to Environment (chgr.mc.vanderbilt.edu/eagle/) study, which utilizes genotypic and phenotypic data from the National Health and Nutrition Examination Surveys (NHANES) from the Centers for Disease Control and Prevention (CDC). The studies of PAGE represent a pool of over 200,000 individuals with genotypic data collected across multiple race/ethnicities, and an extremely diverse collection of phenotypic data. Within PAGE there are numerous analyses and writing groups that focus on specific diseases. Each group selects variables relevant to their disease and harmonizes the variables across studies.

A group within PAGE is investigating a novel approach to genetic association analysis called a Phenome Wide Association Studies (PheWAS) (**?**). This is a different approach compared to the current paradigm of Genome Wide Association Studies (GWAS) (**?**; **?**). GWAS focus on calculating the association between the variation of hundreds of thousands of genotyped single nucleotide polymorphisms (SNPs) and a single or small number of phenotypes. This approach has provided valuable information about the contribution of genetic variation to a wide range of diseases and phenotypes. A common limitation of GWAS is the investigation of a limited phenotypic domain. In contrast, PheWAS utilizes an extensive range of detailed phenotypic measurements including intermediary biomarkers, in addition to prevalent and incident status for multiple common clinical conditions, risk factors, and quantitative traits for comprehensively exploring the association between genetic variations and all PheWAS phenotypes. The investigation of a broad range of phenotypes has the potential to identify pleiotropy, novel mechanistic insights fostering hypothesis generation, and to define a more complete picture of genetic variations and their impact on human diseases.

In order to compare PheWAS results across studies within PAGE to seek replication for significant genotype/phenotype associations, an important step is matching and mapping phenotypes across studies. As the number and range of phenotypes is large across studies, manually matching phenotypes is less than ideal. Therefore, an important step in improving the feasibility of PheWAS studies is to use computational approaches to map phenotypes across studies, effectively matching related phenotypes.

Definition *Phenotype Mapping* is the task of assigning every variable from each participating study to one out of a set of categories. The categories can be defined for a given integrated study or consortium, or can be taken from pre-existing ontologies, such as PhenX (www.phenx.org).

For one example, consider the variable `hypt` from WHI which is described by the text `'Hypertension ever'` and the variable `HAE5A` from the EAGLE study described by the text `'Now taking prescribed medicine for HBP'`. To manually match these phenotypes, a human expert declares these two variables to be relevant to class `'hypertension'`. Table 1 shows additional examples.

The phenotype mapping problem is quite challenging. First, the variable descriptions are quite short (around 10 words, often less). Second, mapping the variables to a category, such as hypertension, may require significant background knowledge (HBP stands for High Blood Pressure, also known as hypertension). Third, there are large numbers of variables, so the solution needs to scale gracefully.

In summary, in order to integrate data from public repositories, such as dbGaP, or from large consortia, such as the PAGE network, a critical task is to understand how the available phenotypes relate to each other. In this paper, we present machine-learning techniques for phenotype mapping that significantly reduce the burden on researchers when integrating data from multiple studies.

2 Related Work

From the perspective of biomedical sciences, phenotype mapping is a pre-requisite and a generalization for the task of *phenotype harmonization* (**?**). In harmonization, a single variable is identified or calculated for each phenotype within each study. This can only be accomplished for a very limited set of variables. There is a need, however, to provide enough information on a much larger set of phenotype variables so that researchers can determine the *common denominator* version of a measure across studies. For example, if a researcher is interested in hypertension status as an outcome, there needs to be an assessment of how hypertension status was ascertained in each study. Different approaches include self-report, clinic-based blood pressure measurement and/or anti-hypertensive medication use. Only after this information is obtained, along with other information, such as at what visit was status assessed and whether the variable is available for the entire cohort or only a portion of it will the researcher be able to determine what to use in analysis and how to interpret the findings. The phenotype mapping task that we address in this paper enables a researcher to rapidly find all the phenotype variables that are related to a given category, which then constitutes the input to the harmonization process.

From the computer science perspective, the task of phenotype mapping can be seen as an instance of the problem of entity linkage, which appears in a variety of forms across many contexts, namely record linkage (**?**), object identification (**?**), duplicate detection (**?**), and coreference (**?**; **?**). That is, the problem of recognizing when multiple objects (in multiple sources) actually correspond to the same entity.

Record linkage generally consists of three phases: (1) blocking, where the number of pairs of objects is reduced, which is critical for large datasets (*e.g.*, (**?**; **?**; **?**)), (2) field similarity, where the attributes of an object are compared (*e.g.*, (**?**; **?**; **?**; **?**; **?**), and (3) record similarity, which weights how different attributes contribute to the similarity of records as a whole (*e.g.*, (**?**; **?**)). Machine learning techniques are used for many of these tasks.

The task of phenotype mapping is related, but differs from previous incarnations of record linkage. In our case, the variables are the objects to be mapped. However, the only attribute of an object is a terse textual description (cf. Table 1). This makes the problem harder since, as we will see, string similarity measures are not enough, and term expansion with additional background knowledge is necessary. We do not consider blocking techniques in this paper, since the number of phenotypes is in the thousands and an exhaustive $O(n^2)$ comparison is still feasible.

In this paper, we define and present an approach to phenotype mapping with good experimental performance, but there are many opportunities for refinement by incorporating additional techniques from the record linkage literature.

3 Phenotype Mapping

For the PAGE PheWAS study, phenotypes were first manually matched, through the creation of 106 phenotype classes, in order to bring together related phenotypes across studies. The following steps were then used: First, the data from different studies were filtered independently for any significant association results with $p < 0.01$. Closely related phenotypes were then matched up between studies and assigned to phenotype classes. Finally, phenotypes from all studies, regardless of association results, were matched up to the already defined phenotype classes. In this way, a phenotype that might not have shown a significant association result for a single study, but that matched a phenotype class, would still be added to the phenotype-class list. To scale up the process it is important to develop a semi or fully automatic approach for the task.

Table 1 shows some example phenotypes and their classification. **Class** labels were assigned when we manually matched the phenotypes. The real ID of a phenotype in a **study** is given in column **ID**. **Description** will be the main clue for automatic matching. These examples were chosen to illustrate unique characteristics that we observed in the manually matched data set and the challenges of the task.

- The descriptions are in a wide variety of forms. They may be a compound term, a phrase, a sentence, or even a question, and usually contain

Class	Study	ID	Description
Allergy	ARIC	MHQA2A	EVER TOLD HAD HAY FEVER
Allergy	ARIC	MHQA2B	STILL HAVE HAY FEVER
Allergy	EAGLEIII	ALPBERFL	Cat - flare length (mm)
Allergy	EAGLEIII	ALPCATWL	Cat - wheal length (mm)
Allergy	EAGLEIII	ALPBERFL	Cat - flare width (mm)
Allergy	EAGLEIII	ALPCATWL	Cat - wheal width (mm)
Allergy	MEC	asthma	History of Asthma, Hayfever, Skin Allergy, Food Allergy or Any Other Allergy from Baseline Questionnaire
CigaretteSmokedPerDay	ARIC	HOM32	NUMBER OF CIGARETTES PER DAY
CigaretteSmokedPerDay	ARIC	HOM35	OVERALL NUM OF CIGARETTES PER DAY
CigaretteSmokedPerDay	CHS	AMOUNT	CIGS SMOKED/DAY
CigaretteSmokedPerDay	WHI	cigsday	Smoke or smoked, cigarettes/day
Hematocrit	ARIC	HMTA01	HEMATOCRIT
Hematocrit	EAGLEIII	HTP	Hematocrit (%)
Hematocrit	WHI	hematocr	Hematocrit (%)
Hypertension	ARIC	HYPERT04	HYPERTENTION, DEFINITION 4
Hypertension	ARIC	HOM10A	HIGH BP EVER DIAGNOSED
Hypertension	CHS	HYPER_1	CALCULATED HTN STATUS
Hypertension	CHS	HYPER_2	CALCULATED HTN STATUS
Hypertension	CHS	HYPER_3	CALCULATED HTN STATUS
Hypertension	CHS	HTNMED06	ANY HYPERTENTION MEDICATION
Hypertension	EAGLEIII	HAE2	Doctor ever told had hypertension/HBP
Hypertension	EAGLEIII	HAE5A	Now taking prescribed medicine for HBP
Hypertension	MEC	q2hibp	History of High Blood Pressure from QX2
Hypertension	MEC	hibp	History of High Blood Pressure from Baseline Questionnaire
Hypertension	WHI	hypt_f30	Hypertension ever
Hypertension	WHI	htntrt_f30	Hypertension
Smoker	ARIC	CURSMK01	CURRENT CIGARETTE SMOKER
Smoker	CHS	PRESSM	PRESENT SMOKER
Smoker	WHI	smoknow	Smoke cigarettes now

Table 1: Example phenotypes and their classification

less than 10 words, so it is difficult to apply sophisticated Natural Language Processing techniques.

- Phenotypes may be related in different ways: subsumption, overlapping, at the same layer of semantic hierarchy, *etc.*

- The granularity of the classes varies. For example, we have classes as specifically defined as `Hematocrit`, the ratio of the volume of red blood cells to the total volume of blood. But the class `Allergy` covers a wide range of allergy sources and symptoms. In Table 1, we show four phenotype variables for allergies against cats with flare and wheal sizes measured. Similar variables include those for allergies of a wide range of sources: alternaria, bermuda grass, german cockroach, mite, peanut, ragweed, rye grass, Russian thistle, and white oak. While in the same class, `MEC` uses a single phenotype `asthma` to cover just about all types of allergies. On the other hand, phenotypes about cigarette smoking are distinctively divided into two categories: cigarettes smoked per day and currently smoking. As we explained earlier, the main criterion here is to maximize the chance to detect unexpected associations, not necessarily to match the most semantically similar phenotypes. As a result, directly applying conventional clustering or topic modeling techniques in Information Retrieval may not be appropriate here.

- Some phenotypes in the same class appear nearly identical. For example, the three hematocrit phenotypes have almost identical descriptions. `HYPER_1`, `2` and `3` of the study `CHS` in the class `Hypertension` have exactly the same descriptions. For those cases, applying string similarity metrics can easily match them together. However, some phenotypes in the same class appear completely different due to the use of synonyms and abbreviations. Again in class `Hypertension`, 'hypertension,' 'HTN,' 'high blood pressure,' 'HBP,' and 'high BP' are keywords appearing in the descriptions of phenotypes. It is possible for an effective string similarity metric to recognize abbreviations like 'HTN' for 'hypertension,' but without additional information there is no way for a string similarity metric to match 'hypertension' and 'high blood pressure.'

4 Methods

We formulate the task as a problem of learning to score the degree of match of a pair of phenotypes based on their descriptions. By setting a threshold of the score for match or not, the problem reduces to a standard binary classification problem in Machine Learning.

We started by performing a pre-processing step of data cleaning to remove redundant phenotypes with no description, then pairing the resulting phenotypes for training and testing in a supervised learning framework. The data is skewed as most pairs are negative.

Studies	5	Phenotypes	733
Classes	106	Total pairs	298378
Positives	10906	Negatives	287472

Table 2: Statistics of Data

Another pre-processing step is tokenization, which was applied to the description of each phenotype before we extracted a set of features from each pairs. The tokenization step includes converting all uppercase letters to lowercase letters, removing punctuations, segmenting the text into tokens, and using Porter's stemmer (?) to stem tokens, removing stop words and digits. For example, `'TRANSIENT ISCHEMIC ATTACK'` will become (`transient`, `ischem`, `attack`). Note that `'ic'` was removed from `'ischemic'` by the stemming process.

The next step is feature extraction. The goal here is to represent each pair of phenotype variables by a set of feature values as the input to a machine-learning model. We considered two types of features. The first type is based on string similarity metrics. The idea is to combine the strength of a variety of string similarity metrics to measure the edit distance between the descriptions of a pair of phenotypes and use the result to determine if they match each other. We chose 16 metrics as shown in Table 3. Some of them are sophisticated and designed for challenging record linkage tasks, such as matching personal records in census data.

Levenshtein Distance	
Needleman-Wunch Distance	
Smith-Waterman Distance	
Smith-Waterman-Gotoh Distance	
Monge Elkan Distance	Q-grams Distance
Jaro Distance	Jaro Winkler
Block Distance	Soundex Distance
Matching Coefficient	Dice's Coefficient
Jaccard Similarity	Overlap Coefficient
Euclidean Distance	Cosine Similarity

Table 3: String similarity metrics

We used the Java implementation provided by SimMetrics[1] to obtain the values of these metrics given a pair of phenotype descriptions. SimMetrics also provides descriptions and references of these string similarity metrics. Each metric is treated as one feature and normalized into a real value between 0 and 1, where 1 indicates that the two strings are identical.

These string similarity metrics, however, treat all words equally but apparently some words are more important than others when we match phenotypes. To assign different weights to different words, we designed a feature set that can be considered as *weighted Jaccard* as follows. Let t be a token or a bi-gram (i.e., pair of consecutive tokens). For each t there are two features in the feature set of the following forms:

- share-t: if t appears in the pre-processed descriptions of both variables, then its value is 1

[1] `staffwww.dcs.shef.ac.uk/people/S.Chapman/simmetrics.html`

and 0 otherwise;

- miss-t: if t appears in the pre-processed description of one variable only, then its value is 1 and 0 otherwise;

For example, suppose we have tokenized variables V_1 = (age, menopause, start), and V_2 = (menopause, start, when), then the features for this pair will be

(miss-'age' : 1,
share-'menopause' : 1,
share-'start' : 1,
miss-'when' : 1,
miss-'age_menopause' : 1,
share-'menopause_start' : 1,
miss-'start_when' : 1).

All other features will have value 0. In this way, each example pair of variables will be represented as a very high-dimensional feature vector of binary values. The dimensionality is proportional to the square of the number of all distinct tokens appearing in the training set.

Now we are ready to train a model by a machine-learning algorithm using the examples represented as feature vectors. The model of our choice is the maximum entropy model (MaxEnt), also known as logistic regression (?). An advantage of this model is that efficient learning algorithms are available for training this model with high-dimensional data and the model not only classifies an example into positive or negative but also gives an estimated probability as its confidence. The basic idea of logistic regression is to search for a weight vector of the same dimension as the feature vector such that this weight vector when applied in the logit function of the probability estimation of the training examples will maximize the likelihood of the positive-negative assignment of the training examples (?). The same model can also be derived from the principle of maximum entropy. We randomly selected half of the pairs as the training examples and the rest as the holdout set for evaluation.

We used the Merriam-Webster Medical Dictionary (?)[2] to augment the descriptions of phenotypes. If there is an entry for a token in the dictionary, then its definition will be included in the description and then the same pre-processing and feature extraction steps will be applied. Pre-processing is also required to remove useless words from the definitions in the dictionary. We chose this dictionary instead of some ontology or phenotype knowledge base for its quality of contents and comprehensive coverage of biomedical terms. The Merriam-Webster Medical Dictionary is also chosen as the only medical dictionary included in the MedlinePlus[3], a Web service produced by the National Library of Medicine for the National Institute of Health to provide reliable and up-to-date information about diseases, conditions and wellness issues to the patients and their families and friends.

5 Results

Table 4 shows the results in terms of precision, recall, and F-score. The first two rows show the use of string similarity metrics as features to train a Naive Bayes model and a MaxEnt model. The F-scores of both models are similar, but Naive Bayes has higher false positives while MaxEnt made more false negative errors. MaxEnt with weighted Jaccard outperforms one with string-similarity features. Augmentation by dictionary lookup ("w/ dictionary") is proved effective by improving recall from 0.59 to 0.82, as more positive mappings were identified for those phenotype pairs described in different terms. One may suspect that the augmentation may increase false positives due to incorrectly associating common words in the descriptions. But remarkably, the false positives also decreased, resulting in the improvement in precision as well.

Table 5 shows a set of selected examples to illustrate the effectiveness of augmentation by dictionary lookup. The first column shows the original descriptions of the phenotype variable pairs. The second and third columns show the classification results (0 for negative, 1 for positive) and the confidence scores by the MaxEnt model without augmentation. The next two columns are their counterparts for the model with augmentation.

For example, the definition of 'Goiter' is 'an enlargement of the thyroid gland.' Therefore, after augmented by dictionary lookup, goi-

[2] www.m-w.com/browse/medical/a.htm

[3] www.nlm.nih.gov/medlineplus

Method / Model	Precision	Recall	F-score
String similarity metrics feature			
NaiveBayes	0.5236	0.6492	0.5797
MaxEnt	0.8092	0.4760	0.5994
Weighted Jaccard			
MaxEnt	0.9655	0.5931	0.7348
w/ dictionary	0.9776	0.8208	0.8924
w/ transitive closure (depth= 1)	0.9138	0.8064	0.8568
w/ both	0.8961	0.9177	0.9068

Table 4: Performance results

Phenotypes	w/o dic	Score	w/ dic	Score
`Goiter ever` `Overactive thyroid ever`	0	0.014562	1	0.996656
`History of High Blood Pressure from Baseline Questionnaire` `Hypertension ever`	0	0.014562	1	0.641408
`DIABETES W/ FASTING GLUCOSE CUTPT.<126` `Insulin shots now`	0	0.014562	1	0.523262
`TIA STATUS AT BASELINE` `Stroke`	0	0.014562	1	0.517444
`NUMBER OF CIGARETTES PER DAY` `CIGS SMOKED/DAY`	0	0.014562	0	0.002509

Table 5: Examples of Mapping Results

ter can be matched with overactive thyroid. Similarly, it is now possible to match `'High Blood Pressure'` with `'hypertension'` and `'TIA'` with `'stroke.'` `'DIABETES'`, `'GLUCOSE'` and `'Insulin'` can also be associated together.

However, terms must be covered in the medical dictionary for this method to work. For example, since `'CIGARETTES'` is not a medical term and even the most sophisticated string similarity metrics cannot match the local abbreviation `'CIGS'` to `'CIGARETTES'`, both models failed to match `'SMOKE'` and `'CIGARETTES'` together.

A solution to this issue is to compute transitive closure of the mapping. For example, if

$$V_1 = (\texttt{SMOKE}) \quad \text{and}$$
$$V_2 = (\texttt{SMOKE CIGARETTES})$$

are matched together by the model because of a shared term `'smoke'` and so are V_2 and

$$V_3 = (\texttt{cigarettes}),$$

but not V_1 and V_3, then transitive closure will infer a match of V_1 and V_3. That will improve recall and F-score further.

Figure 1 shows the performance of applying increasing depths of transitive closure to the results (a) without and (b) with augmentation by dictionary lookup. Transitive closure improves the performance for both models in the beginning but degrades quickly afterward because a phenotype may be assigned to multiple classes. As false positives increase, they will ripple when we infer new positives from false positives. Improvement for the model (a) is more obvious and degradation is not as grave. Applying transitive closure with depth $= 1$ yields the best performance. The exact scores are shown in Table 4 (See "w/ transitive closure" and "w/ both").

The results above were obtained by splitting the set of all pairs by half into training and test sets. It is possible that the model *remembers* phenotype descriptions because they distribute evenly in both training and test sets. To apply the system in practice, the model must generalize to unseen phenotypes. To evaluate the generalization power, instead of splitting the set of pairs, we split the set of vari-

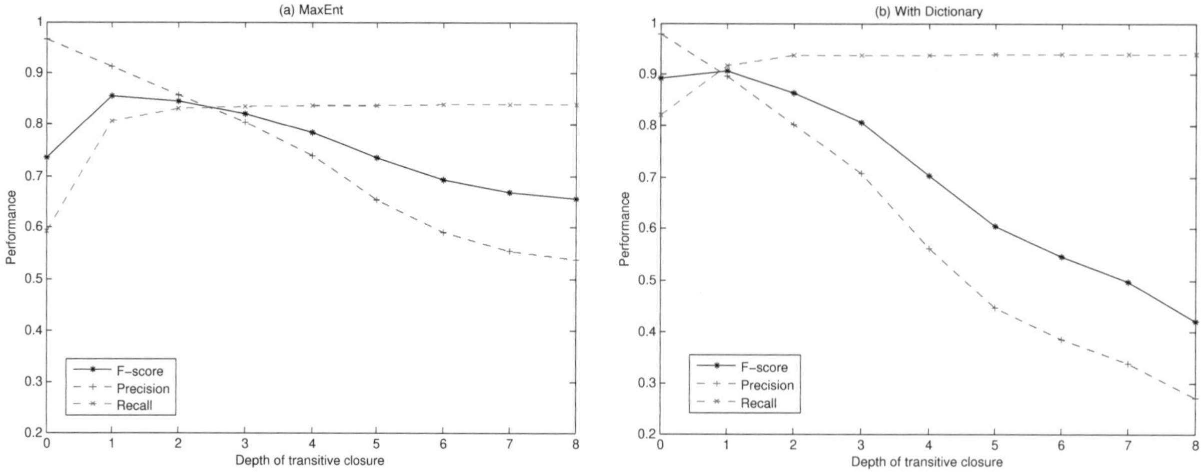

(a) MaxEnt model

(b) MaxEnt model with augmentation by dictionary lookup

Figure 1: Performance with increasing depths of transitive closure

ables by 2 to 1, and used $2/3$ of phenotype variables to generate pairs as the training set and $1/3$ to pair with those in the $2/3$ set as well as with each other for testing. That resulted in 129286 pairs for training and 169092 pairs for testing. In this test set, 6356 pairs are positive.

We used this training set to train MaxEnt models using the weighted Jaccard feature set with and without dictionary augmentation. Table 6 shows the results. Again, dictionary augmentation significantly improves the performance in this case, too, with the F-score reaching 0.81. Though the results degrade slightly from the ones obtained by splitting by pairs, this is expected as the training set is smaller (129286 pairs vs. $149189 = 298378/2$, see Table 2). Consequently, the proposed models can generalize well to unseen phenotypes to some extent.

Method/Model	Precision	Recall	F-score
w/o dictionary	0.9398	0.5817	0.7186
w/ dictionary	0.8213	0.7977	0.8093

Table 6: Performance results of splitting by variables

6 Conclusions and Future Work

In this paper, we define the problem of phenotype mapping and present a solution by learning to score and classify pairs of phenotypes. We evaluate our solution using a data set of manually matched phenotypes from the PAGE PheWAS study. We show that weighted Jaccard features are more effective for this problem than combining string similarity metrics for a MaxEnt model and that dictionary augmentation improves the performance by allowing matching of phenotypes with semantically related but syntactically different descriptions. We show that inferring more positives by depth-one transitive closure fixes those false negatives due to the lack of dictionary definitions. Finally, the evaluation results of splitting-by-variables show that the models generalize well to unseen variables, which is important for the solution to be practical.

Our future work includes to apply blocking as a pre-processing step to keep the number of pairs manageable and to apply active or unsupervised learning to alleviate the burden of generating training corpora by manual matching.

Acknowledgments

This work was supported by NHGRI grant HG004801 to C.-N.H. and J.L.A. and HG004798 to S.A.P. and M.D.R. C.-J.K. was supported by NSC 99-3112-B-001-028, Taiwan. The data were made available by the participating components of the NHGRI PAGE program. The complete list of PAGE members can be found at `www.pagestudy.org`. The contents of this paper are solely the responsibility of the authors and do not necessarily represent the official views of the NIH.

References

Siiri N. Bennett, Neil Caporaso, Annette L. Fitzpatrick, Arpana Agrawal, Kathleen Barnes, Heather A. Boyd, Marilyn C. Cornelis, Nadia N. Hansel, Gerardo Heiss, John A. Heit, Jae Hee Kang, Steven J. Kittner, Peter Kraft, William Lowe, Mary L. Marazita, Kristine R. Monroe, Louis R. Pasquale, Erin M. Ramos, Rob M. van Dam, Jenna Udren, Kayleen Williams, and for the GENEVA Consortium. 2011. Phenotype harmonization and cross-study collaboration in gwas consortia: the GENEVA experience. *Genetic Epidemiology*, 35(3):159–173.

Mikhail Bilenko and Raymond J. Mooney. 2003. Adaptive duplicate detection using learnable string similarity measures. In *Proceedings of the Ninth ACM SIGKDD International Conference on Knowledge Discovery and Data Mining*, pages 39–48, Washington, DC, USA.

William W. Cohen, Pradeep Ravikumar, and Stephen Fienberg. 2003. A comparison of string distance metrics for name-matching tasks. In *Proceedings of IJCAI-03 Workshop on Information Integration on the Web (IIWeb-03)*.

The Wellcome Trust Case Control Consortium. 2007. Genome-wide association study of 14,000 cases of seven common diseases and 3,000 shared controls. *Nature*, 447(7145):661–678, June.

Joshua C. Denny, Marylyn D. Ritchie, Melissa A. Basford, Jill M. Pulley, Lisa Bastarache, Kristin Brown-Gentry, Deede Wang, Dan R. Masys, Dan M. Roden, and Dana C. Crawford. 2010. Phewas: demonstrating the feasibility of a phenome-wide scan to discover gene-disease associations. *Bioinformatics*, 26(9):1205–1210.

Ivan P. Felligi and Alan B. Sunter. 1969. A theory for record linkage. *Journal of the American Statistical Association*, 64(328):1183–1210.

Michael Y. Galperin and Guy R. Cochrane. 2011. The 2011 nucleic acids research database issue and the online molecular biology database collection. *Nucleic Acids Research*, 39(suppl 1):D1–D6.

John Hardy and Andrew Singleton. 2009. Genomewide association studies and human disease. *New England Journal of Medicine*, 360(17):1759–1768.

T. Hastie, R. Tibshirani, and J. Friedmann. 2009. *The Elements of Statistical Learning (2nd Edition)*. Springer-Verlag, New York, NY, USA.

Mauricio A. Hernández and Salvatore J. Stolfo. 1998. Real-world data is dirty: Data cleansing and the merge/purge problem. *Data Mining and Knowledge Discovery*, 2:9–37.

Jerry R. Hobbs. 1979. Coherence and coreference. *Cognitive Science*, 3(1):67–90.

Andrew McCallum, Kamal Nigam, and Lyle Ungar. 2000. Efficient clustering of high-dimensional data sets with application to reference matching. In *Proceedings of the Sixth ACM SIGKDD International Conference on Knowledge Discovery and Data Mining*, pages 169–178.

Merriam-Webster. 2006. *Medical Dictionary*. Merriam-Webster, Springfield, MA, USA.

Matthew Michelson and Craig A. Knoblock. 2006. Learning blocking schemes for record linkage. In *Proceedings of the 21st National Conference on Artificial Intelligence (AAAI-06)*, Boston, MA.

Steven Minton, Claude Nanjo, Craig A. Knoblock, martin Michalowski, and Matthew Michelson. 2005. A heterogeneous field matching method for record linkage. In *Proceedings of the Fifth IEEE International Conference on Data Mining*, Novemeber.

Alvaro Monge and Charles Elkan. 1996. The field matching problem: Algorithms and applications. In *In Proceedings of the Second International Conference on Knowledge Discovery and Data Mining*, pages 267–270.

Felix Naumann and Melanie Herschel. 2010. *An Introduction to Duplicate Detection*. Synthesis Lectures on Data Management. Morgan & Claypool Publishers.

Vincent Ng and Claire Cardie. 2002. Improving machine learning approaches to coreference resolution. In *Proceedings of the 40th Annual Meeting on Association for Computational Linguistics*, pages 104–111, Stroudsburg, PA, USA. Association for Computational Linguistics.

Martin F. Porter. 1980. An algorithm for suffix stripping. *Program*, 14(3):130–137.

Sheila Tejada, Craig A. Knoblock, and Steven Minton. 2001. Learning object identification rules for information integration. *Information Systems*, 26(8).

EVEX: A PubMed-Scale Resource for Homology-Based Generalization of Text Mining Predictions

Sofie Van Landeghem[1,2], Filip Ginter[3], Yves Van de Peer[1,2] and Tapio Salakoski[3,4]
1. Dept. of Plant Systems Biology, VIB, Belgium
2. Dept. of Plant Biotechnology and Genetics, Ghent University, Belgium
3. Dept. of Information Technology, University of Turku, Finland
4. Turku Centre for Computer Science (TUCS), Finland
solan@psb.ugent.be, ginter@cs.utu.fi
yvpee@psb.ugent.be, tapio.salakoski@utu.fi

Abstract

In comparative genomics, functional annotations are transferred from one organism to another relying on sequence similarity. With more than 20 million citations in PubMed, text mining provides the ideal tool for generating additional large-scale homology-based predictions. To this end, we have refined a recent dataset of biomolecular events extracted from text, and integrated these predictions with records from public gene databases. Accounting for lexical variation of gene symbols, we have implemented a disambiguation algorithm that uniquely links the arguments of 11.2 million biomolecular events to well-defined gene families, providing interesting opportunities for query expansion and hypothesis generation. The resulting MySQL database, including all 19.2 million original events as well as their homology-based variants, is publicly available at http://bionlp.utu.fi/.

1 Introduction

Owing to recent advances in high-throughput sequencing technologies, whole genomes are being sequenced at an ever increasing rate (Metzker, 2010). However, for the DNA sequence to truly unravel its secrets, structural annotation is necessary to identify important elements on the genome, such as coding regions and regulatory motifs. Subsequently, functional annotation is crucial to link these structural elements to their biological function.

Functional annotation of genomes often requires extensive *in vivo* experiments. This time-consuming procedure can be expedited by integrating knowledge from closely related species (Fulton et al., 2002; Proost et al., 2009). Over the past few years, homology-based functional annotation has become a widely used technique in the bioinformatics field (Loewenstein et al., 2009).

Unfortunately, many known genotype-phenotype links are still buried in research articles: The largest biomolecular literature database, PubMed, consists of more than 20 million citations[1]. Due to its exponential growth, automated tools have become a necessity to uncover all relevant information.

There exist several text mining efforts focusing on pairwise interactions and co-occurrence links of genes and proteins (Hoffmann and Valencia, 2004; Ohta et al., 2006; Szklarczyk et al., 2011). In this paper, we present the first large-scale text mining resource which both utilizes a detailed event-based representation of biological statements and provides homology-based generalization of the text mining predictions. This resource results from the integration of text mining predictions from nearly 18M PubMed citations with records from public gene databases (Section 2). To enable such integration, it is crucial to first produce canonical forms of the automatically tagged biological entities (Section 3.1). A gene symbol disambiguation algorithm then links these canonical forms to gene families and gene identifiers (Section 3.2). Finally, a MySQL-driven framework aggregates the text-bound event occurrences into generalized events, creating a rich resource of homology-based predictions extracted from text (Section 3.3).

[1] http://www.ncbi.nlm.nih.gov/pubmed/

Proceedings of the 2011 Workshop on Biomedical Natural Language Processing, ACL-HLT 2011, pages 28–37,
Portland, Oregon, USA, June 23-24, 2011.

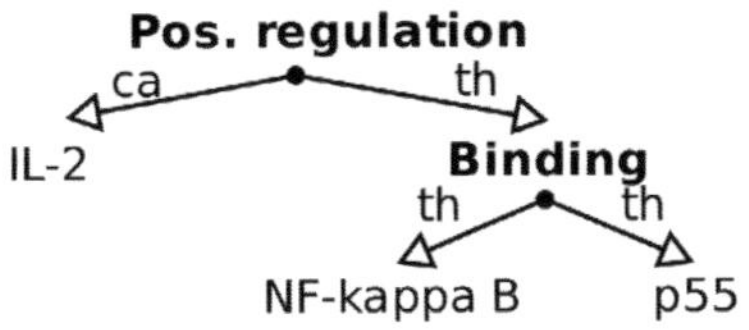

Figure 1: Event representation of the statement *IL-2 acts by enhancing binding activity of NF-kappa B to p55*, illustrating recursive nesting of events where the (th)eme of the *Positive regulation* event is the *Binding* event. The (ca)use argument is the gene symbol *IL-2*.

2 Data

Our integrative approach is based on two types of data: text mining predictions generated for the whole of PubMed (Section 2.1) and publicly available gene database records (Section 2.2).

2.1 Text mining predictions

Björne et al. (2010) have applied to all PubMed abstracts an event extraction pipeline comprising of the BANNER named entity recognizer (Leaman and Gonzalez, 2008) and the Turku Event Extraction System (Björne et al., 2009). The resulting dataset contains 36.5M occurrences of gene / gene product (GGP) entities and 19.2M occurrences of events pertaining to these entities.

The file format and information scheme of the resource correspond to the definition of the BioNLP'09 Shared Task on Event Extraction (Kim et al., 2009). Events are defined as typed relations between arguments that are either entity occurrences or, recursively, other events. There are nine possible event types: *Localization, Binding, Gene expression, Transcription, Protein catabolism, Phosphorylation, Regulation, Positive regulation,* and *Negative regulation.* Further, arguments are assigned a role: *Theme* or *Cause* for the core arguments and *AtLoc, ToLoc, Site,* and *CSite* for auxiliary arguments that define additional information such as cellular location of the event. In addition, each event occurrence may be marked as negative and/or speculative. Figure 1 depicts an example event.

2.2 Database records

During the last few decades, several large-scale databases have been designed to deal with the abundance of data in the field of life sciences. In this study, we are particularly interested in databases of gene symbols and homologous gene groups or gene families. These families are composed by clustering pairwise orthologs, which are genes sharing common ancestry evolved through speciation, often having a similar biological function.

Entrez Gene[2] is the default cross-species gene nomenclature authority, hosted by NCBI (Sayers et al., 2009). It bundles information from species-specific resources as well as from RefSeq records[3]. More than 8M Entrez Gene identifiers were collected from over 8,000 different taxa, all together referring to more than 10M distinct gene symbols, descriptions, abbreviations and synonyms. While Entrez Gene IDs are unique across taxa, gene symbols are highly ambiguous. Section 3 describes how we tackle gene symbol ambiguity across and within species.

The HomoloGene[4] database is also hosted at NCBI and provides the results of automated detection of orthologs in 20 completely sequenced eukaryotic genomes. From this resource, around 43,700 HomoloGene families were extracted, containing about 242,000 distinct genes. A second set of gene families was retrieved from Ensembl (Flicek et al., 2011). More than 13,000 Ensembl clusters were assembled comprising about 220,000 genes.

As a general rule, the functional similarity scores per homologous pair in a gene family are higher when more stringent criteria are used to define the families (Hulsen et al., 2006). While HomoloGene consists of many strict clusters containing true orthologs, bigger Ensembl clusters were obtained by assembling all pairwise orthologous mappings between genes. Ultimately, such clusters may also include paralogs, genes originated by duplication. As an example, consider the *nhr-35* gene from *C. elegans*, which has both *Esr-1* and *Esr-2* as known orthologs, resulting in the two paralogs being assigned to the same final Ensembl cluster. The Ensembl clustering algorithm can thus be seen as a more coarse-grained method while the HomoloGene mapping results in more strictly defined gene families. The implications are discussed on a specific use-case in Section 4.3.1.

[2] `http://www.ncbi.nlm.nih.gov/gene`
[3] `http://www.ncbi.nlm.nih.gov/refseq`
[4] `http://www.ncbi.nlm.nih.gov/homologene`

3 Methods

Widely known biomolecular events occur in many different articles, often mentioning a different gene synonym or lexical variant. Canonicalization of the entity occurrences deals with these lexical variants (Section 3.1), while the disambiguation algorithm then uniquely links canonical forms to a gene families (Section 3.2). In a final step, these links can be used to generalize the text mining events to their homology-based variants (Section 3.3).

3.1 Canonicalization of the entity occurrences

The entity occurrences predicted by BANNER (Section 2.1) follow the guidelines of GENETAG (Tanabe et al., 2005), the corpus it was trained on. These guidelines allow not only gene and gene products, but also related entities such as protein complexes and gene promoters. Furthermore, BANNER frequently tags noun phrases such as *human Esr-1 gene* rather than only the minimal symbol *Esr-1*.

To enable integration of text mining predictions with external databases, it is necessary to refine the entity occurrences to canonical forms that can be linked to gene records such as those in Entrez Gene. To this end, common prefixes and suffixes such as *gene* and *wild-type* should be removed.

In a first step towards canonicalization of the entities, a mapping table was assembled containing common contexts in which a gene symbol appears and where the full noun phrase can be reduced to that embedded symbol for the sake of information retrieval (Table 1). This mapping table was created by matching[5] a list of candidate minimal gene symbols to the extracted BANNER entities.

To define the list of candidate minimal gene symbols, two approaches have been combined. First, a set of around 15,000 likely gene symbols is extracted by looking for single token strings that were tagged by BANNER at least 50% of the times they occur in a PubMed abstract. Secondly, all official gene names are extracted from Entrez Gene. As this latter list also contains common English words such as *was* and *protein*, we have only selected those that were likely to be standalone gene symbols. We calculate this likelihood by $C_s/(C_s + C_n)$ where C_s is the number of times a string is tagged standalone and C_n is the number of times the string occurs in PubMed but is not tagged (neither as standalone, nor as part of a larger entity). This likelihood represents the proportion of standalone occurrences of the string that are tagged. We experimentally set a threshold on this value to be higher than 0.01, excluding a list of 2,865 common English words.

GGP contexts
`-ORG-` `-GGP-` gene
`-GGP-` sequences
mutant `-GGP-` proteins
`-GGP-` homologs
cytoplasmic wild-type `-GGP-`

Table 1: This table lists a few examples of entity occurrences extracted with BANNER that are resolved to the embedded minimal gene symbol (marked as `-GGP-`).

Subsequently, all BANNER entity occurrences are screened and likely minimal gene symbols substituted with `-GGP-`, resulting in generalized contexts. Then, we have matched these contexts with an extensive list of organism names from the Linneaus distribution (Gerner et al., 2010) and a small collection of miscellaneous non-formal organism terms (e.g. *monkey*), replacing all known organisms with an `-ORG-` placeholder. Finally, we have excluded all contexts where the embedded GGP is likely to be functionally too far removed from the embedding noun phrase (e.g. "`-GGP-` inhibitor"), relying on a corpus defining and categorizing such relationships (Ohta et al., 2009). Some of the contexts that were retained after this step, such as "`-GGP-` mutant" or "`-GGP-` promoter" still refer to entities that are distinctly different from the embedded GGP. These results are considered valid, as the goal of the affix stripping algorithm is to increase recall and offer explorative results involving various types of information on gene symbols.

The final list of contexts, generalized with `-GGP-` and `-ORG-` placeholders, is split into two separate lists of prefixes and suffixes, ranked by frequency. Also, numerical affixes as well as those shorter than 3 characters are discarded from these lists.

[5] All string matching steps have been implemented using the *SimString* string retrieval library (Okazaki and Tsujii, 2010).

Each text-bound entity occurrence can then be canonicalized by applying the following algorithm:

1. Replace all organism names with the placeholder `-ORG-`
2. If the string can be matched[6] to a known symbol in Entrez Gene, stop the algorithm
3. Find all occurring affixes and strip the one associated with the highest count
4. Repeat (2-3) until no more affixes match
5. Strip remaining `-ORG-` placeholders and all whitespace and non-alphanumeric characters

For example, the canonicalization of *human anti-inflammatory il-10 gene* proceeds as *-ORG- anti-inflamatory il-10 gene* → *anti-inflammatory il-10 gene* → *anti-inflammatory il-10* → *il-10*, at which point the string *il10* is matched in Entrez Gene, becoming the final canonical form. In the following section, we describe how these canonical forms are assigned to unique gene families.

3.2 Disambiguation of gene symbols

Gene name ambiguity is caused by the lack of community-wide approved standards for assigning gene symbols (Chen et al., 2005). Furthermore, authors often introduce their own lexical variants or abbreviations for specific genes.

From Entrez Gene, we have retrieved 8,034,512 gene identifiers that link to 10,177,542 unique symbols. Some of these symbols are highly ambiguous and uninformative, such as *NEWENTRY*. Others are ambiguous because they are abbreviations. Finally, many symbols can not be linked to one unique gene, but do represent a homologous group of genes sharing a similar function. Often, orthologs with similar functions are assigned similar official gene names.

The first step towards gene symbol disambiguation involves collecting all possible synonyms for each gene family from either Ensembl or HomoloGene. We strip these symbols of all whitespace and non-alphanumeric characters to match the final step in the canonicalization algorithm.

The disambiguation pipeline then synthesizes the ambiguity for all gene symbols by counting their occurrences in the gene families. Each such relation records whether the symbol is registered as an official or default gene symbol, as the gene description, an abbreviation, or a synonym. As an example, Table 2 depicts the intrinsic ambiguity of *esr1*.

Family	Type of symbol	Count
HG:47906	Default symbol	7
HG:99739	Synonym	1
HG:3740	Synonym	1
ECL:10415	Default symbol	12
ECL:8731	Synonym	1
ECL:8226	Synonym	1

Table 2: Intrinsic ambiguity of *esr1*, analysed in both HomoloGene (HG) and Ensembl clusters (ECL).

In a subsequent step, the ambiguity is reduced by applying the following set of rules, relying on a priority list imposed on the type of the symbol, ensuring we choose an official or default symbol over a description or synonym.

1. If one family has the most (or all) hits for a certain symbol and these hits refer to a symbol type having priority over other possibilities, this family is uniquely assigned to that symbol.
2. If a conflict exists between one family having the highest linkage count for a certain symbol, and another family linking that symbol to a higher priority type, the latter is chosen.
3. If two families have equal counts and type priorities for a certain symbol, this symbol can not be unambiguously resolved and is removed from further processing.
4. If the ambiguity is still not resolved, all families with only one hit for a certain symbol are removed, and steps 1-3 repeated.

The above disambiguation rules were applied to the 458,505 gene symbols in HomoloGene. In the third step, 6,891 symbols were deleted, and when the algorithm ends, 555 symbols remained ambiguous. In total, 451,059 gene symbols could thus be uniquely linked to a HomoloGene family (98%). In the *esr1* example depicted in Table 2, only the link to HG:47906 will be retained. The results for Ensembl were very similar, with 342,252 out of 345,906 symbols uniquely resolved (99%).

[6] The comparison is done ignoring whitespace and non-alphanumeric characters.

	All	Ensembl	HomoloGene
No stripping	39.9 / 67.5 / 50.2	62.8 / 70.0 / 66.2	64.2 / 69.2 / 66.6
Affix stripping	48.7 / 82.3 / 61.1	61.7 / 88.0 / 72.5	62.8 / 87.9 / 73.3

Table 3: Influence on precision, recall and F-measure (given as P/R/F) of the affix stripping algorithm on the entity recognition module, as measured across all BioNLP'09 ST entity occurrences and also separately on the subsets which can be uniquely mapped to Ensembl and HomoloGene (77.3% and 75.5% of all occurrences, respectively).

3.3 Homology-based generalization of the text mining events

In order to gain a broader insight into the 19.2M event occurrences obtained by Björne et al. (2010), it is necessary to identify and aggregate multiple occurrences of the same underlying event. This generalization also notably simplifies working with the data, as the number of generalized events is an order of magnitude smaller than the number of event occurrences.

To aggregate event occurrences into generalized events, it is necessary to first define equivalence of two event occurrences: Two event occurrences are equivalent, if they have the same event type, and their core arguments are equivalent and have the same roles. For arguments that are themselves events, the equivalence is applied recursively. The equivalence of arguments that are entities can be established in a number of different ways, affecting the granularity of the event generalization. One approach is to use the string canonicalization described in Section 3.1; two entities are then equivalent if their canonical forms are equal. This, however, does not take symbol synonymy into account. A different approach which we believe to be more powerful, is to disambiguate gene symbols to gene families, as described in Section 3.2. In this latter approach, two entity occurrences are deemed equivalent if their canonical forms can be uniquely resolved to the same gene family. Consequently, two event occurrences are considered equivalent if they pertain to the same gene families.

As both approaches have their merits, three distinct generalization procedures have been implemented: one on top of the canonical gene symbols, and one on top of the gene families defined by HomoloGene and Ensembl, respectively.

4 Results and discussion

4.1 Evaluation of entity canonicalization

The affix stripping step of the canonicalization algorithm described in Section 3.1 often substantially shortens the entity strings and an evaluation of its impact is thus necessary. One of the primary objectives of the canonicalization is to increase the proportion of entity occurrences that can be matched to Entrez Gene identifiers. We evaluate its impact using manually tagged entities from the publicly available BioNLP'09 Shared Task (ST) training set, which specifically aims at identifying entities that are likely to match gene and protein symbol databases (Kim et al., 2009). Further, the ST set comprises of PubMed abstracts and its underlying text is thus covered in our data. Consequently, the ST training set forms a very suitable gold standard for the evaluation.

First, we compare[7] the precision and recall of the BANNER output before and after affix stripping (Table 3, first column). The affix stripping results in a notable gain in both precision and recall. In particular, the nearly 15pp gain on recall clearly demonstrates that the affix stripping results in entity strings more likely to match existing resources.

Second, the effect of affix stripping is evaluated on the subset of entity strings that can be uniquely mapped into Ensembl and HomoloGene (77.3% and 75.5% of the ST entity strings, respectively). This subset is of particular interest, since the generalized events are built on top of the entities that can be found in these resources and any gain on this particular subset is thus likely to be beneficial for the overall quality of the generalized events. Here, affix stripping leads to a substantial increase in recall when compared to no stripping being applied

[7] The comparison is performed on the level of bags of strings from each PubMed abstract, avoiding the complexity of aligning character offsets across different resources.

	Entities	Ent. occ.
Canonical	1.6M (100%)	36.4M (100%)
HomoloGene	64.0K (3.9%)	18.8M (51.7%)
Ensembl	54.6K (3.3%)	18.7M (51.2%)

Table 4: Entity coverage comparison. The *entities* column gives the number of canonical entities, also shown as a percentage of all unique, canonical BANNER entities (1.6M). The *entity occurrences* column shows the number of occurrences for which the generalization could be established, out of the total number of 36.4M extracted BANNER entities.

	Events	Ev. occ.
Canonical	3223K	19.2M (100%)
HomoloGene	614K	10.2M (53%)
Ensembl	505K	10.2M (52.9%)

Table 5: Comparison of the three event generalization methods. The *events* column gives the number of generalized events and the *event occurrences* column shows the number of occurrences for which the generalization could be established, out of the total number of 19.2M text-bound event occurrences.

(around 18pp), which is offset by a comparatively smaller drop in precision (less than 2pp). Global performance increases with about 6.5pp in F-score for both the Ensembl and HomoloGene subsets.

Björne et al. (2010) used a simpler, domain-restricted affix stripping algorithm whereby candidate affixes were extracted only from NP-internal relations in the GENIA corpus (Ohta et al., 2009). This original algorithm affects 11.5% unique entity strings and results in 3.5M unique canonical forms and 4.5M unique events. In comparison, our current affix stripping algorithm results in 1.6M unique canonical forms and 3.2M unique events, thus demonstrating the improved generalization capability of the current affix stripping algorithm.

4.2 Evaluation of homology-based disambiguation

The symbol to gene family disambiguation algorithm succesfully resolves almost all gene symbols in HomoloGene or Ensembl (Section 3.2). However, not all genes are a member of a known gene family, and the event generalization on top of the gene families will thus inevitably discard a significant portion of the text mining predictions.

Table 4 shows that only a small fraction of all unique canonical entities matches the gene families from HomoloGene or Ensembl (3.9% and 3.3%, respectively). However, this small fraction of symbols accounts for approximately half of all entity occurrences in the text mining data (51.7% and 51.2%). The algorithm thus discards a long tail of very infrequent entities. Table 5 shows a similar result for the events and event occurrences. We find that mapping to HomoloGene and Ensembl results in a considerably smaller number of generalized events, yet accounts for more than half of all event occurrences (53% and 52.9%, respectively).

Finally, merging the canonical entities and the corresponding generalized events for both HomoloGene and Ensembl, we can assess the percentage of all text mining predictions that can be linked to at least one homology-based variant: 21.8M (59.8%) of all entity occurrences and 11.2M (58.4%) of all event occurrences can be resolved. Nearly 60% of entity and event occurrences in the original text mining data could thus be uniquely linked to well defined gene families. Also, as shown in Section 4.1, the 60% entities retained are expected to contain proportionally more true positives, compared to the 40% entities that could not be mapped. One might speculate that a similar effect will be seen also among events.

4.3 MySQL database and Use-cases

As the PubMed events extracted by Björne et al. (2010) are purely text-bound and distributed as text files, they can not easily be searched. One important contribution of this paper is the release of all text mining predictions as a MySQL database. During the conversion, all original information is kept, including links to the PubMed IDs and the offsets in text for all entities and triggers, referring to the original strings as they were obtained by BANNER and the event extraction system. This allows for fast retrieval of text mining data on a PubMed-scale.

As described in Section 3.3, three distinct generalization methods have been applied to the original events. On the database level, each generalization is represented by a separate set of tables for the generalized events and their arguments, aggregating important event statistics such as occurrence count and

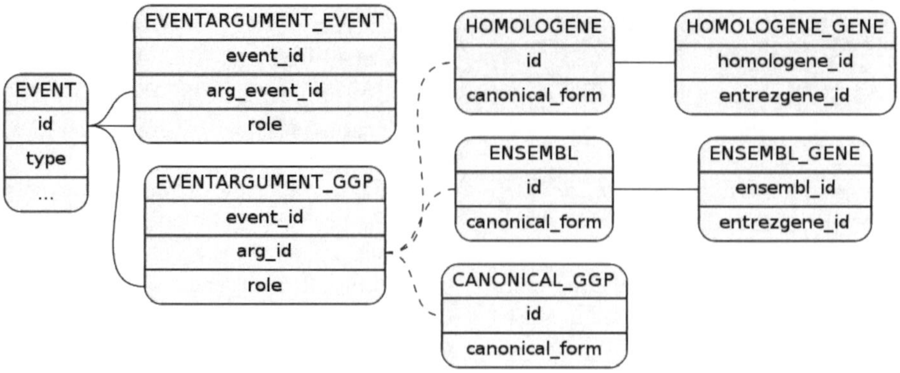

Figure 2: Database scheme of the generalized events. Three instantiations of the general scheme (i.e. the three leftmost tables) exist in the database. Following the dotted lines, each instance links to a different table in which the canonical forms and the gene identifiers can be looked up.

negation/speculation information (Figure 2). Table 5 states general statistics for the three different sets. Finally, a mapping table is provided that links the generalized events to the event occurrences from which they were abstracted. More technical details on the MySQL scheme and example queries can be found at `http://bionlp.utu.fi/`.

4.3.1 Use case: Query expansion

The MySQL database is the ideal resource to retrieve information on a PubMed-scale for a certain gene or set of genes. Suppose there would be an interest in *Esr-1*, then all abstract events on top of the canonical form *esr1* can be retrieved. However, results will display events for both the *Estrogen receptor* as well as for the much less common *Enhancer of shoot regeneration*. Furthermore, it makes no sense to add known synonyms of both genes to the query, as this will generate an incoherent list of synonyms and even more false positive hits.

In such a case, it is to be recommended to use the homology-based generalization of the events. For example, *esr1* hits the HomoloGene family HG:47906, which contains all *Estrogen receptor-alpha* genes across eukaryotic species. Canonical symbols linked to this family include *era*, *estra*, *nr3a1* and *estrogenreceptor1alpha*.

A similar analysis can be done for the Ensembl clustering, where *esr1* links to ECL:10415. However, this more coarse-grained Ensembl family contains all genes from the two closely related subgroups *Estrogen receptor* and *Estrogen related receptor*, both belonging to the *Estrogen Receptor-like* group of the superfamily of nuclear receptors (Zhang et al., 2004). On top of the synonyms mentioned previously, this family thus also includes *erb*, *esr2b*, *errbetagamma* and *similartoesrrbprotein*. By using this list for query expansion, more general text mining predictions can be retrieved.

It is to be noted that both homology-based approaches will also include events mentioning *Esr-1* as the abbreviation for *Enhancer of shoot regeneration*. While this usage is much less common, it will result in a few false positive hits. These false positives may be prevented by taking into account local context such as organism mentions, as the *Enhancer of shoot regeneration* gene is only present in *A. thaliana*. We believe our current homology-based approach could be integrated with existing or future normalization algorithms (Krallinger and Hirschman, 2007; Wermter et al., 2009) to provide such fine-grained resolution. This is regarded as interesting future work.

4.3.2 Use case: Homology-based hypotheses

Consider a newly annotated, protein-coding gene for which no database information currently exists. To generate homology-based text mining hypotheses, the orthologs of this gene first have to be defined by assessing sequence similarity through BLAST (Altschul et al., 1997).

Imagine for example a newly sequenced genome X for which a gene similar to the mouse gene *Esr-1* is identified. This gene will soon be known as "genome X *Esr-1*" and thus related to the *Esr-1* gene family. As described in Section 4.3.1, homology-based query expansion can then be used to retrieve all events involving lexical variants and synonyms of the canonical string *esr1*.

5 Conclusions

We present a large-scale resource for research and application of text mining from biomedical literature. The resource is obtained by integrating text mining predictions in the dataset of Björne et al. (2010) with public databases of gene symbols and gene families: Entrez Gene, Ensembl, and HomoloGene. The integration is performed on the level of gene families, allowing for a number of novel use cases for both text mining and exploratory analysis of the biological statements in PubMed literature. To achieve the linking between text-based event predictions and gene databases, several algorithms are introduced to solve the problems involved.

First, we propose an algorithm for stripping affixes in entity occurrences tagged by the BANNER named entity recognizer, addressing the problem of such entities often including wider context which prevents direct matching against gene symbol databases. Using the BioNLP'09 Shared Task data as gold standard, we show that the algorithm substantially increases both precision and recall of the resulting canonical entities, the gain in recall being particularly pronounced.

Second, we propose an algorithm which assigns to the vast majority of gene symbols found in HomoloGene and Ensembl a single unique gene family, resolving the present intra-organism ambiguity based on symbol occurrence statistics and symbol type information. Matching these disambiguated symbols with the affix-stripped canonical forms of entity occurrences, we were able to assign a unique gene family from either HomoloGene or Ensembl to nearly 60% of all entities in the text, thus linking the text-bound predictions with gene databases.

Finally, we use the resolution of entity occurrences to unique gene families to generalize the events in the text mining data, aggregating together event occurrences whose arguments are equivalent with respect to their gene family. Depending on whether HomoloGene or Ensembl is used for the gene family definition, this generalization process results in 500K-600K generalized events, which together aggregate over 11.2M (58.4%) of all event occurrences in the text mining data. Being able to link the literature-based events with well-defined gene families opens a number of interesting new use-cases for biomedical text mining, such as the ability to use the homology information to search for events relevant to newly discovered sequences. The remaining 41.6% of event occurrences not generalizable to gene families can still be retrieved through an additional generalization on the level of entity canonical forms.

All relevant data, namely all original events and entities together with their canonical forms, the generalizations of events based on canonical entity forms and gene families, as well as the gene symbol to unique family mapping are made publicly available as records in a MySQL database. We also provide detailed online documentation of the database scheme and example queries. Finally, we release the affix lists used in the canonicalization algorithm.

We believe this resource to be very valuable for explorative analysis of text mining results and homology-based hypothesis generation, as well as for supporting future research on data integration and biomedical text mining.

One important future work direction is a further disambiguation of canonical gene symbols to unique gene identifiers rather than entire gene families, which would allow for more fine-grained event generalization. There is an ongoing active, community-wide research focusing on this challenge and the current resource could be integrated as an additional source of information. Another future work direction is to create a visualization method and a web interface which would allow simple, user-friendly access to the data for researchers outside of the BioNLP research community itself.

Acknowledgments

The authors would like to thank Sampo Pyysalo (University of Tokyo) and Jari Björne (University of Turku) for their contribution. SVL would like to thank the Research Foundation Flanders (FWO) for funding her research and a travel grant to Turku. This work was partly funded by the Academy of Finland and the computational resources were provided by CSC IT Center for Science Ltd., Espoo, Finland.

References

Stephen F. Altschul, Thomas L. Madden, Alejandro A. Schäffer, Jinghui Zhang, Zheng Zhang, Webb Miller, and David J. Lipman. 1997. Gapped BLAST and PSI-BLAST: a new generation of protein database search programs. *Nucleic acids research*, 25(17):3389–3402, September.

Jari Björne, Juho Heimonen, Filip Ginter, Antti Airola, Tapio Pahikkala, and Tapio Salakoski. 2009. Extracting complex biological events with rich graph-based feature sets. In *BioNLP '09: Proceedings of the Workshop on BioNLP*, pages 10–18, Morristown, NJ, USA. Association for Computational Linguistics.

Jari Björne, Filip Ginter, Sampo Pyysalo, Jun'ichi Tsujii, and Tapio Salakoski. 2010. Scaling up biomedical event extraction to the entire PubMed. In *Proceedings of the 2010 Workshop on Biomedical Natural Language Processing*, BioNLP '10, pages 28–36, Stroudsburg, PA, USA. Association for Computational Linguistics.

Lifeng Chen, Hongfang Liu, and Carol Friedman. 2005. Gene name ambiguity of eukaryotic nomenclatures. *Bioinformatics*, 21:248–256, January.

Paul Flicek, M. Ridwan Amode, Daniel Barrell, Kathryn Beal, Simon Brent, Yuan Chen, Peter Clapham, Guy Coates, Susan Fairley, Stephen Fitzgerald, Leo Gordon, Maurice Hendrix, Thibaut Hourlier, Nathan Johnson, Andreas Kähäri, Damian Keefe, Stephen Keenan, Rhoda Kinsella, Felix Kokocinski, Eugene Kulesha, Pontus Larsson, Ian Longden, William McLaren, Bert Overduin, Bethan Pritchard, Harpreet Singh S. Riat, Daniel Rios, Graham R. Ritchie, Magali Ruffier, Michael Schuster, Daniel Sobral, Giulietta Spudich, Y. Amy Tang, Stephen Trevanion, Jana Vandrovcova, Albert J. Vilella, Simon White, Steven P. Wilder, Amonida Zadissa, Jorge Zamora, Bronwen L. Aken, Ewan Birney, Fiona Cunningham, Ian Dunham, Richard Durbin, Xosé M. Fernández-Suarez, Javier Herrero, Tim J. Hubbard, Anne Parker, Glenn Proctor, Jan Vogel, and Stephen M. Searle. 2011. Ensembl 2011. *Nucleic acids research*, 39(Database issue), January.

Theresa M. Fulton, Rutger Van der Hoeven, Nancy T. Eannetta, and Steven D. Tanksley. 2002. Identification, analysis, and utilization of conserved ortholog set markers for comparative genomics in higher plants. *Plant Cell*, 14(5):1457–1467.

Martin Gerner, Goran Nenadic, and Casey M. Bergman. 2010. LINNAEUS: a species name identification system for biomedical literature. *BMC bioinformatics*, 11(1):85+, February.

Robert Hoffmann and Alfonso Valencia. 2004. A gene network for navigating the literature. *Nat Genet*, 36(7):664, Jul.

Tim Hulsen, Martijn Huynen, Jacob de Vlieg, and Peter Groenen. 2006. Benchmarking ortholog identification methods using functional genomics data. *Genome Biology*, 7(4):R31+, April.

Jin-Dong Kim, Tomoko Ohta, Sampo Pyysalo, Yoshinobu Kano, and Jun'ichi Tsujii. 2009. Overview of BioNLP'09 shared task on event extraction. In *Proceedings of the BioNLP 2009 Workshop Companion Volume for Shared Task*, pages 1–9, Boulder, Colorado, June. Association for Computational Linguistics.

Martin Krallinger and Lynette Hirschman, editors. 2007. *Proceedings of the Second BioCreative Challenge Evaluation Workshop*, Madrid, April.

Robert Leaman and Graciela Gonzalez. 2008. BANNER: an executable survey of advances in biomedical named entity recognition. *Pacific Symposium on Biocomputing. Pacific Symposium on Biocomputing*, pages 652–663.

Yaniv Loewenstein, Domenico Raimondo, Oliver C. Redfern, James Watson, Dmitrij Frishman, Michal Linial, Christine Orengo, Janet Thornton, and Anna Tramontano. 2009. Protein function annotation by homology-based inference. *Genome biology*, 10(2):207, February.

Michael L. Metzker. 2010. Sequencing technologies - the next generation. *Nature reviews. Genetics*, 11(1):31–46, January.

Tomoko Ohta, Yusuke Miyao, Takashi Ninomiya, Yoshimasa Tsuruoka, Akane Yakushiji, Katsuya Masuda, Jumpei Takeuchi, Kazuhiro Yoshida, Tadayoshi Hara, Jin-Dong Kim, Yuka Tateisi, and Jun'ichi Tsujii. 2006. An intelligent search engine and GUI-based efficient MEDLINE search tool based on deep syntactic parsing. In *Proceedings of the COLING/ACL 2006 Interactive Presentation Sessions*, pages 17–20, Sydney, Australia, July. Association for Computational Linguistics.

Tomoko Ohta, Sampo Pyysalo, Kim Jin-Dong, and Jun'ichi Tsujii. 2009. A re-evaluation of biomedical named entity - term relations. In *Proceedings of LBM'09*.

Naoaki Okazaki and Jun'ichi Tsujii. 2010. Simple and efficient algorithm for approximate dictionary matching. In *Proceedings of the 23rd International Conference on Computational Linguistics (Coling 2010)*, pages 851–859, Beijing, China, August.

Sebastian Proost, Michiel Van Bel, Lieven Sterck, Kenny Billiau, Thomas Van Parys, Yves Van de Peer, and Klaas Vandepoele. 2009. PLAZA: A comparative genomics resource to study gene and genome evolution in plants. *Plant Cell*, 21(12):3718–3731, December.

Eric W. W. Sayers, Tanya Barrett, Dennis A. A. Benson, Stephen H. H. Bryant, Kathi Canese, Vyacheslav Chetvernin, Deanna M. M. Church, Michael Dicuccio, Ron Edgar, Scott Federhen, Michael Feolo, Lewis Y. Y. Geer, Wolfgang Helmberg, Yuri Kapustin, David Landsman, David J. J. Lipman, Thomas L. L. Madden, Donna R. R. Maglott, Vadim Miller, Ilene Mizrachi, James Ostell, Kim D. D. Pruitt, Gregory D. D. Schuler, Edwin Sequeira, Stephen T. T. Sherry, Martin Shumway, Karl Sirotkin, Alexandre Souvorov, Grigory Starchenko, Tatiana A. A. Tatusova, Lukas Wagner, Eugene Yaschenko, and Jian Ye. 2009. Database resources of the National Center for Biotechnology Information. *Nucleic Acids Research*, 37(Database issue):D5–15, January.

Damian Szklarczyk, Andrea Franceschini, Michael Kuhn, Milan Simonovic, Alexander Roth, Pablo Minguez, Tobias Doerks, Manuel Stark, Jean Muller, Peer Bork, Lars J. Jensen, and Christian von Mering. 2011. The STRING database in 2011: functional interaction networks of proteins, globally integrated and scored. *Nucleic acids research*, 39(Database issue):D561–D568, January.

Lorraine Tanabe, Natalie Xie, Lynne H. Thom, Wayne Matten, and W. John Wilbur. 2005. GENETAG: a tagged corpus for gene/protein named entity recognition. *BMC bioinformatics*, 6 Suppl 1.

Joachim Wermter, Katrin Tomanek, and Udo Hahn. 2009. High-performance gene name normalization with GENO. *Bioinformatics*, 25(6):815–821.

Zhengdong Zhang, Paula E. Burch, Austin J. Cooney, Rainer B. Lanz, Fred A. Pereira, Jiaqian Wu, Richard A. Gibbs, George Weinstock, and David A. Wheeler. 2004. Genomic analysis of the nuclear receptor family: New insights into structure, regulation, and evolution from the rat genome. *Genome Research*, 14(4):580–590, April.

Fast and simple semantic class assignment for biomedical text

K. Bretonnel Cohen
Computational Bioscience Program
U. Colorado School of Medicine
and
Department of Linguistics
U. of Colorado at Boulder
kevin.cohen@gmail.com

Tom Christiansen
Comput. Bioscience Prog.
U. Colorado Sch. of Medicine
tchrist@perl.com

William A. Baumgartner Jr.
Computational Bioscience Program
U. Colorado School of Medicine
william.baumgartner@ucdenver.edu

Karin Verspoor
Computational Bioscience Program
U. Colorado School of Medicine
karin.verspoor@ucdenver.edu

Lawrence E. Hunter
Computational Bioscience Program
U. Colorado School of Medicine
larry.hunter@ucdenver.edu

Abstract

A simple and accurate method for assigning broad semantic classes to text strings is presented. The method is to map text strings to terms in ontologies based on a pipeline of exact matches, normalized strings, headword matching, and stemming headwords. The results of three experiments evaluating the technique are given. Five semantic classes are evaluated against the CRAFT corpus of full-text journal articles. Twenty semantic classes are evaluated against the corresponding full ontologies, i.e. by reflexive matching. One semantic class is evaluated against a structured test suite. Precision, recall, and F-measure on the corpus when evaluating against only the ontologies in the corpus is micro-averaged 67.06/78.49/72.32 and macro-averaged 69.84/83.12/75.31. Accuracy on the corpus when evaluating against all twenty semantic classes ranges from 77.12% to 95.73%. Reflexive matching is generally successful, but reveals a small number of errors in the implementation. Evaluation with the structured test suite reveals a number of characteristics of the performance of the approach.

1 Introduction

Broad semantic class assignment is useful for a number of language processing tasks, including coreference resolution (Hobbs, 1978), document classification (Caporaso et al., 2005), and information extraction (Baumgartner Jr. et al., 2008). A limited number of semantic classes have been studied extensively, such as assigning text strings to the category gene or protein (Yeh et al., 2005; Smith et al., 2008), or the PERSON, ORGANIZATION, and LOCATION categories introduced in the Message Understanding Conferences (Chinchor, 1998). A larger number of semantic classes have received smaller amounts of attention, e.g. the classes in the GENIA ontology (Kim et al., 2004), various event types derived from the Gene Ontology (Kim et al., 2009), and diseases (Leaman and Gonzalez, 2008). However, many semantic types have not been studied at all. In addition, where ontologies are concerned, although there has been work on finding mentions or evidence of specific terms in text (Blaschke et al., 2005; Stoica and Hearst, 2006; Davis et al., 2006; Shah et al., 2009), there has been no work specifically addressing assigning multiple very broad semantic classes with potential overlap. In particular, this paper examines the problem of taking a set of ontologies and a text string (typically, but not necessarily, a noun phrase) as input and determining which ontology defines the semantic class that that text string refers to. We make an equivalence here between the notion of belonging to the domain of an ontology and belonging to a specific semantic class. For example, if a string in text refers to something in the domain of the Gene Ontology, we take it as belonging to a Gene Ontology semantic class (using the name of the ontology only for convenience); if a string in text refers to something belonging to the domain of the Sequence Ontology, we take it as belonging to a Sequence Ontology semantic class. We focus especially on rapid, simple methods for making such a determination.

The problem is most closely related to multi-class

Proceedings of the 2011 Workshop on Biomedical Natural Language Processing, ACL-HLT 2011, pages 38–45,
Portland, Oregon, USA, June 23-24, 2011. ©2011 Association for Computational Linguistics

classification, where in the case of this study we are including an unusually large number of categories, with possible overlap between them. A text string might refer to something that legitimately belongs to the domain of more than one ontology. For example, it might belong to the semantic classes of both the Gene Ontology and the Gene Regulation Ontology; regulation is an important and frequent concept in the Gene Ontology. This fact has consequences for defining the notion of a false positive class assignment; we return to this issue in the *Results* section.

2 Methods

2.1 Target semantic classes

The following ontologies were used to define semantic classes:

- Gene Ontology
- Sequence Ontology
- Foundational Model of Anatomy
- NCBI Taxonomy
- Chemical Entities of Biological Interest
- Phenotypic Quality
- BRENDA Tissue/Enzyme Source
- Cell Type Ontology
- Gene Regulation Ontology
- Homology Ontology
- Human Disease Ontology
- Human Phenotype Ontology
- Mammalian Phenotype Ontology
- Molecule Role Ontology
- Mouse Adult Gross Anatomy Ontology
- Mouse Pathology Ontology
- Protein Modification Ontology
- Protein-Protein Interaction Ontology
- Sample Processing and Separation Techniques Ontology
- Suggested Ontology for Pharmacogenomics

2.2 Methodology for assigning semantic class

We applied four simple techniques for attempting to match a text string to an ontology. They are arranged in order of decreasing stringency. That is, each subsequent method has looser requirements for a match. This both allows us to evaluate the contribution of each component more easily and, at run time, allows the user to set a stringency level, if the default is not desired.

2.2.1 Exact match

The first and most stringent technique is exact match. (This is essentially the only technique used by the NCBO (National Center for Biomedical Ontology) Annotator (Jonquet et al., 2009), although it can also do substring matching.) We normalize terms in the ontology and text strings in the input for case and look for a match.

2.2.2 Stripping

All non-alphanumeric characters, including whitespace, are deleted from the terms in the ontology and from text strings in the input (e.g. *cadmium-binding* and *cadmium binding* both become *cadmiumbinding*) and look for a match.

2.2.3 Head nouns

This method involves a lightweight linguistic analysis. We traversed each ontology and determined the head noun (see method below) of each term and synonym in the ontology. We then prepared a dictionary mapping from head nouns to lists of ontologies in which those head nouns were found.

Head nouns were determined by two simple heuristics (cf. (Collins, 1999)). For terms fitting the pattern *X of...* (where *of* represents any preposition) the term *X* was taken as the head noun. For all other terms, the rightmost word was taken as the head noun. These two heuristics were applied in sequence when applicable, so that for example *positive regulation of growth* (GO:0045927) becomes *positive regulation* by application of the first heuristic and *regulation* by application of the second heuristic. In the case of some ontologies, very limited pre-processing was necessary—for example, it was necessary to delete double quotes that appeared around synonyms, and in some ontologies we had to delete strings like *[EXACT SYNONYM]* from some terms before extracting the head noun.

2.2.4 Stemming head nouns

In this technique, the headwords obtained by the previous step were stemmed with the Porter stemmer.

2.3 Corpus and other materials

We made use of three sources in our evaluation. One is the CRAFT (Colorado Richly Annotated Full Text) corpus (Verspoor et al., 2009; Cohen et al., 2010a). This is a collection of 97 full-text journal articles, comprising about 597,000 words, each of which has been used as evidence for at least one annotation by the Mouse Genome Informatics group. It has been annotated with a number of ontologies and database identifiers, including:

- Gene Ontology
- Sequence Ontology
- Cell Type Ontology
- NCBI Taxonomy
- Chemical Entities of Biological Interest (ChEBI)

In total, there are over 119,783 annotations. (For the breakdown across semantic categories, see Table 1.) All of these annotations were done by biological scientists and have been double-annotated with inter-annotator agreement in the nineties for most categories.

The second source is the full sets of terms from the twenty ontologies listed in the Introduction. All of the twenty ontologies that we used were obtained from the OBO portal. Version numbers are omitted here due to space limitations, but are available from the authors on request.

The third source is a structured test suite based on the Gene Ontology (Cohen et al., 2010b). Structured test suites are developed to test the performance of a system on specific categories of input types. This test set was especially designed to test difficult cases that do not correspond to exact matches of Gene Ontology terms, as well as the full range of types of terms. The test suite includes 300 concepts from GO, as well as a number of transformations of their terms, such as *cells migrated* derived from the term *cell migration* and *migration of cells* derived from *cell migration*, classified according to a number of linguistic attributes, such as length, whether or not punctuation is included in the term, whether or not it includes function (stop) words, etc. This test suite determines at least one semantic category that should be returned for each term. Unlike using the entire ontologies, this evaluation method made detailed error analysis possible. This test suite has been used by other groups for broad characterizations of successes and failures of concept recognizers, and to tune the parameters of concept recognition systems.

2.4 Evaluation

We did three separate evaluations. In one, we compared the output of our system against manually-generated gold-standard annotations in the CRAFT corpus (op. cit.). This was possible only for the ontologies that have been annotated in CRAFT, which are listed above.

In the second evaluation, we used the entire ontologies themselves as inputs. In this method, all responses should be the same—for example, every term from the Gene Ontology should be classified as belonging to the GO semantic class.

In the third, we utilized the structured test suite described above.

2.4.1 Baselines

Two baselines are possible, but neither is optimal. The first would be to use MetaMap (Aronson, 2001), the industry standard for semantic category assignment. (Note that MetaMap assigns specific categories, not broad ones.) However, MetaMap outputs only semantic classes that are elements of the UMLS, which of the ontologies that we looked at, includes only the Gene Ontology. The other is the NCBO Annotator. The NCBO Annotator detects only exact matches (or substring matches) to ontology terms, so it is not clear that it is a strong enough baseline to allow for a stringent analysis of our approach.

3 Results

We present our results in three sections:

- For the CRAFT corpus
- For the ontologies themselves
- For the Gene Ontology test suite

3.1 Corpus results

Table 1 (see next page) shows the results on the CRAFT corpus if only the five ontologies that were actually annotated in CRAFT are used as inputs. The results are given for stemmed heads. Performance on the four techniques that make up the approach is cumulative, and results for stemmed heads reflects the application of all four techniques. In this case, where we evaluate against the corpus, it is possible to determine false positives, so we can give precision, recall, and F-measures for each semantic class, as well as for the corpus as a whole. Micro-averaged results were 67.06 precision, 78.49 recall, and 72.32 F-measure. Macro-averaged results were 69.84 precision, 83.12 recall, and 75.31 F-measure.

Table 2 (see next page) shows the results for the CRAFT corpus when all twenty ontologies are matched against the corpus data, including the many ontologies that are not annotated in the data. We give results for just the five annotated ontologies below. Rather than calculating precision, recall, and F-measure, we calculate only accuracy. This is because when classes other than the gold standard class is returned, we have no way of knowing if they are incorrect without manually examining them—that is, we have no way to identify false positives. If the set of classes returned included the gold standard class, a correct answer was counted. If the classifier returned zero or more classes and none of them was the gold standard, an incorrect answer was counted. Results are given separately for each of the four techniques. This allows us to evaluate the contribution of each technique to the overall results; the value in each column is cumulative, so the value for *Stemmed head* includes the contribution of all four of the techniques that make up the general approach. Accuracies of 77.12% to 95.73% were achieved, depending on the ontology. We see that the linguistic technique of locating the head noun makes a contribution to all categories, but makes an especially strong contribution to the Gene Ontology and Cell Type Ontology classes. Stemming of headwords is also effective for all five categories. We see that exact match is effective only for those semantic classes for which terminology is relatively fixed, i.e. the NCBI taxonomy and chemical names. In some of the others, matching natural language text is very difficult by any technique. For example, of the 8,665 Sequence Ontology false negatives in the data reflected in the P/R/F values in Table 1, a full 2,050 are due to the single character +, which does not appear in any of the twenty ontologies that we examined and that was marked by the annotators as a Sequence Ontology term, *wild_type* (SO:0000817).

3.2 Ontology results

As the second form of evaluation, we used the terms from the ontologies themselves as the inputs to which we attempted to assign a semantic class. In this case, no annotation is required, and it is straightforwardly the case that each term in a given ontology should be assigned the semantic class of that ontology. We used only the head noun technique. We did *not* use the exact match or stripping heuristics, since they are guaranteed to return the correct answer, nor did we use stemming. Thus, this section of the evaluation gives us a good indication of the performance of the head noun approach.

As might be expected, almost all twenty ontologies returned results in the 97-100% correct rate. However, we noted much lower performance in two ontologies, the Sequence Ontology and the Molecule Role Ontology. This lower performance reflects a number of preprocessing errors or omissions. The fact that we were able to detect these low-performing ontologies indicates that our evaluation technique in this experiment—trying to match terms from an ontology against that ontology itself—is a robust evaluation technique and should be used in similar studies.

3.2.1 Structured test suite results

The third approach to evaluation involved use of the structured test suite. The structured test suite revealed a number of trends in the performance of the system.

Ontology	Annotations	Precision	Recall	F-measure
Gene Ontology	39,626	66.31	73.06	69.52
Sequence Ontology	40,692	63.00	72.21	67.29
Cell Type Ontology	8,383	53.58	87.27	66.40
NCBI Taxonomy	11,775	96.24	92.51	94.34
ChEBI	19,307	70.07	90.53	79.00
Total (micro-averaged)	119,783	67.06	78.49	72.32
Total (macro-averaged)		69.84	83.12	75.31

Table 1: Results on the CRAFT corpus when only the CRAFT ontologies are used as input. Results are for stemmed heads. Precision, recall, and F-measure are given for each semantic category in the corpus. *Totals* are micro-averaged (over all tokens) and macro-averaged (over all categories), respectively. P/R/F are cumulative, so that the results for stemmed heads reflect the application of all four techniques.

Ontology	Exact	Stripped	Head noun	Stemmed head
Gene Ontology	24.26	24.68	59.18	77.12
Sequence Ontology	44.28	47.63	56.63	73.33
Cell Type Ontology	25.26	25.80	70.09	88.38
NCBI Taxonomy	84.67	84.71	90.97	95.73
ChEBI	86.93	87.44	92.43	95.49

Table 2: Results on the CRAFT corpus when all twenty ontologies are used as input. Accuracy is given for each technique. Accuracy is cumulative, so that accuracy in the final column reflects the application of all four techniques.

- The headword technique works very well for recognizing syntactic variants. For example, if the GO term *induction of apoptosis* is written as *apoptosis induction*, the headword technique allows it to be picked up.

- The headword technique works in situations where text has been inserted into a term. For example, if the GO term *ensheathment of neurons* appears as *ensheathment of some neurons*, the headword technique will allow it to be picked up. If the GO term *regulation of growth* shows up as *regulation of vascular growth*, the headword technique will allow it to be picked up.

- The headword stemming technique allows us to pick up many verb phrases, which is important for event detection and event coreference. For example, if the GO term *cell migration* appears in text as *cells migrate*, the technique will detect it. The test suite also showed that failures to recognize verb phrases still occur when the morphological relationship between the nominal term and the verb are irregular, as for example between the GO term *growth* and the verb *grows*.

- The technique's ability to handle coordination is very dependent on the type of coordination. For example, simple coordination (e.g. *cell migration and proliferation*) is handled well, but complex coordination (e.g. *cell migration, proliferation and adhesion*) is handled poorly.

- Stemming is necessary for recognition of plurals, regardless of the length of the term in words.

- The approach currently fails on irregular plurals, due to failure of the Porter stemmer to handle plurals like *nuclei* and *nucleoli* well.

- The approach handles classification of terms that others have characterized as "ungrammatical," such as *transposition, DNA-mediated* (GO:0006313). This is important, because exact matches will always fail on these terms.

4 Discussion

4.1 Related work

We are not aware of similar work that tries to assign a large set of broad semantic categories to individual text strings. There is a body of work on selecting a single ontology for a domain or text. (Martínez-Romero et al., 2010) proposes a method for selecting an ontology given a list of terms, all of which must appear in the ontology. (Jonquet et al., 2009) describes an ontology recommender that first annotates terms in a text with the Open Biomedical Annotator service, then uses the sum of the scores of the individual annotations to recommend a single ontology for the domain as a whole.

4.2 Possible alternate approaches

Three possible alternative approaches exist, all of which would have as their goal the returning of a single best semantic class for every input. However, for the use cases that we have identified—coreference resolution, document classification, information extraction, and curator assistance—we are more interested in wide coverage of a broad range of semantic classes, so these approaches are not evaluated here. However, we describe them for completeness and for the use of researchers who might be interested in pursuing single-class assignment.

4.2.1 Frequent words

One alternative approach would be to use simple word frequencies. For example, for each ontology, one could determine the *N* most frequent words, filtering out stop words. At run time, check the words in each noun phrase in the text against the lists of frequent words. For every word from the text that appeared in the list of frequent words from some ontology, assign a score to each ontology in which it was found, weighting it according to its position in the list of frequent words. In theory, this could accommodate for the non-uniqueness of word-to-ontology mappings, i.e. the fact that a single word might appear in the lists for multiple ontologies. However, we found the technique to perform very poorly for differentiating between ontologies and do not recommend it.

4.2.2 Measuring informativeness

If the system is desired to return only one single semantic class per text string, then one approach would be to determine the informativeness of each word in each ontology. That is, we want to find the maximal probability of an ontology given a word from that ontology. This approach is very difficult to normalize for the wide variability in size of the many ontologies that we wanted to be able to deal with.

4.2.3 Combining scores

Finally, one could conceivably combine scores for matches obtained by the different strategies, weighting them according to their stringency, i.e. exact match receiving a higher weight than head noun match, which in turn would receive a higher weight than stemmed head noun match. This weighting might also include informativeness, as described above.

4.3 Why the linguistic method works

As pointed out above, the lightweight linguistic method makes a large contribution to the performance of the approach for some ontologies, particularly those for which the exact match and stripping techniques do not perform well. It works for two reasons, one related to the approach itself and one related to the nature of the OBO ontologies. From a methodological perspective, the approach is effective because headwords are a good reflection of the semantic content of the noun phrase and they are relatively easy to access via simple heuristics. Of course simple heuristics will fail, as we can observe most obviously in the cases where we failed to identify members of the ontologies in the second evaluation step. However, overall the approach works well enough to constitute a viable tool for coreference systems and other applications that benefit from the ability to assign broad semantic classes to text strings.

The approach is also able to succeed because of the nature of the OBO ontologies. OBO ontologies are meant to be orthogonal (Smith et al., 2007). A distributional analysis of the distribution of terms and words between the ontologies (data not shown here, although some of it is discussed below), as well as the false positives found in the corpus study, suggests that orthogonality between the OBO ontologies is by no means complete. However, it holds often enough for the headword method to be effective.

4.4 Additional error analysis

In the section on the results for the structured test suite, we give a number of observations on contributions to errors, primarily related either to the characteristics of individual words or to particular syntactic instantiations of terms. Here, we discuss some aspects of the distribution of lexical items and of the corpus that contributed to errors.

- The ten most common headwords appear in from 6-16 of the twenty ontologies. However, they typically appear in one ontology at a frequency many orders of magnitude greater than their frequency in the other ontologies. Taking this frequency data into account for just these ten headwords would likely decrease false positives quite significantly.

- More than 50% of Gene Ontology terms share one of only ten headwords. Many of our Gene Ontology false negatives on the corpus are because the annotated text string does not contain a word such as *process* or *complex* that is the head word of the canonical term.

4.5 Future work

The heuristics that we implemented for extracting headwords from OBO terms were very simple, in keeping with our initial goal of developing an easy, fast method for semantic class assignment. However, it is clear that we could achieve substantial performance improvements from improving the heuristics. We may pursue this track, if it becomes clear that coreference performance would benefit from this when we incorporate the semantic classification approach into a coreference system.

On acceptance of the paper, we will make Perl and Java versions of the semantic class assigner publicly available on SourceForge.

4.6 Conclusion

The goal of this paper was to develop a simple approach to assigning text strings to an unprecedentedly large range of semantic classes, where membership in a semantic class is equated with belonging to the semantic domain of a specific ontology. The approach described in this paper is able to do that at a micro-averaged F-measure of 72.32 and macro-averaged F-measure of 75.31 as evaluated on a manually annotated corpus where false positives can be determined, and with an accuracy of 77.12-95.73% when only true positives and false negatives can be determined.

References

A. Aronson. 2001. Effective mapping of biomedical text to the UMLS Metathesaurus: The MetaMap program. In *Proc AMIA 2001*, pages 17–21.

William A. Baumgartner Jr., Zhiyong Lu, Helen L. Johnson, J. Gregory Caporaso, Jesse Paquette, Anna Lindemann, Elizabeth K. White, Olga Medvedeva, K. Bretonnel Cohen, and Lawrence Hunter. 2008. Concept recognition for extracting protein interaction relations from biomedical text. *Genome Biology*, 9.

Christian Blaschke, Eduardo A. Leon, Martin Krallinger, and Alfonso Valencia. 2005. Evaluation of BioCreative assessment of task 2. *BMC Bioinformatics*, 6 Suppl 1.

J. Gregory Caporaso, William A. Baumgartner Jr.., K. Bretonnel Cohen, Helen L. Johnson, Jesse Paquette, and Lawrence Hunter. 2005. Concept recognition and the TREC Genomics tasks. In *The Fourteenth Text REtrieval Conference (TREC 2005) Proceedings*.

Nancy A. Chinchor. 1998. Overview of MUC-7/MET-2.

K. Bretonnel Cohen, Helen L. Johnson, Karin Verspoor, Christophe Roeder, and Lawrence E. Hunter. 2010a. The structural and content aspects of abstracts versus bodies of full text journal articles are different. *BMC Bioinformatics*, 11(492).

K. Bretonnel Cohen, Christophe Roeder, William A. Baumgartner Jr., Lawrence Hunter, and Karin Verspoor. 2010b. Test suite design for biomedical ontology concept recognition systems. In *Proceedings of the Language Resources and Evaluation Conference*.

Michael Collins. 1999. *Head-driven statistical models for natural language parsing*. Ph.D. thesis, University of Pennsylvania.

N. Davis, H. Harkema, R. Gaizauskas, Y. K. Guo, M. Ghanem, T. Barnwell, Y. Guo, and J. Ratcliffe. 2006. Three approaches to GO-tagging biomedical abstracts. In *Proceedings of the Second International Symposium on Semantic Mining in Biomedicine*, pages 21–28, Jena, Germany.

Jerry R. Hobbs. 1978. Resolving pronoun references. *Lingua*, 44:311–338.

C. Jonquet, N.H. Shah, and M.A. Musen. 2009. Prototyping a biomedical ontology recommender service. In *Bio-Ontologies: Knowledge in Biology, ISMB/ECCB SIG*.

Jin-Dong Kim, Tomoko Ohta, Yoshimasa Tsuruoka, Yuka Tateisi, and Nigel Collier. 2004. Introduction to the bio-entity recognition task at JNLPBA. In *Proceedings of the international joint workshop on natural language processing in biomedicine and its applications*, pages 70–75.

Jin-Dong Kim, Tomoko Ohta, Sampo Pyysalo, Yoshinobu Kano, and Jun'ichi Tsujii. 2009. Overview of BioNLP'09 shared task on event extraction. In *BioNLP 2009 Companion Volume: Shared Task on Entity Extraction*, pages 1–9.

Robert Leaman and Graciela Gonzalez. 2008. BANNER: An executable survey of advances in biomedical named entity recognition. In *Pac Symp Biocomput*.

Marcos Martínez-Romero, José Vázquez-Naya, Cristian R. Munteanu, Javier Pereira, and Alejandro Pazos. 2010. An approach for the automatic recommendation of ontologies using collaborative knowledge. In *KES 2010, Part II, LNAI 6277*, pages 74–81.

Nigam H. Shah, Nipun Bhatia, Clement Jonquet, Daniel Rubin, Annie P. Chiang, and Mark A. Musen. 2009. Comparison of concept recognizers for building the Open Biomedical Annotator. *BMC Bioinformatics*, 10.

Barry Smith, Michael Ashburner, Cornelius Rosse, Jonathan Bard, William Bug, Werner Ceusters, Louis J. Goldberg, Karen Eilbeck, Amelia Ireland, Christopher J. Mungall, The OBI Consortium, Neocles Leontis, Philippe Rocca-Serra, Alan Ruttenberg, Susanna-Assunta Sansone, Richard H. Scheuermann, Nigam Shah, Patricia L. Whetzel, and Suzanna Lewis. 2007. The OBO Foundry: coordinated evolution of ontologies to support biomedical data integration. *Nature Biotechnology*, 25:1251–1255.

Larry Smith, Lorraine Tanabe, Rie Johnson nee Ando, Cheng-Ju Kuo, I-Fang Chung, Chun-Nan Hsu, Yu-Shi Lin, Roman Klinger, Christof Friedrich, Kuzman Ganchev, Manabu Torii, Hongfang Liu, Barry Haddow, Craig Struble, Richard Povinelli, Andreas Vlachos, William Baumgartner, Jr., Lawrence Hunter, Bob Carpenter, Richard Tzong-Han Tsai, Hong-Jie Dai, Feng Liu, Yifei Chen, Chengjie Sun, Sophia Katrenko, Pieter Adriaans, Christian Blaschke, Rafael Torres Perez, Mariana Neves, Preslav Nakov, Anna Divoli, Manuel Mana, Jacinto Mata-Vazquez, and W. John Wilbur. 2008. Overview of BioCreative II gene mention recognition. *Genome Biology*.

E. Stoica and M. Hearst. 2006. Predicting gene functions from text using a cross-species approach. In *Proceedings of the 11th Pacific Symposium on Biocomputing*.

Karin Verspoor, K. Bretonnel Cohen, and Lawrence Hunter. 2009. The textual characteristics of traditional and Open Access scientific journals are similar. *BMC Bioinformatics*, 10.

A. Yeh, A. Morgan, M. Colosimo, and L. Hirschman. 2005. BioCreatve task 1A: gene mention finding evaluation. *BMC Bioinformatics*, 6(Suppl. 1).

The Role of Information Extraction in the Design of a Document Triage Application for Biocuration

Sandeep Pokkunuri
School of Computing
University of Utah
Salt Lake City, UT
sandeepp@cs.utah.edu

Cartic Ramakrishnan
Information Sciences Institute
Univ. of Southern California
Marina del Rey, CA
cartic@isi.edu

Ellen Riloff
School of Computing
University of Utah
Salt Lake City, UT
riloff@cs.utah.edu

Eduard Hovy
Information Sciences Institute
Univ. of Southern California
Marina del Rey, CA
hovy@isi.edu

Gully APC Burns
Information Sciences Institute
Univ. of Southern California
Marina del Rey, CA
burns@isi.edu

Abstract

Traditionally, automated triage of papers is performed using lexical (unigram, bigram, and sometimes trigram) features. This paper explores the use of information extraction (IE) techniques to create richer linguistic features than traditional bag-of-words models. Our classifier includes lexico-syntactic patterns and more-complex features that represent a pattern coupled with its extracted noun, represented both as a lexical term and as a semantic category. Our experimental results show that the IE-based features can improve performance over unigram and bigram features alone. We present intrinsic evaluation results of full-text document classification experiments to determine automatically whether a paper should be considered of interest to biologists at the Mouse Genome Informatics (MGI) system at the Jackson Laboratories. We also further discuss issues relating to design and deployment of our classifiers as an application to support scientific knowledge curation at MGI.

1 Introduction

A long-standing promise of Biomedical Natural Language Processing is to accelerate the process of literature-based 'biocuration', where published information must be carefully and appropriately translated into the knowledge architecture of a biomedical database. Typically, biocuration is a manual activity, performed by specialists with expertise in both biomedicine and the computational representation of the target database. It is widely acknowledged as a vital lynch-pin of biomedical informatics (Bourne and McEntyre, 2006).

A key step in biocuration is the initial triage of documents in order to direct to specialists only the documents appropriate for them. This classification (Cohen and Hersh, 2006)(Hersh W, 2005) can be followed by a step in which desired information is extracted and appropriately standardized and formalized for entry into the database. Both these steps can be enhanced by suitably powerful Natural Language Processing (NLP) technology. In this paper, we address text mining as a step within the broader context of developing both infrastructure and tools for biocuration support within the Mouse Genome Informatics (MGI) system at the Jackson Laboratories. We previously identified 'document triage' as a crucial bottleneck (Ramakrishnan et al., 2010) within MGI's biocuration workflow.

Our research explores the use of information extraction (IE) techniques to create richer linguistic features than traditional bag-of-words models. These features are employed by a classifier to perform the triage step. The features include lexico-syntactic patterns as well as more-complex features, such as a pattern coupled with its extracted noun, where the noun is represented both as a lexical term and by its semantic category. Our experimental results show that the IE-based enhanced features can improve performance over unigram and bigram features alone.

Proceedings of the 2011 Workshop on Biomedical Natural Language Processing, ACL-HLT 2011, pages 46–55,
Portland, Oregon, USA, June 23-24, 2011.

Evaluating the performance of BioNLP tools is not trivial. So-called *intrinsic* metrics measure the performance of a tool against some gold standard of performance, while *extrinsic* ones (Alex et al., 2008) measure how much the overall biocuration process is benefited. Such metrics necessarily involve the deployment of the software in-house for testing by biocurators, and require a large-scale software-engineering infrastructure effort. In this paper, we present intrinsic evaluation results of full-text document classification experiments to determine automatically whether a paper should be considered of interest to MGI curators. We plan in-house deployment and extrinsic evaluation in near-term work.

Our work should be considered as the first step of a broader process within which (a) the features used in this particular classification approach will be re-engineered so that they may be dynamically recreated in any new domain by a reusable component, (b) this component is deployed into reusable infrastructure that also includes document-, annotation- and feature-storage capabilities that support scaling and reuse, and (c) the overall functionality can then be delivered as a software application to biocurators themselves for extrinsic evaluation in any domain they choose. Within the 'SciKnowMine' project, we are constructing such a framework (Ramakrishnan et al., 2010), and this work reported here forms a prototype component that we plan to incorporate into a live application. We describe the underlying NLP research here, and provide context for the work by describing the overall design and implementation of the SciKnowMine infrastructure.

1.1 Motivation

MGI's biocurators use very specific guidelines for triage that continuously evolve. These guidelines are tailored to specific subcategories within MGI's triage task (phenotype, Gene Ontology[1] (GO) term, gene expression, tumor biology and chromosomal location mapping). They help biocurators decide whether a paper is relevant to one or more subcategories. As an example, consider the guideline for the phenotype category shown in Table 1.

This example makes clear that it is not sufficient to match on relevant words like 'transgene' alone.

[1]http://www.geneontology.org/

'Select paper **If**: it is about transgenes where a gene from any species is inserted in mice **and** this results in a phenotype. **Except**: if the paper uses transgenes to examine promoter function'.

Table 1: Sample triage guideline used by MGI biocurators

To identify a paper as being *'within-scope'* or *'out-of-scope'* requires that a biocurator understand the context of the experiment described in the paper. To check this we examined two sample papers; one that matches the precondition of the above rule and another that matches its exception. The first paper (Sjögren et al., 2009) is about a transgene insertion causing a pheotype and is a positive example of the category phenotype, while the second paper (Bouatia-Naji et al., 2010) is about the use of transgenes to study promoter function and is a negative example for the same category.

Inspection of the negative-example paper illustrates the following issues concerning the language used: (1) This paper is about transgene-use in studying promoter function. Understanding this requires the following background knowledge: (a) the two genes mentioned in the title are transgenes; (b) the phrase 'elevation of fasting glucose levels' in the title represents an up-regulation phenotype event. (2) Note that the word 'transgene' never occurs in the entire negative-example paper. This suggests that recognizing that a paper involves the use of transgenes requires annotation of domain-specific entities and a richer representation than that offered by a simple bag-of-words model.

Similar inspection of the positive-example paper reveals that (3) the paper contains experimental evidence showing the phenotype resulting from the transgene insertion. (4) The 'Materials and Methods' section of the positive-example paper clearly identifies the construction of the transgene and the 'Results' section describes the development of the transgenic mouse model used in the study. (3) and (4) above suggest that domain knowledge about complex biological phenomena (events) such as phenotype and experimental protocol may be helpful for the triage task.

Together, points (1)–(4) suggest that different sections of a paper contain additional important context-specific clues. The example highlights the complex nature of the triage task facing the MGI biocurators. At present, this level of nuanced 'understanding' of content semantics is extremely hard for machines to replicate. Nonetheless, merely treating the papers as a bag-of-words is unlikely to make nuanced distinctions between positive and negative examples with the level of precision and recall required in MGI's triage task.

In this paper we therefore describe: (1) the design and performance of a classifier that is enriched with three types of features, all derived from information extraction: (a) lexico-syntactic patterns, (b) patterns coupled with lexical extractions, and (c) patterns coupled with semantic extractions. We compare the enriched classifier against classifiers that use only unigram and bigram features; (2) the design of a biocuration application for MGI along with the first prototype system where we emphasize the infrastructure necessary to support the engineering of domain-specific features of the kind described in the examples above. Our application is based on Unstructured Information Management Architecture (UIMA) (Ferrucci and Lally, 2004), which is a pipeline-based framework for the development of software systems that analyze large volumes of unstructured information.

2 Information Extraction for Triage Classification

In this section, we present the information extraction techniques that we used as the basis for our IE-based features, and we describe the three types of IE features that we incorporated into the triage classifier.

2.1 Information Extraction Techniques

Information extraction (IE) includes a variety of techniques for extracting factual information from text. We focus on pattern-based IE methods that were originally designed for event extraction. Event extraction systems identify the role fillers associated with events. For example, consider the task of extracting information from disease outbreak reports, such as ProMed-mail articles (http://www.promedmail.org/). In contrast to a named entity recognizer, which should identify all mentions of diseases and people, an event extraction system should only extract the diseases involved in an outbreak incident and the people who were the victims. Other mentions of diseases (*e.g.*, in historical discussions) or people (*e.g.*, doctors or scientists) should be discarded.

We utilized the Sundance/AutoSlog software package (Riloff and Phillips, 2004), which is freely available for research. Sundance is an information extraction engine that applies lexico-syntactic patterns to extract noun phrases from specific linguistic contexts. Sundance performs its own syntactic analysis, which includes morphological analysis, shallow parsing, clause segmentation, and syntactic role assignment (*i.e.*, identifying subjects and direct objects of verb phrases). Sundance labels verb phrases with respect to active/passive voice, which is important for event role labelling. For example, *"Tom Smith was diagnosed with bird flu"* means that Tom Smith is a victim, but *"Tom Smith diagnosed the elderly man with bird flu"* means that the elderly man is the victim.

Sundance's information extraction engine can apply lexico-syntactic patterns to extract noun phrases that participate in syntactic relations. Each pattern represents a linguistic expression, and extracts a noun phrase (NP) argument from one of three syntactic positions: Subject, Direct Object, or Prepositional Phrase. Patterns may be defined manually, or they can be generated by the AutoSlog pattern generator (Riloff, 1993), which automatically generates patterns from a domain-specific text corpus. AutoSlog uses 17 syntactic 'templates' that are matched against the text. Lexico-syntactic patterns are generated by instantiating the matching words in the text with the syntactic template. For example, five of AutoSlog's syntactic templates are shown in Table 2:

(a) <SUBJ> PassVP
(b) PassVP Prep <NP>
(c) <SUBJ> ActVP
(d) ActVP Prep <NP>
(e) Subject PassVP Prep <NP>

Table 2: Five example syntactic templates (PassVP means passive voice verb phrase, ActVP means active voice verb phrase)

Pattern (a) matches any verb phrase (VP) in a passive voice construction and extracts the Subject of the VP. Pattern (b) matches passive voice VPs that are followed by a prepositional phrase. The NP in the prepositional phrase is extracted. Pattern (c) matches any active voice VP and extracts its Subject, while Pattern (d) matches active voice VPs followed by a prepositional phrase. Pattern (e) is a more complex pattern that requires a specific Subject[2], passive voice VP, and a prepositional phrase. We applied the AutoSlog pattern generator to our corpus (described in Section 3.1) to exhaustively generate every pattern that occurs in the corpus.

As an example, consider the following sentence, taken from an article in PLoS Genetics:

> *USP14 is endogenously expressed in HEK293 cells and in kidney tissue derived from wt mice.*

<SUBJ> PassVP(expressed)
<SUBJ> ActVP(derived)
PassVP(expressed) Prep(in) <NP>
ActVP(derived) Prep(from) <NP>
Subject(USP14) PassVP(expressed) Prep(in) <NP>

Table 3: Lexico-syntactic patterns for the PLoS Genetics sentence shown above.

AutoSlog generates five patterns from this sentence, which are shown in Table 3:

The first pattern matches passive voice instances of the verb *'expressed'*, and the second pattern matches active voice instances of the verb *'derived'*.[3] These patterns rely on syntactic analysis, so they will match any syntactically appropriate construction. For example, the first pattern would match *'was expressed'*, *'were expressed'*, *'have been expressed'* and *'was very clearly expressed'*. The third and fourth patterns represent the same two VPs but also require the presence of a specific prepositional phrase. The prepositional phrase does not need to be adjacent to the VP, so long as it is attached to the VP syntactically. The last pattern is very specific and will only match passive voice instances of *'expressed'* that also have a Subject with a particular head noun (*'USP14'*) and an attached prepositional phrase with the preposition *'in'*.

The example sentence contains four noun phrases, which are underlined. When the patterns generated by AutoSlog are applied to the sentence, they produce the following NP extractions (shown in **boldface** in Table 4):

<**USP14**> PassVP(expressed)
<**kidney tissue**> ActVP(derived)
PassVP(expressed) Prep(in) <**HEK293 cells**>
ActVP(derived) Prep(from) <**wt mice**>
Subject(USP14) PassVP(expressed) Prep(in) <**HEK293 cells**>

Table 4: Noun phrase extractions produced by Sundance for the sample sentence.

In the next section, we explain how we use the information extraction system to produce rich linguistic features for our triage classifier.

2.2 IE Pattern Features

For the triage classification task, we experimented with four types of IE-based features: *Patterns*, *Lexical Extractions*, and *Semantic Extractions*.

The *Pattern* features are the lexico-syntactic IE patterns. Intuitively, each pattern represents a phrase or expression that could potentially capture contexts associated with mouse genomics better than isolated words (unigrams). We ran the AutoSlog pattern generator over the training set to exhaustively generate every pattern that appeared in the corpus. We then defined one feature for each pattern and gave it a binary feature value (*i.e.*, 1 if the pattern occurred anywhere in the document, 0 otherwise).

We also created features that capture not just the pattern expression, but also its argument. The *Lexical Extraction* features represent a pattern paired with the head noun of its extracted noun phrase. Table 5 shows the Lexical Extraction features that would be generated for the sample sentence shown earlier. Our hypothesis was that these features could help to distinguish between contexts where an activity is relevant (or irrelevant) to MGI because of the combination of an activity and its argument.

The Lexical Extraction features are very specific, requiring the presence of multiple terms. So we

[2] Only the head nouns must match.

[3] Actually, the second clause is in reduced passive voice (*i.e.*, tissue that was derived from mice), but the parser misidentifies it as an active voice construction.

PassVP(expressed), **USP14**
ActVP(derived), **tissue**
PassVP(expressed) Prep(in), **cells**
ActVP(derived) Prep(from), **mice**
Subject(USP14) PassVP(expressed) Prep(in), **cells**

Table 5: Lexical Extraction features

also experimented with generalizing the extracted nouns by replacing them with a semantic category. To generate a semantic dictionary for the mouse genomics domain, we used the Basilisk bootstrapping algorithm (Thelen and Riloff, 2002). Basilisk has been used previously to create semantic lexicons for terrorist events (Thelen and Riloff, 2002) and sentiment analysis (Riloff et al., 2003), and recent work has shown good results for bioNLP domains using similar bootstrapping algorithms (McIntosh, 2010; McIntosh and Curran, 2009).

As input, Basilisk requires a domain-specific text corpus (unannotated) and a handful of seed nouns for each semantic category to be learned. A bootstrapping algorithm then iteratively hypothesizes additional words that belong to each semantic category based on their association with the seed words in pattern contexts. The output is a lexicon of nouns paired with their corresponding semantic class. (e.g., *liver :* BODY PART).

We used Basilisk to create a lexicon for eight semantic categories associated with mouse genomics: BIOLOGICAL PROCESS, BODY PART, CELL TYPE, CELLULAR LOCATION, BIOLOGICAL SUBSTANCE, EXPERIMENTAL REAGENT, RESEARCH SUBJECT, TUMOR. To choose the seed nouns, we parsed the training corpus, ranked all of the nouns by frequency[4], and selected the 10 most frequent, unambiguous nouns belonging to each semantic category. The seed words that we used for each semantic category are shown in Table 6.

Finally, we defined *Semantic Extraction* features as a pair consisting of a pattern coupled with the semantic category of the noun that it extracted. If the noun was not present in the semantic lexicons, then no feature was created. The Basilisk-generated lexicons are not perfect, so some entries will be incorrect. But our hope was that replacing the lexical terms with semantic categories might help the classifier learn more general associations. For example, "PassVP(expressed) Prep(in), CELLULAR LOCATION" will apply much more broadly than the corresponding lexical extraction with just one specific cellular location (e.g., 'mitochondria').

[4]We only used nouns that occurred as the head of a NP.

BIOLOGICAL PROCESS: expression, activity, activation, development, function, production, differentiation, regulation, reduction, proliferation
BODY PART: brain, muscle, thymus, cortex, retina, skin, spleen, heart, lung, pancreas
CELL TYPE: neurons, macrophages, thymocytes, splenocytes, fibroblasts, lymphocytes, oocytes, monocytes, hepatocytes, spermatocytes
CELLULAR LOCATION: receptor, nuclei, axons, chromosome, membrane, nucleus, chromatin, peroxisome, mitochondria, cilia
BIOLOGICAL SUBSTANCE: antibody, lysates, kinase, cytokines, peptide, antigen, insulin, ligands, peptides, enzyme
EXPERIMENTAL REAGENT: buffer, primers, glucose, acid, nacl, water, saline, ethanol, reagents, paraffin
RESEARCH SUBJECT: mice, embryos, animals, mouse, mutants, patients, littermates, females, males, individuals
TUMOR: tumors, tumor, lymphomas, tumours, carcinomas, malignancies, melanoma, adenocarcinomas, gliomas, sarcoma

Table 6: Seed words given to Basilisk

Information extraction patterns and their arguments have been used for text classification in previous work (Riloff and Lehnert, 1994; Riloff and Lorenzen, 1999), but the patterns and arguments were represented separately and the semantic features came from a hand-crafted dictionary. In contrast, our work couples each pattern with its extracted argument as a single feature, uses an automatically generated semantic lexicon, and is the first application of these techniques to the biocuration triage task.

3 Results

3.1 Data Set

For our experiments in this paper we use articles within the PubMed Central (PMC) Open Access Subset[5]. From this subset we select all articles that are published in journals of interest to biocurators at MGI. This results in a total of 14,827 documents out of which 981 have been selected manually by MGI biocurators as relevant (referred to as **IN** documents). This leaves 13,846 that are presumably out of scope (referred to as **OUT** documents), although it was not guaranteed that all of them had been manually reviewed so some relevant documents could be included as well. (We plan eventually to present to the biocurators those papers not included by them but nonetheless selected by our tools as **IN** with high confidence, for possible reclassification. Such changes will improve the system's evaluated score.)

As preprocessing for the NLP tools, we split the input text into sentences using the Lingua::EN::Sentence perl package. We trimmed non-alpha-numerics attached before and after words. We also removed stop words using the Lingua::EN::StopWords package.

3.2 Classifier

We used SVM Light[6](Joachims, 1999) for all of our experiments. We used a linear kernel and a tolerance value of 0.1 for QP solver termination. In preliminary experiments, we observed that the cost factor (C value) made a big difference in performance. In SVMs, the cost factor represents the importance of penalizing errors on the training instances in comparison to the complexity (generalization) of the model. We observed that higher values of C produced increased recall, though at the expense of some precision. We used a tuning set to experiment with different values of C, trying a wide range of powers of 2. We found that C=1024 generally produced the best balance of recall and precision, so we used that value throughout our experiments.

[5] `http://www.ncbi.nlm.nih.gov/pmc/about/openftlist.html`

[6] `http://svmlight.joachims.org/`

3.3 Experiments

We randomly partitioned our text corpus into 5 subsets of 2,965 documents each.[7] We used the first 4 subsets as the training set, and reserved the fifth subset as a blind test set.

In preliminary experiments, we found that the classifiers consistently benefitted from feature selection when we discarded low-frequency features. This helps to keep the classifier from overfitting to the training data. For each type of feature, we set a frequency threshold θ and discarded any features that occurred fewer than θ times in the training set. We chose these θ values empirically by performing 4-fold cross-validation on the training set. We evaluated θ values ranging from 1 to 50, and chose the value that produced the highest F score. The θ values that were selected are: 7 for unigrams, 50 for bigrams, 35 for patterns, 50 for lexical extractions, and 5 for semantic extractions.

Finally, we trained an SVM classifier on the entire training set and evaluated the classifier on the test set. We computed Precision (P), Recall (R), and the F score, which is the harmonic mean of precision and recall. Precision and recall were equally weighted, so this is sometimes called an F1 score.

Table 7 shows the results obtained by using each of the features in isolation. The lexical extraction features are shown as 'lexExts' and the semantic extraction features are shown as 'semExts'. We also experimented with using a hybrid extraction feature, 'hybridExts', which replaced a lexical extraction noun with its semantic category when one was available but left the noun as the extraction term when no semantic category was known.

Table 7 shows that the bigram features produced the best Recall (65.87%) and F-Score (74.05%), while the hybrid extraction features produced the best Precision (85.52%) but could not match the bigrams in terms of recall. This is not surprising because the extraction features on their own are quite specific, often requiring 3-4 words to match.

Next, we experimented with adding the IE-based features to the bigram features to allow the classifier to choose among both feature sets and get the best of both worlds. Combining bigrams with IE-based

[7] Our 5-way random split left 2 documents aside, which we ignored for our experiments.

Feature	P	R	F
unigrams	79.75	60.58	68.85
bigrams	84.57	**65.87**	**74.05**
patterns	78.98	59.62	67.95
lexExts	76.54	59.62	67.03
semExts	72.39	46.63	56.73
hybridExts	**85.52**	59.62	70.25
bigrams + patterns	84.87	62.02	71.67
bigrams + lexExts	85.28	**66.83**	**74.93**
bigrams + semExts	85.43	62.02	71.87
bigrams + hybridExts	**87.10**	64.90	74.38

Table 7: Triage classifier performance using different sets of features.

features did in fact yield the best results. Using bigrams and lexical extraction features achieved both the highest recall (66.83%) and the highest F score (74.93%). In terms of overall F score, we see a relatively modest gain of about 1% by adding the lexical extraction features to the bigram features, which is primarily due to the 1% gain in recall.

However, precision is of paramount importance for many applications because users don't want to wade through incorrect predictions. So it is worth noting that adding the hybrid extraction features to the bigram features produced a 2.5% increase in precision (84.57% → 87.10%) with just a 1% drop in recall. This recall/precision trade-off is likely to be worthwhile for many real-world application settings, including biocuration.

4 Biocuration Application for MGI

Developing an application that supports MGI biocurators necessitates an application design that minimally alters existing curation workflows while maintaining high classification F-scores (intrinsic measures) and speeding up the curation process (extrinsic measures). We seek improvements with respect to intrinsic measures by engineering context-specific features and seek extrinsic evaluations by instrumenting the deployed triage application to record usage statistics that serve as input to extrinsic evaluation measures.

4.1 Software Architecture

As stated earlier, one of our major goals is to build, deploy, and extrinsically evaluate an NLP-assisted curation application (Alex et al., 2008) for triage at MGI. By definition, an extrinsic evaluation of our triage application requires its deployment and subsequent tuning to obtain optimal performance with respect to extrinsic evaluation criteria. We anticipate that features, learning parameters, and training data distributions may all need to be adjusted during a tuning process. Cognizant of these future needs, we have designed the SciKnowMine system so as to integrate the various components and algorithms using the UIMA infrastructure. Figure 1 shows a schematic of SciKnowMine's overall architecture.

4.1.1 Building configurable & reusable UIMA pipelines

The experiments we have presented in this paper have been conducted using third party implementations of a variety of algorithms implemented on a wide variety of platforms. We use SVMLight to train a triage classifier on features that were produced by AutoSlog and Sundance on sentences identified by the perl package Lingua::EN::Sentence. Each of these types of components has either been reimplemented or wrapped as a component reusable in UIMA pipelines within the SciKnowMine infrastructure. We hope that building such a library of reusable components will help galvanize the BioNLP community towards standardization of an interoperable and open-access set of NLP components. Such a standardization effort is likely to lower the barrier-of-entry for NLP researchers interested in applying their algorithms to knowledge engineering problems in Biology (such as biocuration).

4.1.2 Storage infrastructure for annotations & features

As we develop richer section-specific and context-specific features we anticipate the need for provenance pertaining to classification decisions for a given paper. We have therefore built an Annotation Store and a Feature Store collectively referred to as the Classification Metadata Store[8] in Figure 1. Figure 1 also shows parallel pre-processing populating the annotation store. We are working on developing parallel UIMA pipelines that extract expensive (resource & time intensive) features (such as depen-

[8] Our classification metadata store has been implemented using Solr `http://lucene.apache.org/solr/`

dency parses).The annotation store holds features produced by pre-processing pipelines. The annotation store has been designed to support query-based composition of feature sets specific to a classification run. These feature sets can be asserted to the feature store and reused later by any pipeline. This design provides us with the flexibility necessary to experiment with a wide variety of features and tune our classifiers in response to feedback from biocurators.

5 Discussions & Conclusions

In this paper we have argued the need for richer semantic features for the MGI biocuration task. Our results show that simple lexical and semantic features used to augment bigram features can yield higher classification performance with respect to intrinsic metrics (such as F-Score). It is noteworthy that using a hybrid of lexical and semantic features results in the highest precision of 87%.

In our motivating example, we have proposed the need for sectional-zoning of articles and have demonstrated that certain zones like the 'Materials and Methods' section can contain contextual features that might increase classification performance. It is clear from the samples of MGI manual classification guidelines that biocurators do, in fact, use zone-specific features in triage. It therefore seems likely that section specific feature extraction might result in better classification performance in the triage task. Our preliminary analysis of the MGI biocuration guidelines suggests that experimental procedures described in the 'Materials and Methods' seem to be a good source of triage clues. We therefore propose to investigate zone and context specific features and the explicit use of domain models of experimental procedure as features for document triage.

We have also identified infrastructure needs arising within the construction of a biocuration application. In response we have constructed preliminary versions of metadata stores and UIMA pipelines to support MGI's biocuration. Our next step is to deploy a prototype assisted-curation application that uses a classifier trained on the best performing features discussed in this paper. This application will be instrumented to record usage statistics for use in extrinsic evaluations (Alex et al., 2008). We hope that construction on such an application will also engender the creation of an open environment for NLP scientists to apply their algorithms to biomedical corpora in addressing biomedical knowledge engineering challenges.

6 Acknowledgements

This research is funded by the U.S. National Science Foundation under grant #0849977 for the SciKnowMine project (`http://sciknowmine.isi.edu/`). We wish to acknowledge Kevin Cohen for helping us collect the seed terms for Basilisk and Karin Verspoor for discussions regarding feature engineering.

References

[Alex et al.2008] Beatrice Alex, Claire Grover, Barry Haddow, Mijail Kabadjov, Ewan Klein, Michael Matthews, Stuart Roebuck, Richard Tobin, and Xinglong Wang. 2008. Assisted curation: does text mining really help? *Pacific Symposium On Biocomputing*, 567:556–567.

[Bouatia-Naji et al.2010] Nabila Bouatia-Naji, Amélie Bonnefond, Devin A Baerenwald, Marion Marchand, Marco Bugliani, Piero Marchetti, François Pattou, Richard L Printz, Brian P Flemming, Obi C Umunakwe, Nicholas L Conley, Martine Vaxillaire, Olivier Lantieri, Beverley Balkau, Michel Marre, Claire Lévy-Marchal, Paul Elliott, Marjo-Riitta Jarvelin, David Meyre, Christian Dina, James K Oeser, Philippe Froguel, and Richard M O'Brien. 2010. Genetic and functional assessment of the role of the rs13431652-A and rs573225-A alleles in the G6PC2 promoter that are strongly associated with elevated fasting glucose levels. *Diabetes*, 59(10):2662–2671.

[Bourne and McEntyre2006] Philip E Bourne and Johanna McEntyre. 2006. Biocurators: Contributors to the World of Science. *PLoS Computational Biology*, 2(10):1.

[Cohen and Hersh2006] Aaron M Cohen and William R Hersh. 2006. The TREC 2004 genomics track categorization task: classifying full text biomedical documents. *Journal of Biomedical Discovery and Collaboration*, 1:4.

[Ferrucci and Lally2004] D Ferrucci and A Lally. 2004. Building an example application with the Unstructured Information Management Architecture. *IBM Systems Journal*, 43(3):455–475.

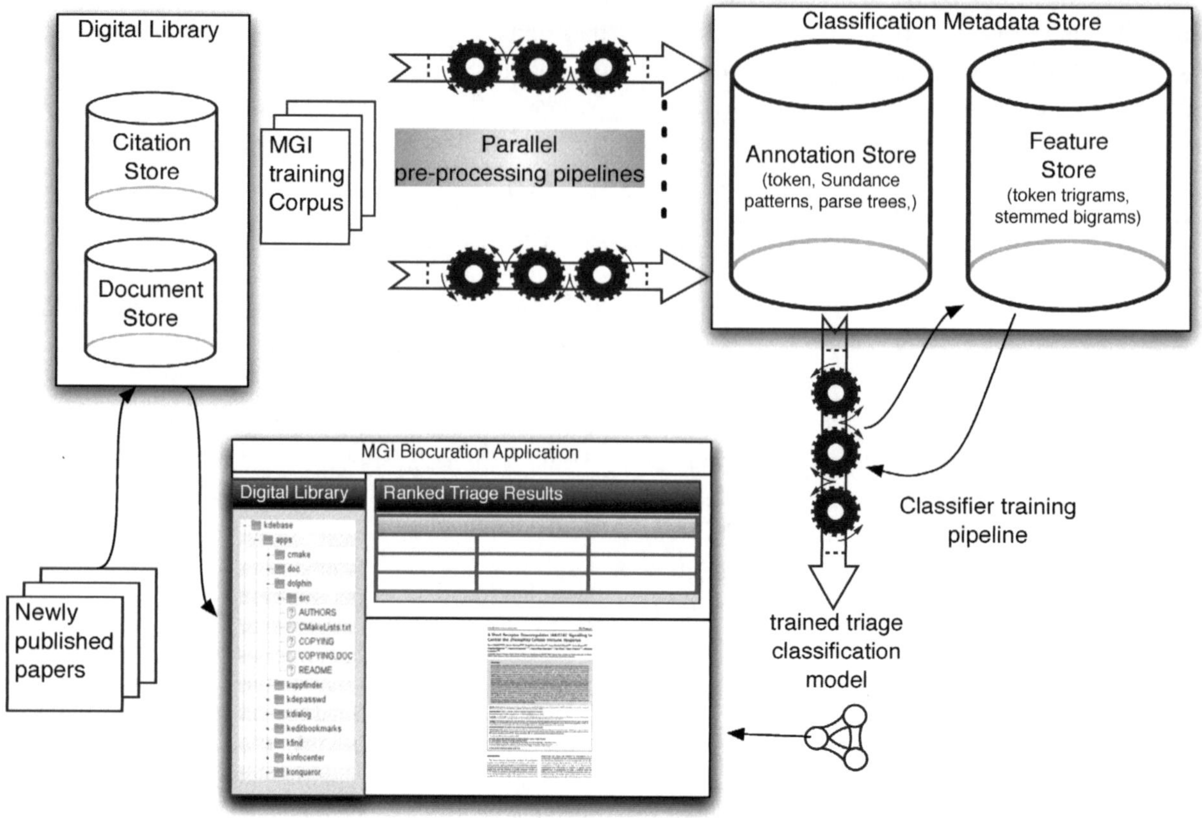

Figure 1: Design schematic of the MGI biocuration application. The components of the application are: (A) Digital Library composed of a citation store and document store. (B) Pre-processing UIMA pipelines which are a mechanism to pre-extract standard features such as parse trees, tokenizations *etc.* (C) Classification Metadata Store which is composed of an Annotation Store for the pre-extracted standard features from (B), and a Feature Store to hold derived features constructed from the standard ones in the Annotation Store. (D) Classifier training pipeline. (E) MGI Biocuration Application.

[Hersh W2005] Yang J Bhupatiraju RT Roberts P M. Hearst M Hersh W, Cohen AM. 2005. TREC 2005 genomics track overview. In *The Fourteenth Text Retrieval Conference.*

[Joachims1999] Thorsten Joachims. 1999. Making Large-Scale SVM Learning Practical. *Advances in Kernel Methods Support Vector Learning*, pages 169–184.

[McIntosh and Curran2009] T. McIntosh and J. Curran. 2009. Reducing Semantic Drift with Bagging and Distributional Similarity. In *Proceedings of the 47th Annual Meeting of the Association for Computational Linguistics.*

[McIntosh2010] Tara McIntosh. 2010. Unsupervised discovery of negative categories in lexicon bootstrapping. In *Proceedings of the 2010 Conference on Empirical Methods in Natural Language Processing*, number October, pages 356–365. Association for Computational Linguistics.

[Ramakrishnan et al.2010] Cartic Ramakrishnan, William A Baumgartner Jr, Judith A Blake, Gully A P C Burns, K Bretonnel Cohen, Harold Drabkin, Janan Eppig, Eduard Hovy, Chun-Nan Hsu, Lawrence E Hunter, Tommy Ingulfsen, Hiroaki Rocky Onda, Sandeep Pokkunuri, Ellen Riloff, and Karin Verspoor. 2010. Building the Scientific Knowledge Mine (SciKnowMine 1): a community- driven framework for text mining tools in direct service to biocuration. *In proceeding of Workshop "New Challenges for NLP Frameworks" collocated with The seventh international conference on Language Resources and Evaluation (LREC) 2010.*

[Riloff and Lehnert1994] E. Riloff and W. Lehnert. 1994. Information Extraction as a Basis for High-Precision Text Classification. *ACM Transactions on Information*

Systems, 12(3):296–333, July.

[Riloff and Lorenzen1999] E. Riloff and J. Lorenzen. 1999. Extraction-based text categorization: Generating domain-specific role relationships automatically. In Tomek Strzalkowski, editor, *Natural Language Information Retrieval*. Kluwer Academic Publishers.

[Riloff and Phillips2004] E. Riloff and W. Phillips. 2004. An Introduction to the Sundance and AutoSlog Systems. Technical Report UUCS-04-015, School of Computing, University of Utah.

[Riloff et al.2003] E. Riloff, J. Wiebe, and T. Wilson. 2003. Learning Subjective Nouns using Extraction Pattern Bootstrapping. In *Proceedings of the Seventh Conference on Natural Language Learning (CoNLL-2003)*, pages 25–32.

[Riloff1993] E. Riloff. 1993. Automatically Constructing a Dictionary for Information Extraction Tasks. In *Proceedings of the 11th National Conference on Artificial Intelligence*.

[Sjögren et al.2009] Klara Sjögren, Marie Lagerquist, Sofia Moverare-Skrtic, Niklas Andersson, Sara H Windahl, Charlotte Swanson, Subburaman Mohan, Matti Poutanen, and Claes Ohlsson. 2009. Elevated aromatase expression in osteoblasts leads to increased bone mass without systemic adverse effects. *Journal of bone and mineral research the official journal of the American Society for Bone and Mineral Research*, 24(7):1263–1270.

[Thelen and Riloff2002] M. Thelen and E. Riloff. 2002. A Bootstrapping Method for Learning Semantic Lexicons Using Extraction Pa ttern Contexts. In *Proceedings of the 2002 Conference on Empirical Methods in Natural Language Processing*, pages 214–221.

Medical Entity Recognition: A Comparison of Semantic and Statistical Methods

Asma Ben Abacha
LIMSI-CNRS
BP 133, 91403 Orsay Cedex, France
asma.benabacha@limsi.fr

Pierre Zweigenbaum
LIMSI-CNRS
BP 133, 91403 Orsay Cedex, France
pz@limsi.fr

Abstract

Medical Entity Recognition is a crucial step towards efficient medical texts analysis. In this paper we present and compare three methods based on domain-knowledge and machine-learning techniques. We study two research directions through these approaches: (i) a first direction where noun phrases are extracted in a first step with a chunker before the final classification step and (ii) a second direction where machine learning techniques are used to identify simultaneously entities boundaries and categories. Each of the presented approaches is tested on a standard corpus of clinical texts. The obtained results show that the hybrid approach based on both machine learning and domain knowledge obtains the best performance.

1 Introduction

Medical Entity Recognition (MER) consists in two main steps: (i) detection and delimitation of phrasal information referring to medical entities in textual corpora (e.g. *pyogenic liver abscess*, *infection of biliary system*) and (ii) identification of the semantic category of located entities (e.g. Medical Problem, Test). Example 1 shows the result of MER on a sentence where the located entity and its category are marked with *treatment* and *problem* tags.

(1) *<treatment> Adrenal-sparing surgery </treatment> is safe and effective , and may become the treatment of choice in patients with <problem> hereditary phaeochromocytoma </problem>.*

This task is very important for many applications such as Question-Answering where MER is used in the question analysis step (to determine the expected answers' type, the question focus, etc.) and in the offline text tagging or annotation.

One of the most important obstacles to identifying medical entities is the high terminological variation in the medical domain (e.g. *Diabetes mellitus type 1*, *Type 1 diabetes*, *IDDM*, or *juvenile diabetes* all express the same concept). Other aspects also have incidence on MER processes such as the evolution of entity naming (e.g. new abbreviations, names for new drugs or diseases). These obstacles limit the scalability of methods relying on dictionaries and/or gazetteers. Thus, it is often the case that other types of approaches are developed by exploiting not only domain knowledge but also domain-independent techniques such as machine learning and natural language processing tools.

In this paper, we study MER with three different methods: (i) a semantic method relying on MetaMap (Aronson, 2001) (a state-of-the-art tool for MER) (ii) chunker-based noun phrase extraction and SVM classification and (iii) a last method using supervised learning with Conditional Random Fields (CRF), which is then combined with the semantic method. With these methods we particularly study two processing directions: (i) pre-extraction of noun phrases with specialized tools, followed by a medical classification step and (ii) exploitation of machine-learning techniques to detect simultaneously entity boundaries and their categories.

We also present a comparative study of the performance of different noun phrase chunkers on medical

Proceedings of the 2011 Workshop on Biomedical Natural Language Processing, ACL-HLT 2011, pages 56–64,
Portland, Oregon, USA, June 23-24, 2011.

texts: Treetagger-chunker, OpenNLP and MetaMap. The best chunker was then used to feed some of the proposed MER approaches. All three methods were experimented on the i2b2/VA 2010 challenge corpus of clinical texts (Uzuner, 2010). Our study shows that hybrid methods achieve the best performance w.r.t machine learning approaches or domain knowledge-based approaches if applied separately.

After a review of related work (Section 2), we describe the chunker comparison and the three MER methods (Section 3). We present experiments on clinical texts (Section 4), followed by a discussion and variant experiments on literature abstracts (Section 5), then conclude and draw some perspectives for further work (Section 6).

2 Related Work

Several teams have tackled named entity recognition in the medical domain. (Rindflesch et al., 2000) presented the EDGAR system which extracts information about drugs and genes related to a given cancer from biomedical texts. The system exploits the MEDLINE database and the UMLS. Protein name extraction has also been studied through several approaches (e.g. (Liang and Shih, 2005; Wang, 2007)). (Embarek and Ferret, 2008) proposed an approach relying on linguistic patterns and canonical entities for the extraction of medical entities belonging to five categories: Disease, Treatment, Drug, Test, and Symptom. Another kind of approach uses domain-specific tools such as MetaMap (Aronson, 2001). MetaMap recognizes and categorizes medical terms by associating them to concepts and semantic types of the UMLS Metathesaurus and Semantic Network. (Shadow and MacDonald, 2003) presented an approach based on MetaMap for the extraction of medical entities of 20 medical classes from pathologist reports. (Meystre and Haug, 2005) obtained 89.9% recall and 75.5% precision for the extraction of medical problems with an approach based on MetaMap Transfer (MMTx) and the NegEx negation detection algorithm.

In contrast with semantic approaches which require rich domain-knowledge for rule or pattern construction, statistical approaches are more scalable. Several approaches used classifiers such as decision trees or SVMs (Isozaki and Kazawa, 2002). Markov models-based methods are also frequently used (e.g. Hidden Markov Models, or CRFs (He and Kayaalp, 2008)). However, the performance achieved by such supervised algorithms depends on the availability of a well-annotated training corpus and on the selection of a relevant feature set.

Hybrid approaches aim to combine the advantages of semantic and statistical approaches and to bypass some of their weaknesses (e.g. scalability of rule-based approaches, performance of statistical methods with small training corpora). (Proux et al., 1998) proposed a hybrid approach for the extraction of gene symbols and names. The presented system processed unknown words with lexical rules in order to obtain candidate categories which were then disambiguated with Markov models. (Liang and Shih, 2005) developed a similar approach using empirical rules and a statistical method for protein-name recognition.

3 Medical Entity Recognition Approaches

Named entity recognition from medical texts involves two main tasks: (i) identification of entity boundaries in the sentences and (ii) entity categorization. We address these tasks through three main approaches which are listed in Table 1.

3.1 Noun Phrase Chunking

Although noun phrase segmentation is an important task for MER, few comparative studies on available tools have been published. A recent study (Kang et al., 2010), which claims to be the first to do such comparative experiments, tested six state-of-the-art chunkers on a biomedical corpus: GATE chunker, Genia Tagger, Lingpipe, MetaMap, OpenNLP, and Yamcha. This study encompassed sentence splitting, tokenization and part-of-speech tagging and showed that for both noun-phrase chunking and verb-phrase chunking, OpenNLP performed best (F-scores 89.7% and 95.7%, respectively), but differences with Genia Tagger and Yamcha were small.

With a similar objective, we compared the performance of three different noun-phrase chunkers in the medical domain: (i) Treetagger-chunker[1], a state-of-the-art open-domain tool, (ii) OpenNLP[2] and (iii)

[1] http://www.ims.uni-stuttgart.de/projekte/corplex/TreeTagger

[2] http://incubator.apache.org/opennlp

	Medical Entity Recognition	
	1. Boundary identification	**2. Type categorization (with *n* medical entity categories)**
Method 1 (MetaMap+)	Noun phrase segmentation	- Rule-based method, - *Noun phrase* classification, - Number of classes = $n + 1$
Method 2 (TT-SVM)	Noun phrase segmentation	- Statistical method with a SVM classifier, - *Noun phrase* classification, - Number of classes = $n + 1$
Method 3 (BIO-CRF)	- Statistical method with a CRF classifier, - and the BIO format, - *word-level* classification, - Number of classes = $2n + 1$	

Table 1: Proposed MER methods

	Corpus of clinical texts (i2b2)			Corpus of scientific abstracts (Berkeley)		
	MetaMap	**TreeTagger**	**OpenNLP**	**MetaMap**	**TreeTagger**	**OpenNLP**
Reference entities	58115	58115	58115	3371	3371	3371
Correct entities	6532	35314	26862	151	2106	1874
Found entities	212227	129912	122131	22334	19796	18850
Recall	11.14%	**60.06%**	46.62%	4.48%	**62.27%**	55.59%

Table 2: NP Segmentation Results

MetaMap. Regardless of the differences in corpora with (Kang et al., 2010) we chose these particular tools to compare medical-domain specific tools with open domain tools and to highlight the lower performance of MetaMap for noun-phrase chunking compared to other tools. This last point led us to introduce the MetaMap+ approach for MER (Ben Abacha and Zweigenbaum, 2011) in order to take advantage of MetaMap's domain-knowledge approach while increasing performance by relying on external tools for noun-phrase chunking.

We evaluate these tools on the subset of noun phrases referring to medical entities in our corpora (cf. Section 4.1 for a description of the i2b2 corpus and Section 5 for the Berkeley corpus). We consider that a noun phrase is correctly extracted if it corresponds exactly to an annotated medical entity from the reference corpora. Also, as our corpora are not fully annotated (only entities of the targeted types are annotated), we do not evaluate "extra noun-phrases" corresponding to non-annotated entities. The retrieved noun phrases are heterogeneous: many of them are not all relevant to the medical field and therefore not relevant to the MER task. Our goal is to obtain the maximal number of correct noun phrases and leave it to the next step to filter out those that are irrelevant. We therefore wish to maximize recall at this stage.

Table 2 shows that in this framework, Treetagger-chunker obtains the best recall. We thus used it for noun-phrase segmentation in the experimented MER approaches (cf. Sections 3.2 and 3.3).

3.2 Semantic and Rule-Based Method: MM+

MetaMap is a reference tool which uses the UMLS to map noun phrases in raw texts to the best matching UMLS concepts according to matching scores. MetaMap leads however to some residual problems, which we can arrange into three classes: (i) noun phrase chunking is not at the same level of performance as some specialized NLP tools, (ii) medical entity detection often retrieves general words and verbs which are not medical entities and (iii) some ambiguity is left in entity categorization since MetaMap can provide several concepts for the same term as well as several semantic types for the same concept. Several "term/concept/type" combinations are then possible.

To improve MetaMap output, we therefore use an external noun phrase chunker (cf. Section 3.1) and stop-list based filtering to recover frequent/noticeable errors. MetaMap can propose different UMLS semantic types for the same noun phrase, thus leading to different categories for the same entity. In such cases we apply a voting procedure. For instance, if the process retrieves three UMLS semantic types for one noun phrase where two are associated to the target category "Problem" and one is associated to "Treatment", the "Problem" category is chosen as the entity's category. In case of a tie, we rely on the order output by MetaMap and take the first returned type.

More precisely, our rule-based method, which we call MetaMap+ (MM+), can be decomposed into the following steps:

1. Chunker-based noun phrase extraction. We use Treetagger-chunker according to the above-mentioned test (cf. Table 2).

2. Noun phrase filtering with a stop-word list.

3. Search for candidate terms in specialized lists of medical problems, treatments and tests gathered from the training corpus, Wikipedia, Health on the Net and Biomedical Entity Network.

4. Use MetaMap to annotate medical entities (which were not retrieved in the specialized lists) with UMLS concepts and semantic types.

5. Finally, filter MetaMap results with (i) a list of common/noticeable errors and (ii) the selection of only a subset of semantic types to look for (e.g. Quantitative Concept, Functional Concept, Qualitative Concept are too general semantic types and produce noise in the extraction process).

3.3 Statistical Method: TT-SVM

The second presented approach uses Treetagger-chunker to extract noun phrases followed by a machine learning step to categorize medical entities (e.g. Treatment, Problem, Test). The problem is then modeled as a supervised classification task with $n + 1$ categories (n is the number of entity categories). We chose an SVM classifier.

As noted by (Ekbal and Bandyopadhyay, 2010), SVMs (Support Vector Machines) have advantages over conventional statistical learning algorithms, such as Decision Trees or Hidden Markov Models, in the following two aspects: (1) SVMs have high generalization performance independent of the dimension of feature vectors, and (2) SVMs allow learning with all feature combinations without increasing computational complexity, by introducing kernel functions.

In our experiments we use the libSVM (Chang and Lin, 2001) implementation of the SVM classifier. We chose the following feature set to describe each noun phrase (NP):

1. Word features:
   - words of the NP
   - number of the NP words
   - lemmas of the NP words
   - 3 words and their lemmas before the NP
   - 3 words and their lemmas after the NP

2. Orthographic features (some examples):
   - first letter capitalized for the first word, one word or all words
   - all letters uppercase for the first word, one word or all words
   - all letters lowercase for the first word, one word or all words
   - NP is or contains an abbreviation
   - word of NP contains a single uppercase, digits, hyphen, plus sign, ampersand, slash, etc.

3. Part-of-speech tags: POS tags of the NP words, of the 3 previous and 3 next words.

3.4 Statistical Method: BIO-CRF

We conducted MER with a CRF in one single step by determining medical categories and entity boundaries at the same time. We used the BIO format: B (beginning), I (inside), O (outside) which represents entity tagging by individual word-level tagging. For instance, a problem-tagged entity is represented as a first word tagged B-P (begin problem) and other (following) words tagged I-P (inside a problem). A problem entity comprising one single word will be tagged B-P. Words outside entities are tagged with the letter 'O'.

If we have n categories (e.g. Problem, Treatment, Test), we then have n classes of type B-, n classes of type I- (e.g. P-B and P-I classes associated to the *problem* category) and one class of type 'O'. Figure 1 shows an example sentence tagged with the BIO format. As a result, the classification task consists in a word classification task (instead of a noun-phrase classification task) into $2n + 1$ target classes, where n is the number of categories. As a consequence, relying on a chunker is no longer necessary.

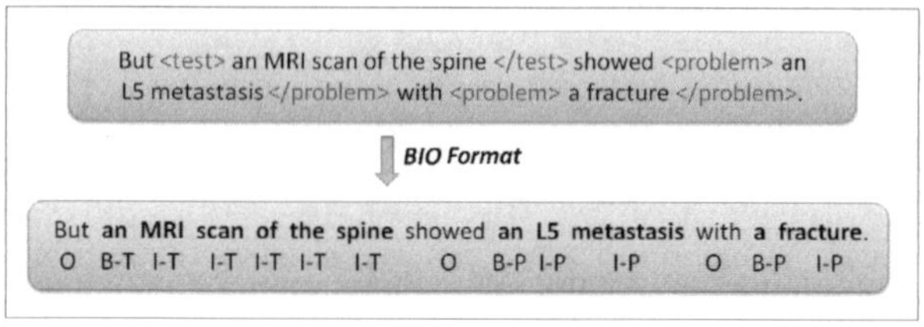

Figure 1: BIO Format (T = Test, P = Problem)

Words in a sentence form a sequence, and the decision on a word's category can be influenced by the decision on the category of the preceding word. This dependency is taken into account in sequential models such as Hidden Markov Models (HMMs) or Conditional Random Fields (CRF). In contrast with HMMs, CRF learning maximizes the conditional probability of classes w.r.t. observations rather than their joint probability. This makes it possible to use any number of features which may be related to all aspects of the input sequence of words. These properties are assets of CRFs for several natural language processing tasks, such as POS tagging, noun phrase chunking, or named entity recognition (see (Tellier and Tommasi, 2010) for a survey).

In our experiments we used the CRF++[3] implementation of CRFs. CRF++ eases feature description through *feature templates*. We list hereafter some of our main features. We instructed CRF++ to model the dependency of successive categories (instruction `B` in feature template).

For each word we use the following features:

1. Word features: The word itself, two words before and three words after, with their lemmas.
2. Morphosyntactic features: POS tag of the word itself, two words before and three words after.
3. Orthographic features (some examples):
   - The word contains hyphen, plus sign, ampersand, slash, etc.
   - The word is a number, a letter, a punctuation sign or a symbol.
   - The word is in uppercase, capitalized, in lowercase (AA, Aa, aa)
   - Prefixes of different lengths (from 1 to 4)
   - Suffixes of different lengths (from 1 to 4)
4. Semantic features: semantic category of the word (provided by MetaMap+)
5. Other features: next verb, next noun, word length over a threshold, etc.

Additionally, we tested semantic features constructed from MM+ results. More detail on these last features is given in Section 5.3.

4 Experiments on Clinical Texts

We performed MER experiments on English clinical texts.

4.1 Corpus

The i2b2 corpus was built for the i2b2/VA 2010 challenge[4] in Natural Language Processing for Clinical Data (Uzuner, 2010). The data for this challenge includes discharge summaries from Partners HealthCare and from Beth Israel Deaconess Medical Center (MIMIC II Database), as well as discharge summaries and progress notes from University of Pittsburgh Medical Center. All records have been fully de-identified and manually annotated for concept, assertion, and relation information. The corpus contains entities of three different categories: Problem, Treatment and Test, 76,665 sentences and 663,476 words with a mean of 8.7 words per sentence. Example 2 shows an annotated sentence from the i2b2 corpus.

[3] `http://crfpp.sourceforge.net/`

[4] `http://www.i2b2.org/NLP/Relations/`

(2) *<problem>CAD</problem> s/p <treatment>3v-CABG </treatment> 2003 and subsequent <treatment>stenting </treatment> of <treatment>SVG</treatment> and LIMA.*

Table 3 presents the number of training and test sentences.

i2b2 Corpus	Sentences	Words
Training Corpus	31 238	267 304
Test Corpus	44 927	396 172

Table 3: Number of training and test sentences

4.2 Experimental Settings

We tested the above-described five configurations (see Table 1):

1. MM: MetaMap is applied as a baseline method
2. MM+: MetaMap Plus (semantic and rule-based method)
3. TT-SVM: Statistical method, chunking with Treetagger and Categorization with a SVM classifier
4. BIO-CRF: Statistical method, BIO format with a CRF classifier
5. BIO-CRF-H: Hybrid method combining semantic and statistical methods (BIO-CRF with semantic features constructed from MM+ results)

We evaluate the usual metrics of Recall (proportion of correctly detected entities among the reference entities), Precision (proportion of correctly detected entities among those output by the system), and F-measure (harmonic means of Recall and Precision).

4.3 Results

Table 4 presents the results obtained by each configuration. BIO-CRF and BIO-CRF-H obtained the best precision, recall and F-measures. MM+ comes next, followed by TT-SVM and MetaMap alone.

Table 5 presents the obtained results per each medical category (i.e. Treatment, Problem and Test) for three configurations. Again, BIO-CRF-H obtains the best results for all metrics and all categories.

Setting	P	R	F
MM	15.52	16.10	15.80
MM+	48.68	56.46	52.28
TT-SVM	43.65	47.16	45.33
BIO-CRF	70.15	83.31	76.17
BIO-CRF-H	**72.18**	**83.78**	**77.55**

Table 4: Results per setting on the i2b2 corpus. R = recall, P = precision, F = F-measure

Setting	Category	P	R	F
MM+	Problem	60.84	53.04	56.67
	Treatment	51.99	61.93	56.53
	Test	56.67	28.48	37.91
TT-SVM	Problem	48.25	43.16	45.56
	Treatment	42.45	50.86	46.28
	Test	57.37	35.76	44.06
BIO-CRF-H	Problem	82.05	73.14	77.45
	Treatment	83.18	73.33	78.12
	Test	87.50	68.69	77.07

Table 5: Results per setting and per category on the i2b2 corpus

5 Discussion and Further Experiments

We presented three different methods for MER: MM+, TT-SVM, and BIO-CRF (with variant BIO-CRF-H). In this section we quickly present supplementary results obtained on a second corpus with the same methods, and discuss differences in results when corpora and methods vary.

5.1 Corpora

Different kinds of corpora exist in the biomedical domain (Zweigenbaum et al., 2001). Among the most recurring ones we may cite (i) clinical texts and (ii) scientific literature (Friedman et al., 2002). Clinical texts have motivated a long stream of research (e.g. (Sager et al., 1995), (Meystre et al., 2008)), and more recently international challenges such as i2b2 2010 (Uzuner, 2010). The scientific literature has also been the subject of much research (e.g. (Rindflesch et al., 2000)), especially in genomics for more than a decade, e.g. through the BioCreative challenge (Yeh et al., 2005).

Section 4 presented experiments in MER on English clinical texts. To have a complementary view on the performance of our methods, we performed additional experiments on the Berkeley corpus (Rosario and Hearst, 2004) of scientific literature abstracts and titles extracted from MEDLINE. The original aim of this corpus was to study the extraction of semantic relationships between problems and treatments (e.g. *cures*, *prevents*, and *side effect*). In our context, we only use its annotation of medical entities. The corpus contains two categories of medical entities: problems (1,660 entities) and treatments (1,179 entities) in 3,654 sentences (74,754 words) with a mean of 20.05 words per sentence. We divided the corpus into 1,462 sentences for training and 2,193 for testing.

We tested the MetaMap (MM), MetaMap+ (MM+) and BIO-CRF methods on the Berkeley corpus. Table 6 presents the results. BIO-CRF again obtain the best results, but it is not much better than MM+ in this case.

		P	**R**	**F**
MM	Problem	5.32	7.63	6.27
	Treatment	6.37	18.84	9.52
	Total	5.35	12.34	7.46
MM+	Problem	34.47	44.97	39.02
	Treatment	18.11	39.36	24.81
	Total	23.43	42.47	30.20
BIO-CRF	Problem	41.88	38.88	40.32
	Treatment	29.85	23.86	26.52
	Total	36.94	32.13	34.37

Table 6: Results on the Berkeley Corpus

We constructed three different models for the BIO-CRF method: a first model constructed from the Berkeley training corpus, a second model constructed from the i2b2 corpus and a third model constructed from a combination of the former two. We obtained the best results with the last model: F=34.37% (F=22.97% for the first model and F=30.08% for the second model). These results were obtained with a feature set with which we obtained 76.17% F-measure on the i2b2 corpus (i.e. words, lemmas, morphosyntactic categories, orthographic features, suffixes and prefixes, cf. set A4 in Table 7).

The results obtained on the two corpora are not on the same scale of performance. This is mainly due to the characteristics of each corpus. For instance, the i2b2 2010 corpus has an average words-per-sentence ratio of 8.7 while the Berkeley corpus has a ratio of 20.45 words per sentence. Besides, the i2b2 corpus uses a quite specific vocabulary such as conventional abbreviations of medical terms (e.g. *k/p* for *kidney pancreas* and *d&c* for *dilation and curettage*) and abbreviations of domain-independent words (e.g. *w/o* for *without* and *y/o* for *year old*).

However, according to our observations, the most important characteristic which may explain these results may be the quality of annotation. The i2b2 corpus was annotated according to well-specified criteria to be relevant for the challenge, while the Berkeley corpus was annotated with different rules and less control measures. We evaluated a random sample of 200 annotated medical entities in the Berkeley corpus, using the i2b2 annotation criteria, and found that 20% did not adhere to these criteria.

5.2 Semantic Methods

The semantic methods have the advantage of being reproducible on all types of corpora without a preprocessing or learning step. However, their dependency to knowledge reduces their performance w.r.t. machine learning approaches. Also the development of their knowledge bases is a relatively slow process if we compare it with the time which is necessary for machine learning approaches to build new extraction and categorization models.

On the other hand, a clear advantage of semantic approaches is that they facilitate semantic access to the extracted information through conventional semantics (e.g. the UMLS Semantic Network).

In our experiments we did not obtain good results when applying MetaMap alone. This is mainly due to the detection of entity boundaries (e.g. *"no pericardial effusion_"* instead of *"pericardial effusion"* and *"(Warfarin"* instead of *"Warfarin"*).

We were able to enhance the overall performance of MetaMap for this task by applying several input and output filtering primitives, among which the use of an external chunker to obtain the noun phrases. Our observation is that the final results are limited by chunker performance. Nevertheless, the approach provided the correct categories for 52.28% correctly extracted entities while the total ratio of the retrieved entities with correct boundaries is 60.76%.

5.3 Machine Learning Methods

We performed several tests with semantic features with the BIO-CRF method. For instance, applying MM+ on each word and using the obtained medical category as an input feature for CRF decreased performance from 76.17% F-measure to 76.01%. The same effect was observed by using the UMLS semantic type instead of the final category for each word, with an F-measure decrease from 76.17% to 73.55%. This can be explained by a reduction in feature value space size (22 UMLS types instead of 3 final categories) but also by the reduced performance of MetaMap if it is applied at the word level.

The best solution was obtained by transforming the output of the MM+ approach into BIO format tags and feeding them to the learning process as features for each word. Thus, each word in an entity tagged by MM+ has an input feature value corresponding to one of the following: B-problem, I-problem, B-treatment, I-treatment, B-test and I-test. Words outside entities tagged by MM+ received an 'O' feature value.

With these semantic features we were able to increase the F-measure from 76.19% to 77.55%. Table 7 presents the contribution of each feature category to the BIO-CRF method on the i2b2 corpus.

Features	**P**	**R**	**F**
A1: Words/Lemmas/POS	62.81	82.25	71.23
A2: A1 + orthographic features	63.72	82.19	71.78
A3: A2 + suffixes	67.91	82.89	74.65
A4: A3 + prefixes	70.15	83.31	76.17
A5: A4 + other features	70.22	83.28	76.19
A6: A5 + semantic features	**72.18**	**83.78**	**77.55**

Table 7: Contribution of each feature category (BIO-CRF method) on the i2b2 corpus

6 Conclusion

We presented and compared three different approaches to MER. Our experiments show that performing the identification of entity boundaries with a chunker in a first step limits the overall performance, even though categorization can be performed efficiently in a second step. Using machine learning methods for joint boundary and category identification allowed us to bypass such limits. We obtained the best results with a hybrid approach combining machine learning and domain knowledge. More precisely, the best performance was obtained with a CRF classifier using the BIO format with lexical and morphosyntactic features combined with semantic features obtained from a domain-knowledge based method using MetaMap.

Future work will tackle French corpora with both a semantic method and the BIO-CRF approach. We also plan to exploit these techniques to build a cross-language question answering system. Finally, it would be interesting to try ensemble methods to combine the set of MER methods tested in this paper.

Acknowledgments

This work has been partially supported by OSEO under the Quaero program.

References

Alan R. Aronson. 2001. Effective mapping of biomedical text to the UMLS metathesaurus: the MetaMap program. In *AMIA Annu Symp Proc*, pages 17–21.

Asma Ben Abacha and Pierre Zweigenbaum. 2011. Automatic extraction of semantic relations between medical entities: a rule based approach. *Journal of Biomedical Semantics*. In Press.

Chih-Chung Chang and Chih-Jen Lin, 2001. *LIBSVM: a library for support vector machines*. Software available at `http://www.csie.ntu.edu.tw/~cjlin/libsvm`.

Asif Ekbal and Sivaji Bandyopadhyay. 2010. Named entity recognition using support vector machine: A language independent approach. *International Journal of Electrical and Electronics Engineering*, 4(2):155–170.

Mehdi Embarek and Olivier Ferret. 2008. Learning patterns for building resources about semantic relations in the medical domain. In *LREC'08*, May.

Carol Friedman, Pauline Kra, and Andrey Rzhetsky. 2002. Two biomedical sublanguages: a description based on the theories of Zellig Harris. *Journal of Biomedical Informatics*, 35:222–235.

Ying He and Mehmet Kayaalp. 2008. Biological entity recognition with Conditional Random Fields. In *AMIA Annu Symp Proc*, pages 293–297.

Hideki Isozaki and Hideto Kazawa. 2002. Efficient support vector classifiers for named entity recognition. In *Proceedings of COLING-2002*, pages 390–396.

N Kang, EM van Mulligen, and JA Kors. 2010. Comparing and combining chunkers of biomedical text. *J Biomed Inform*, 44(2):354–360, nov.

Tyne Liang and Ping-Ke Shih. 2005. Empirical textual mining to protein entities recognition from PubMed corpus. In *NLDB'05*, pages 56–66.

Stéphane M. Meystre and Peter J. Haug. 2005. Comparing natural language processing tools to extract medical problems from narrative text. In *AMIA Annu Symp Proc*, pages 525–529.

S M Meystre, G K Savova, K C Kipper-Schuler, and J F Hurdle. 2008. Extracting information from textual documents in the electronic health record: a review of recent research. *Yearb Med Inform*, 35:128–44.

Denys Proux, François Rechenmann, Laurent Julliard, Violaine Pillet, and Bernard Jacq. 1998. Detecting gene symbols and names in biological texts : A first step toward pertinent information extraction. In *Proceedings of Genome Informatics*, pages 72–80, Tokyo, Japan : Universal Academy Press.

Thomas C. Rindflesch, Lorraine Tanabe, John N. Weinstein, and Lawrence Hunter. 2000. Edgar: Extraction of drugs, genes and relations from the biomedical literature. In *Proceedings of Pacific Symposium on Biocomputing*, pages 517–528.

Barbara Rosario and Marti A. Hearst. 2004. Classifying semantic relations in bioscience text. In *Proceedings of the 42nd Annual Meeting of the Association for Computational Linguistics (ACL 2004)*, Barcelona, July.

N Sager, M Lyman, N T Nhàn, and L J Tick. 1995. Medical language processing: applications to patient data representation and automatic encoding. *Meth Inform Med*, 34(1–2):140–6.

G Shadow and C MacDonald. 2003. Extracting structured information from free text pathology reports. In *AMIA Annu Symp Proc*, Washington, DC.

Isabelle Tellier and Marc Tommasi. 2010. Champs Markoviens Conditionnels pour l'extraction d'information. In Éric Gaussier and François Yvon, editors, *Modèles probabilistes pour l'accès à l'information textuelle*. Hermès, Paris.

Özlem Uzuner, editor. 2010. *Working papers of i2b2 Medication Extraction Challenge Workshop*. i2b2.

Xinglong Wang. 2007. Rule-based protein term identification with help from automatic species tagging. In *Proceedings of CICLING 2007*, pages 288–298.

Alexander Yeh, Alexander Morgan, Marc Colosimo, and Lynette Hirschman. 2005. BioCreAtIvE task 1A: gene mention finding evaluation. *BMC Bioinformatics*, 6 Suppl 1.

Pierre Zweigenbaum, Pierre Jacquemart, Natalia Grabar, and Benoît Habert. 2001. Building a text corpus for representing the variety of medical language. In V. L. Patel, R. Rogers, and R. Haux, editors, *Proceedings of Medinfo 2001*, pages 290–294, Londres.

Automatic Acquisition of Huge Training Data for Bio-Medical Named Entity Recognition

Yu Usami[*] [†] **Han-Cheol Cho**[†] **Naoaki Okazaki**[‡] **and Jun'ichi Tsujii**[§]
[*] Aizawa Laboratory, Department of Computer Science, The University of Tokyo, Tokyo, Japan
[†] Tsujii Laboratory, Department of Computer Science, The University of Tokyo, Tokyo, Japan
[‡] Inui Laboratory, Department of System Information Sciences, Tohoku University, Sendai, Japan
[§] Microsoft Research Asia, Beijing, China
{yusmi, hccho}@is.s.u-tokyo.ac.jp
okazaki@ecei.tohoku.ac.jp
jtsujii@microsoft.com

Abstract

Named Entity Recognition (NER) is an important first step for BioNLP tasks, e.g., gene normalization and event extraction. Employing supervised machine learning techniques for achieving high performance recent NER systems require a manually annotated corpus in which every mention of the desired semantic types in a text is annotated. However, great amounts of human effort is necessary to build and maintain an annotated corpus. This study explores a method to build a high-performance NER without a manually annotated corpus, but using a comprehensible lexical database that stores numerous expressions of semantic types and with huge amount of unannotated texts. We underscore the effectiveness of our approach by comparing the performance of NERs trained on an automatically acquired training data and on a manually annotated corpus.

1 Introduction

Named Entity Recognition (NER) is the task widely used to detect various semantic classes such as genes (Yeh et al., 2005), proteins (Tanabe and Wilbur, 2002), and diseases in the biomedical field.

A naïve approach to NER handles the task as a dictionary-matching problem: Prepare a dictionary (gazetteer) containing textual expressions of named entities of specific semantic types. Scan an input text, and recognize a text span as a named entity if the dictionary includes the expression of the span.

Although this approach seemingly works well, it presents some critical issues. First, the dictionary must be comprehensive so that every NE mention can be found in the dictionary. This requirement for dictionaries is stringent because new terminology is being produced continuously, especially in the biomedical field. Second, this approach might suffer from an ambiguity problem in which a dictionary includes an expression as entries for multiple semantic types. For this reason, we must use the context information of an expression to make sure that the expression stands for the target semantic type.

Nadeau and Sekine (2007) reported that a strong trend exists recently in applying machine learning (ML) techniques such as Support Vector Machine (SVM) (Kazama et al., 2002; Isozaki and Kazawa, 2002) and Conditional Random Field (CRF) (Settles, 2004) to NER, which can address these issues. In this approach, NER is formalized as a classification problem in which a given expression is classified into a semantic class or other (non-NE) expressions. Because the classification problem is usually modeled using supervised learning methods, we need a manually annotated corpus for training NER classifier. However, preparing manually annotated corpus for a target domain of text and semantic types is cost-intensive and time-consuming because human experts are needed to reliably annotate NEs in text. For this reason, manually annotated corpora for NER are often limited to a specific domain and covers a small amount of text.

In this paper we propose a novel method for automatically acquiring training data for NER from a comprehensible lexical database and huge amounts of unlabeled text. This paper presents four contribu-

Proceedings of the 2011 Workshop on Biomedical Natural Language Processing, ACL-HLT 2011, pages 65–73,
Portland, Oregon, USA, June 23-24, 2011.

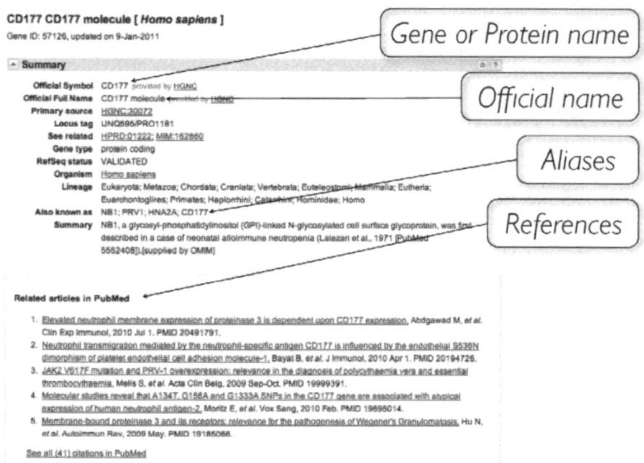

Figure 1: Example of an Entrez Gene record.

tions:

1. We show the ineffectiveness of a naíve dictionary-matching for acquiring a training data automatically and the significance of the quality of training data for supervised NERs

2. We explore the use of reference information that bridges the lexical database and unlabeled text for acquiring high-precision and low-recall training data

3. We develop two strategies for expanding NE annotations, which improves the recall of the training data

4. The proposed method acquires a large amount of high-quality training data rapidly, decreasing the necessity of human efforts

2 Proposed method

The proposed method requires two resources to acquire training data automatically: a comprehensive lexical database and unlabeled texts for a target domain. We chose Entrez Gene (National Library of Medicine, 2005) as the lexical database because it provides rich information for lexical entries and because genes and proteins constitute an important semantic classes for Bio NLP. Entrez Gene consists of more than six million gene or protein records, each of which has various information such as the official gene (protein) name, synonyms, organism, description, and human created references. Figure 1 presents an example of an Entrez Gene record. We created a dictionary by collecting official gene (protein) names and their synonyms from the Entrez Gene records. For unlabeled text, we use the all 2009 release MEDLINE (National Library of Medicine, 2009) data. MEDLINE consists of about ten million abstracts covering various fields of biomedicine and health. In our study, we focused on recognizing gene and protein names within biomedical text.

Our process to construct a NER classifier is as follows: We apply the GENIA tagger (Tsuruoka et al., 2005) to split the training data into tokens and to attach part of speech (POS) tags and chunk tags. In this work, tokenization is performed by an external program that separates tokens by a space, hyphen, comma, period, semicolon, or colon character. Part of speech tags present grammatical roles of tokens, e.g. verbs, nouns, and prepositions. Chunk tags compose tokens into syntactically correlated segments, e.g. verb phrases, noun phrases, and prepositional phrases. We use the IOBES notation (Ratinov and Roth, 2009) to represent NE mentions with label sequences, thereby NER is formalized as a multi-class classification problem in which a given token is classified into IOBES labels. To classify labels of tokens, we use a linear kernel SVM which applies the one-vs.-the-rest method (Weston and Watkins, 1999) to extend binary classification to multi-class classification. Given the t-th token x_t in a sentence, we predict the label y_t,

$$y_t = \underset{y}{\operatorname{argmax}}\, s(y|x_t, y_{t-1}).$$

In this equation, $s(y|x_t, y_{t-1})$ presents the score (sum of feature weights) when the token x_t is labeled y. We use y_{t-1} (the label of the previous token) to predict y_t, expecting that this feature behaves as a label bigram feature (also called translation feature) in CRF. If the sentence consists of x_1 to x_T, we repeat prediction of labels sequentially from the beginning (y_1) to the end (y_T) of a sentence. We used LIBLINEAR (Fan et al., 2008) as an SVM implementation.

Table 1 lists the features used in the classifier modeled by SVM. For each token ("Human" in the example of Table 1), we created several features including: token itself (w), lowercase token (wl), part of speech (pos), chunk tag (chk), character pattern of

Name	Description	Example Value
w	token	Human
wl	token in small letters	human
pos	part of speech	NNP
chk	chunk tag	B-NP
shape	entity pattern	ULLLL
shaped	entity pattern 2	UL
type	token type	InitCap
$p_n(n = 1...4)$	prefix n characters	(H,Hu,Hum,Huma)
$s_n(n = 1...4)$	suffix n characters	(n,an,man,uman)

Table 1: Example of features used in machine learning process.

token (shape), character pattern designated (shaped), token type (type), prefixes of length n (p_n), and suffixes of length n (s_n). More precisely, the character pattern of token (shape) replaces each character in the token with either an uppercase letter (U), a lowercase letter (L), or a digit (D). The character pattern designated (shaped) is similar to a shape feature, but the consecutive character types are reduced to one symbol, for example, "ULLLL" (shape) is represented with "UL" (shaped) in the example of Table 1). The token type (type) represents whether the token satisfies some conditions such as "begins with a capital letter", "written in all capitals", "written only with digits", or "contains symbols". We created unigram features and bigram features (excluding wl, p_n, s_n) from the prior 2 to the subsequent 2 tokens of the current position.

2.1 Preliminary Experiment

As a preliminary experiment, we acquired training data using a naíve dictionary-matching approach. We obtained the training data from all 2009 MEDLINE abstracts with an all gene and protein dictionary in Entrez Gene. The training data consisted of nine hundred million tokens. We constructed a NER classifier using only four million tokens of the training data because of memory limitations. For evaluation, we used the Epigenetics and Post-translational Modification (EPI) corpus BioNLP 2011 Shared Task (SIGBioMed, 2011). Only development data and training data are released as the EPI corpus at present, we used both of the data sets for evaluation in this experiment. Named entities in the corpus are annotated exhaustively and belong to a single semantic class, Gene or Gene Product (GGP) (Ohta et al., 2009). We evaluated the performance of the NER on four measures: Accuracy (a), Precision (P), Recall (R), and F1-measure (F1). We used the strict matching criterion that a predicted named entity is correct if and only if the left and the right boundaries are both correct.

Method	A	P	R	F1
dictionary matching	92.09	39.03	42.69	40.78
trained on acquired data	85.76	10.18	23.83	14.27

Table 2: Results of the preliminary experiment.

(a) It is clear that in culture media of ***AM***, *cystatin C* and *cathepsin B* are present as proteinase–antiproteinase complexes.

(b) Temperature in the puerperium is higher in ***AM***, and lower in PM.

Figure 2: Dictionary-based gene name annotating example (annotated words are shown in italic typeface).

Table 2 presents the evaluation results of this experiment. The first model "dictionary matching" performs exact dictionary-matching on the test corpus. It achieves a 40.78 F1-score. The second model "trained on acquired data" uses the training data acquired automatically for constructing NER classifier. It scores very low-performance (14.27 F1-score), even compared with the simple dictionary-matching NER. Exploring the annotated training data, we investigate why this machine learning approach shows extremely low performance.

Figure 2 presents an example of the acquired training data. The word "AM" in the example (a) is correct because it is gene name, although "AM" in the example (b) is incorrect because "AM" in (b) is the abbreviation of *ante meridiem*, which means before noon. This is a very common problem, especially with abbreviations and acronyms. If we use this noisy training data for learning, then the result of NER might be low because of such ambiguity. It is very difficult to resolve errors in the training data even with the help of machine learning methods.

2.2 Using Reference Information

To obtain high-precision data, we used reference information included with each record in Entrez Gene. Figure 3 portrays a simple example of reference information. It shows the reference information of the

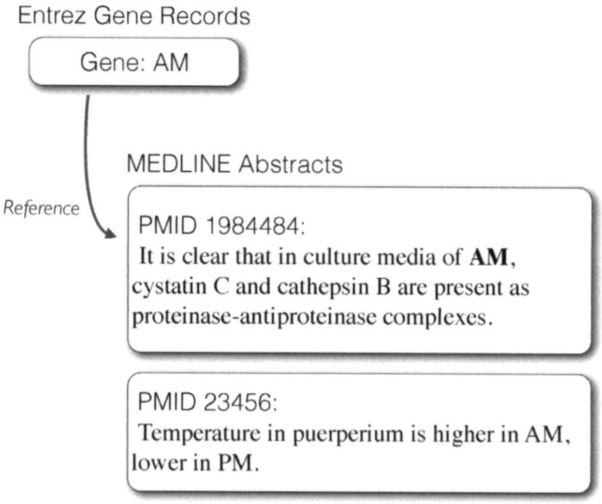

Figure 3: Reference to MEDLINE abstract example.

Entrez Gene record which describes that the gene "AM". The reference information indicates PMIDs in which the gene or protein is described.

We applied the rule whereby we annotated a dictionary-matching in each MEDLINE abstract only if they were referred by the Entrez Gene records. Figure 3 shows that the gene "AM" has reference to the MEDLINE abstract #1984484 only. Using this reference information between the Entrez Gene record "AM" and the MEDLINE abstract #1984484, we can annotate the expansion "AM" in MEDLINE abstract #1984484 only. In this way, we can avoid incorrect annotation such as example b in Figure 2.

We acquired training data automatically using reference information, as follows:

1. Construct a gene and protein dictionary including official names, synonyms and reference information in Entrez Gene
2. Apply a dictionary-matching on the all MEDLINE abstracts with the dictionary
3. Annotate the MEDLINE abstract only if it was referred by the Entrez Gene records which describe the matched expressions

We obtained about 48,000,000 tokens of training data automatically by using this process using all the 2009 MEDLINE data. This training data includes about 3,000,000 gene mentions.

- ... in the following order: *tna*, *gltC*, *gltS*, *pyrE*; *gltR* is located near ...
- The three genes concerned (designated *entA*, *entB* and *entC*) ...
- Within the hypoglossal nucleus large amounts of *acetylcholinesterase* (*AChE*) activity are ...

Figure 4: False negative examples.

2.3 Training Data Expansion

In the previous section, we were able to obtain training data with high-precision by exploiting reference information in the Entrez Gene. However, the resulting data include many false negatives (low-recall), meaning that correct gene names in the data are unannotated. Figure 4 presents an example of missing annotation. In this figure, all gene mentions are shown in italic typeface. The underlined entities were annotated by using the method in Section 2.2, because they were in the Entrez Gene dictionary and this MEDLINE abstract was referred by these entities. However, the entities in italic typeface with no underline were not annotated, because these gene names in Entrez Gene have no link to this MEDLINE abstract. Those expressions became false negatives and became noise for learning. This low-recall problem occurred because no guarantee exists of exhaustiveness in Entrez Gene reference information.

To improve the low-recall while maintaining high-precision, we focused on coordination structures. We assumed that coordinated noun phrases belong to the same semantic class. Figure 5 portrays the algorithm for the annotation expansion based on coordination analysis. We expanded training data annotation using this coordination analysis algorithm to improve annotation recall. This algorithm analyzes whether the words are reachable or not through coordinate tokens such as ",", ".", or "and" from initially annotated entities. If the words are reachable and their entities are in the Entrez Gene records (ignoring reference information), then they are annotated.

Input: Sequence of sentence tokens S, Set of symbols and conjunctions C, Dictionary without reference D, Set of annotated tokens A
Output: Set of Annotated tokens A

begin
for $i = 1$ to $|S|$ **do**
  if $S[i] \in A$ **then**
    $j \leftarrow i - 2$
    while $1 \leq j \leq |S| \wedge S[j] \in D \wedge S[j] \notin A \wedge S[j+1] \in C$ **do**
      $A \leftarrow A \cap \{S[j]\}$
      $j \leftarrow j - 2$
    end while
    $j \leftarrow i + 2$
    while $1 \leq j \leq |S| \wedge S[j] \in D \wedge S[j] \notin A \wedge S[j-1] \in C$ **do**
      $A \leftarrow A \cap \{S[j]\}$
      $j \leftarrow j + 2$
    end while
  end if
end for
Output A
end

Figure 5: Coordination analysis algorithm.

2.4 Self-training

The method described in Section 2.3 reduces false negatives based on coordination structures. However, the training data contain numerous false negatives that cannot be solved through coordination analysis. Therefore, we used a self-training algorithm to automatically correct the training data. In general, a self-training algorithm obtains training data with a small amount of annotated data (seed) and a vast amount of unlabeled text, iterating this process (Zadeh Kaljahi, 2010):

1. Construct a classification model from a seed, then apply the model on the unlabeled text.
2. Annotate recognized expressions as NEs.
3. Add the sentences which contain newly annotated expressions to the seed.

In this way, a self-training algorithm obtains a huge amount of training data.

Input: Labeled training data D, Machine learning algorithm A, Iteration times n, Threshold θ
Output: Training data T_n

begin
$T_0 \leftarrow$ A seed data from D
$i \leftarrow 0$
$D \leftarrow D \backslash T_0$
while $i \neq n$ **do**
  $M_i \leftarrow$ Construct model with T_i
  $U \leftarrow$ Sample some amount of data from D
  $L \leftarrow$ Annotate U with model M_i
  $U_{new} \leftarrow$ Merge U with L if their confidence values are larger than θ
  $T_{i+1} \leftarrow T_i \cup U_{new}$
  $D \leftarrow D \backslash U$
  $i \leftarrow i + 1$
end while
Output T_n
end

Figure 6: Self-training algorithm.

In contrast, our case is that we have a large amount of training data with numerous false negatives. Therefore, we adapt a self-training algorithm to revise the training data obtained using the method described in Section 2.3. Figure 6 shows the algorithm. We split the data set (D) obtained in Section 2.3 into a seed set (T_0) and remaining set ($D \backslash T_0$). Then, we iterate the cycle ($0 \leq i \leq n$):

1. Construct a classification model (M_i) trained on the training data (T_i).
2. Sample some amount of data (U) from the remaining set (D).
3. Apply the model (M_i) on the sampled data (U).
4. Annotate entities (L) recognized by this model.
5. Merge newly annotated expressions (L) with expressions annotated in Section 2.3 (U) if their confidence values are larger than a threshold (θ).
6. Add the merged data (U_{new}) to the training data (T_i).

In this study, we prepared seed data of 683,000 tokens (T_0 in Figure 6). In each step, 227,000 tokens were sampled from the remaining set (U).

Because the remaining set U has high precision and low recall, we need not revise NEs that were annotated in Section 2.3. It might lower the quality of the training data to merge annotated entities, thus we used confidence values (Huang and Riloff, 2010) to revise annotations. Therefore, we retain the NE annotations of the remaining set U and overwrite a span of a non-NE annotation only if the current model predicts the span as an NE with high confidence. We compute the confidence of the prediction ($f(x)$) which a token x is predicted as label y as,

$$f(x) = s(x, y) - max(\forall_{z \neq y} s(x, z)).$$

Here, $s(x, y)$ denotes the score (the sum of feature weights) computed using the SVM model described in the beginning of Section 2. A confidence score presents the difference of scores between the predicted (the best) label and the second-best label. The confidence value is computed for each token label prediction. If the confidence value is greater than a threshold (θ) and predicted as an NE of length 1 token (label `S` in IOBES notation), then we revise the NE annotation. When a new NE with multiple tokens (label `B`, `I`, or `E` in IOBES notation) is predicted, we revise the NE annotation if the average of confidence values is larger than a threshold (θ). If a prediction suggests a new entity with multiple tokens x_i, ..., x_j, then we calculate the average of confidence values as

$$f(x_i, ..., x_j) = \frac{1}{j - i + 1} \sum_{k=i}^{j} f(x_k).$$

The feature set presented in the beginning of Section 2 uses information of the tokens themselves. These features might overfit the noisy seed set, even if we use regularization in training. Therefore, when we use the algorithm of Figure 6, we do not generate token (w) features from tokens themselves but only from tokens surrounding the current token. In other words, we hide information from the tokens of an entity, and learn models using information from surrounding words.

Method	A	P	R	F1
dictionary matching	92.09	39.03	42.69	40.78
svm	85.76	10.18	23.83	14.27
+ reference	93.74	**69.25**	39.12	50.00
+ coordination	93.97	66.79	47.44	55.47
+ self-training	**93.98**	63.72	**51.18**	**56.77**

Table 3: Evaluation results.

3 Experiment

The training data automatically generated using the proposed method have about 48,000,000 tokens and 3,000,000 gene mentions. However, we used only about 10% of this data because of the computational cost. For evaluation, we chose to use the BioNLP 2011 Shared Task EPI corpus and evaluation measures described in Section 2.1.

3.1 Evaluation of Proposed Methods

In the previous section, we proposed three methods for automatic training data acquisition. We first investigate the effect of these methods on the performance of NER. Table 3 presents evaluation results.

The first method "dictionary matching" simply performs exact string matching with the Entrez Gene dictionary on the evaluation corpus. It achieves a 40.78 F1-measure; this F1-measure will be used as the baseline performance. The second method, as described in Section 2.1, "svm" uses training data generated automatically from the Entrez Gene and unlabeled texts without reference information of the Entrez Gene. The third method, "+ reference" exploits the reference information of the Entrez Gene. This method drastically improves the performance. As shown in Table 3, this model achieves the highest precision (69.25%) with comparable recall (39.12%) to the baseline model with a 50.00 F1-measure. The fourth method, "+ coordination", uses coordination analysis results to expand the initial automatic annotation. Compared to the "+ reference" model, the annotation expansion based on coordination analysis greatly improves the recall (+8.32%) with only a slight decrease of the precision (-2.46%). The last method "+ self-training" applies a self-training technique to improve the performance further. This model achieves the highest recall (51.18%) among all models with a reasonable cost in the precision.

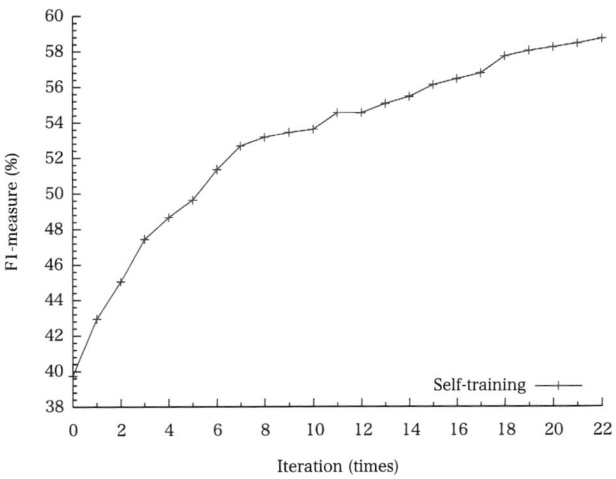

Figure 7: Results of self-training.

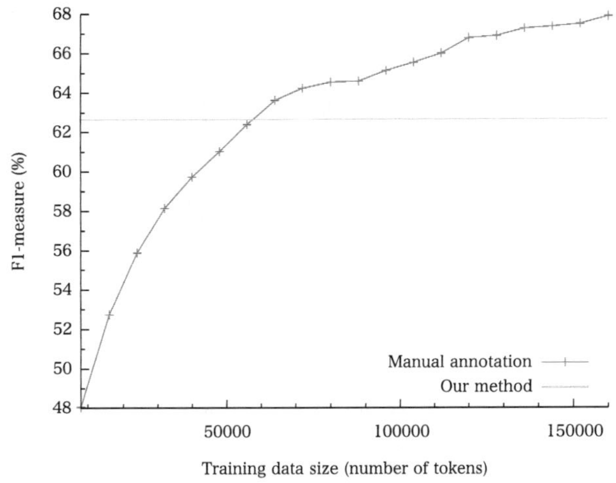

Figure 8: Manual annotation vs. our method.

To analyze the effect of self-training, we evaluated the performance of this model for each iteration. Figure 7 shows the F1-measure of the model as iterations increase. The performance improved gradually. It did not converge even for the last iteration. The size of the training data at the 17th iteration was used in Table 3 experiment. It is the same to the size of the training data for other methods.

3.2 Comparison with a Manually Annotated Corpus

NER systems achieving state-of-the-art performance are based mostly on supervised machine learning trained on manually annotated corpus. In this section, we present a comparison of our best-performing NER model with a NER model trained on manually annotated corpus. In addition to the performance comparison, we investigate how much manually annotated data is necessary to outperform our best-performing system. In this experiment, we used only the development data for evaluation because the training data are used for training the NER model.

We split the training data of EPI corpus randomly into 20 pieces and evaluated the performance of the conventional NER system as the size of manually annotated corpus increases. Figure 8 presents the evaluation results. The performance of our our best-performing NER is a 62.66 F1-measure; this is shown as horizontal line in Figure 8. The NER model trained on the all training data of EPI corpus achieves a 67.89 F1-measure. The result shows that our best-performing models achieve comparable performance to that of the NER model when using about 40% (60,000 tokens, 2,000 sentences) of the manually annotated corpus.

3.3 Discussion

Although the proposed methods help us to obtain training data automatically with reasonably high quality, we found some shortcomings in these methods. For example, the annotation expansion method based on coordination analysis might find new entities in the training data precisely. However, it was insufficient in the following case.

> *tna* loci, in the following order: *tna*, *gltC*, *gltS*, *pyrE*; *gltR* is located near ...

In this example, all gene mentions are shown in italic typeface. The words with underline were initial annotation with reference information. The surrounding words represented in italic typeface are annotated by annotation expansion with coordination analysis. Here, the first word "tna" shown in italic typeface in this example is not annotated, although its second mention is annotated at the annotation expansion step. We might apply the one sense per discourse (Gale et al., 1992) heuristic to label this case.

Second, the improvement of self-training techniques elicited less than a 1.0 F1-measure. To ascertain the reason for this small improvement, we analyzed the distribution of entity length both origi-

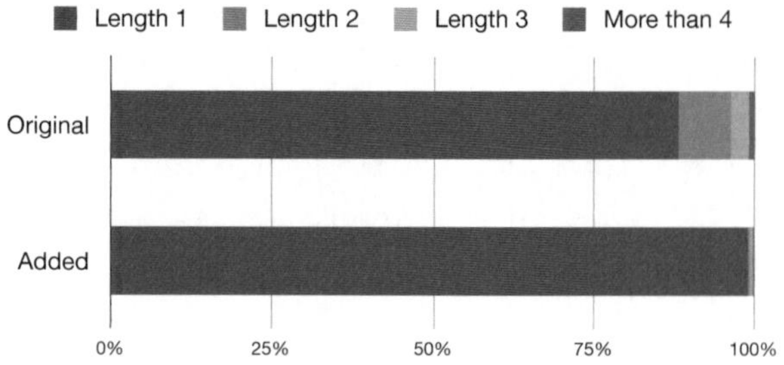

Figure 9: Distribution of entity length.

nally included entities and newly added entities during self-training, as shown in Figure 9. They represent the ratio of entity length to the number of total entities. Figure 9 shows the added distribution of entity length (Added) differs from the original one (Original). Results of this analysis show that self-training mainly annotates entities of the length one and barely recognizes entities of the length two or more. It might be necessary to devise a means to follow the corpus statistics of the ratio among the number of entities of different length as the self-training iteration proceeds.

4 Related Work

Our study focuses mainly on achieving high performance NER without manual annotation. Several previous studies aimed at reducing the cost of manual annotations.

Vlachos and Gasperin (2006) obtained noisy training data from FlyBase[1] with few manually annotated abstracts from FlyBase. This study suggested the possibility of acquiring high-quality training data from noisy training data. It used a bootstrapping method and a highly context-based classifiers to increase the number of NE mentions in the training data. Even though the method achieved a high-performance NER in the biomedical domain, it requires curated seed data.

Whitelaw et al. (2008) attempted to create extremely huge training data from the Web using a seed set of entities and relations. In generating training data automatically, this study used context-based tagging. They reported that quite a few good resources (e.g., Wikipedia[2]) listed entities for obtaining training data automatically.

Muramoto et al. (2010) attempted to create training data from Wikipedia as a lexical database and blogs as unlabeled text. It collected about one million entities from these sources, but they did not report the performance of the NER in their paper.

5 Conclusions

This paper described an approach to the acquisition of huge amounts of training data for high-performance Bio NER automatically from a lexical database and unlabeled text. The results demonstrated that the proposed method outperformed dictionary-based NER. Utilization of reference information greatly improved its precision. Using coordination analysis to expand annotation increased recall with slightly decreased precision. Moreover, self-training techniques raised recall. All strategies presented in the paper contributed greatly to the NER performance.

We showed that the self-training algorithm skewed the length distribution of NEs. We plan to improve the criteria for adding NEs during self-training. Although we obtained a huge amount of training data by using the proposed method, we could not utilize all of acquired training data because they did not fit into the main memory. A future direction for avoiding this limitation is to employ an online learning algorithm (Tong and Koller, 2002; Langford et al., 2009), where updates of feature weights are done for each training instance. The necessity of coordination handling and self-training originates from the insufficiency of reference information in the lexical database, which was not designed to be comprehensive. Therefore, establishing missing reference information from a lexical database to unlabeled texts may provide another solution for improving the recall of the training data.

References

Rong-En Fan, Kai-Wei Chang, Cho-Jui Hsieh, Xiang-Rui Wang, and Chih-Jen Lin. 2008. LIBLINEAR: A Library for Large Linear Classification. *Journal of Machine Learning Research*, 9:1871–1874.

William A. Gale, Kenneth W. Church, and David Yarowsky. 1992. One sense per discourse. In *Proceedings of the workshop on Speech and Natural Language*, pages 233–237.

[1] `http://flybase.org/`

[2] `http://www.wikipedia.org/`

Ruihong Huang and Ellen Riloff. 2010. Inducing domain-specific semantic class taggers from (almost) nothing. In *Proceedings of the 48th Annual Meeting of the Association for Computational Linguistics*, pages 275–285.

Hideki Isozaki and Hideto Kazawa. 2002. Efficient support vector classifiers for named entity recognition. In *Proceedings of the 19th international conference on Computational linguistics - Volume 1*, pages 1–7.

Jun'ichi Kazama, Takaki Makino, Yoshihiro Ohta, and Jun'ichi Tsujii. 2002. Tuning support vector machines for biomedical named entity recognition. In *Proceedings of the ACL-02 workshop on Natural language processing in the biomedical domain - Volume 3*, pages 1–8.

John Langford, Lihong Li, and Tong Zhang. 2009. Sparse online learning via truncated gradient. *J. Mach. Learn. Res.*, 10:777–801.

Hideki Muramoto, Nobuhiro Kaji, Naoki Suenaga, and Masaru Kitsuregawa. 2010. Learning semantic category tagger from unlabeled data. In *The Fifth NLP Symposium for Yung Researchers*. (in Japanese).

David Nadeau and Satoshi Sekine. 2007. A survey of named entity recognition and classification. *Lingvisticae Investigationes*, 30(1):3–26.

National Library of Medicine. 2005. Entrez Gene. available at http://www.ncbi.nlm.nih.gov/gene.

National Library of Medicine. 2009. MEDLINE. available at http://www.ncbi.nlm.nih.gov/.

Tomoko Ohta, Jin-Dong Kim, Sampo Pyysalo, Yue Wang, and Jun'ichi Tsujii. 2009. Incorporating genetag-style annotation to genia corpus. In *Proceedings of the Workshop on Current Trends in Biomedical Natural Language Processing*, pages 106–107.

Lev Ratinov and Dan Roth. 2009. Design challenges and misconceptions in named entity recognition. In *Proceedings of the Thirteenth Conference on Computational Natural Language Learning*, pages 147–155.

Burr Settles. 2004. Biomedical named entity recognition using conditional random fields and rich feature sets. In *Proceedings of the International Joint Workshop on Natural Language Processing in Biomedicine and its Applications*, pages 104–107.

SIGBioMed. 2011. BioNLP 2011 Shared Task. http://sites.google.com/site/bionlpst/.

Lorraine K. Tanabe and W. John Wilbur. 2002. Tagging gene and protein names in biomedical text. *Bioinformatics/computer Applications in The Biosciences*, 18:1124–1132.

Simon Tong and Daphne Koller. 2002. Support vector machine active learning with applications to text classification. *J. Mach. Learn. Res.*, 2:45–66.

Yoshimasa Tsuruoka, Yuka Tateishi, Jin-Dong Kim, Tomoko Ohta, John McNaught, Sophia Ananiadou, and Jun ' ichi Tsujii. 2005. Developing a robust part-of-speech tagger for biomedical text. In *Advances in Informatics*, volume 3746, pages 382–392.

Andreas Vlachos and Caroline Gasperin. 2006. Bootstrapping and evaluating named entity recognition in the biomedical domain. In *Proceedings of the HLT-NAACL BioNLP Workshop on Linking Natural Language and Biology*, pages 138–145.

Jason Weston and Chris Watkins. 1999. Support vector machines for multi-class pattern recognition. In *ESANN'99*, pages 219–224.

Casey Whitelaw, Alex Kehlenbeck, Nemanja Petrovic, and Lyle Ungar. 2008. Web-scale named entity recognition. In *Proceeding of the 17th ACM conference on Information and knowledge management*, pages 123–132.

Alexander Yeh, Alexander Morgan, Marc Colosimo, and Lynette Hirschman. 2005. Biocreative task 1a: gene mention finding evaluation. *BMC Bioinformatics*, 6(1):S2.

Rasoul Samad Zadeh Kaljahi. 2010. Adapting self-training for semantic role labeling. In *Proceedings of the ACL 2010 Student Research Workshop*, pages 91–96.

Building frame-based corpus on the basis of ontological domain knowledge

He Tan
Institutionen för
datavetenskap
Linköpings universitet
Sweden
he.tan@liu.se

Rajaram Kaliyaperumal
Institutionen för
medicinsk teknik
Linköpings universitet
Sweden
rajka625

Nirupama Benis
Institutionen för
medicinsk teknik
Linköpings universitet
Sweden
nirbe455@student.liu.se

Abstract

Semantic Role Labeling (SRL) plays a key role in many NLP applications. The development of SRL systems for the biomedical domain is frustrated by the lack of large domain-specific corpora that are labeled with semantic roles. Corpus development has been very expensive and time-consuming. In this paper we propose a method for building frame-based corpus on the basis of domain knowledge provided by ontologies. We believe that ontologies, as a structured and semantic representation of domain knowledge, can instruct and ease the tasks in building the corpora. In the paper we present a corpus built by using the method. We compared it to BioFrameNet, and examined the gaps between the semantic classification of the target words in the domain-specific corpus and in FrameNet and PropBank/VerbNet.

1 Introduction

The sentence-level semantic analysis of text is concerned with the characterization of events, such as determining "who" did "what" to "whom", "where", "when" and "how". It is believed to play a key role in NLP applications such as Information Extraction, Question Answering and Summarization. Semantic Role Labeling (SRL) is a process that, for each predicate in a sentence, indicates what semantic relations hold among the predicate and other sentence constituents that express the participants in the event (such as who and where). The relations are described by using a list of pre-defined possible semantic roles for that predicate (or class of predicates). Recently, large corpora have been manually annotated with semantic roles in FrameNet (Fillmore et al., 2001) and PropBank (Palmer et al., 2005). With the advent of resources, SRL has become a well-defined task with a substantial body of work and comparative evaluation. Most of the work has been trained and evaluated on newswire text (see (Màrquez et al., 2008)).

Biomedical text considerably differs from the newswire text, both in the style of the written text and the predicates involved. Predicates in newswire text are typically verbs, biomedical text often prefers nominalizations, gerunds, and relational nouns (Kilicoglu et al., 2010). Predicates like *endocytosis*, *exocytosis* and *translocate*, though common in biomedical text, are absent from both the FrameNet and PropBank data (Bethard et al., 2008). Predicates like *block*, *generate* and *transform*, have been used in biomedical documents with different semantic senses and require different number of semantic roles compared to FrameNet (Tan, 2010) and PropBank (Wattarujeekrit et al., 2004). The development of SRL systems for the biomedical domain is frustrated by the lack of large domain-specific corpora that are labeled with semantic roles.

The projects, PASBio (Wattarujeekrit et al., 2004), BioProp (Tsai et al., 2006) and BioFrameNet (Dolbey et al., 2006), have made efforts on building PropBank-like and FrameNet-like corpora for processing biomedical text. Up until recently, these corpora are relatively small. Further, no general methodology exists to support domain-specific corpus construction. The difficulties include, how to discover and define

Proceedings of the 2011 Workshop on Biomedical Natural Language Processing, ACL-HLT 2011, pages 74–82,
Portland, Oregon, USA, June 23-24, 2011. ©2011 Association for Computational Linguistics

semantic frames together with associated semantic roles within the domain? how to collect and group domain-specific predicates to each semantic frame? and how to select example sentences from publication databases, such as the PubMed/MEDLINE database containing over 20 million articles? In this paper, we propose that building frame-based lexicon for the domain can be strongly instructed by domain knowledge provided by ontologies. We believe that ontologies, as a structured and semantic representation of domain-specific knowledge, can instruct and ease all the above tasks.

The paper proceeds as follows: first we explain our method how ontological domain knowledge instructs the main tasks in building a frame-based lexicon. This is followed by the related work. In section 4, we present a "study case" of the method. We built a frame *Protein Transport* containing text annotated with semantic roles. The construction is carried out completely under the supervision of the domain knowledge from the Gene Ontology (GO) (Ashburner et al., 2000). We evaluated it to the frame *Protein_transport* in the BioFrameNet and examined the gaps between the semantic classification of the target words in the domain-specific corpus and in FrameNet and PropBank/VerbNet. Finally, we conclude our work.

2 The Method

The FrameNet project is the application of the theory of *Frames Semantics* (Fillmore et al., 1985) in computational lexicography. Frame semantics begins with the assumption that in order to understand the meanings of the words in a language, we must first have knowledge of the background and motivation for their existence in the language and for their use in discourse. The knowledge is provided by the conceptual structures, or *semantic frames*. In FrameNet, a semantic frame describes an event, a situation or a object, together with the participants (called frame elements (FE)) involved in it. A word evokes the frame, when its sense is based on the frame. The relations between frames include *is-a*, *using* and *subframe*.

Ontology is a formal representation of knowledge of a domain of interest. It has concepts that represent sets or classes of entities within a domain. It defines different types of relations among concepts. Intuitively, ontological concepts and their relations can be used as the frame-semantic descriptions imposed on a lexicon.

A large number of ontologies have been developed in the domain of biomedicine. Many of them contain concepts that comprehensively describe a certain domain of interest, such as GO. GO biological process ontology, containing 20,368 concepts, provides the structured knowledge of biological processes that are recognized series of events or molecular functions. For example, the concept GO:0015031 protein transport defines the scenario, "*the directed movement of proteins into, out of or within a cell, or between cells, by means of some agent such as a transporter or pore*". It is a subclass of GO:0006810:transport and GO:0045184:establishment of protein localization. The class has 177 descendant classes in *is-a* hierarchies. A *Protein Transport* frame can be effectively described by using these classes and relations between them.

In many cases ontological terms can be seen as phrases that exhibit underlying compositional structures (McCray et al., 2002; Ogren et al., 2005). Figure 1 presents the compositional structures of 9 direct subclasses describing various types of protein transport. They provide that `translocation`, `import`, `recycling`, `secretion` and `transport` are the possible predicates, evoking the protein transport event. The more complex expressions, e.g. translocation of peptides or proteins into other organism involved in symbiotic interaction (GO:0051808), express participants involved in the event, i.e. the entity (`peptides or proteins`), destination (`into other organism`) and condition (`involved in symbiotic interaction`) of the event.

So far, we, using these classes and relations between them, have partly defined the semantic frame *Protein Transport*, decided the participants involved in the event, and listed the domain-specific words evoking the frame. The complete frame description can be given after studying all the related classes and their relations. Lastly, collecting example sentences will be based on knowledge based search engine for biomedical text, like GoPubMed (Doms and Schroeder, 2005). As such, domain knowledge

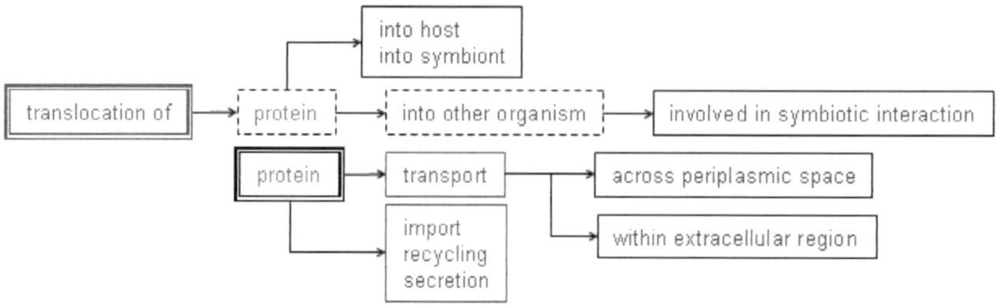

Figure 1: A concise view of 9 GO terms describing *Protein Transport*. We use the modified finite state automaton (FSA) representation given in (Ogren et al., 2005). Any path that begins at a start state, represented by double solid borders, and ends in an end state, represented by a single solid border, corresponds to a term. The nodes with a dashed border are neither start states nor end states.

provided by ontologies, such as GO biological process ontology and molecular function ontology, and pathway ontologies, can instruct us in building large frame-based corpora for the domain.

We outline the aspects of how ontologies instruct building a frame-based corpus:

1. The structure and semantics of domain knowledge in ontologies constrain the frame semantics analysis, i.e. decide the coverage of semantic frames and the relations between them;

2. Ontological terms can comprehensively describe the characteristics of events/scenarios in the domain, so domain-specific semantic roles can be determined based on terms;

3. Ontological terms provide a list of domain-specific predicates, so the semantic senses of the predicates in the domain are determined;

4. The collection and selection of example sentences can be based on knowledge-based search engine for biomedical text.

3 Related Work

The PropBank project is to add a semantic layer on the Penn Treebank (Marcus et al., 1994). For each unique verb sense, a set of semantic roles is defined at its accompanying syntactic realizations. The VerbNet project (Kipper et al., 2000) systematically creates English verb entries in a lexicon with syntactic and semantic information, referring to Levin verb classes. It made efforts to classify individual verbs in PropBank into VerbNet classes, based on patterns of usage (Kingsbury and Kipper, 2003).

The FrameNet project collects and analyzes the corpus (the British National Corpus) attestations of target words with semantic overlapping. The attestations are divided into semantic groups, noting especially the semantic roles of each target words, and then these small groups are combined into frames.

Ontologies have been put under the spotlight for providing the framework for semantic representation of textual information, and thus a basis for text mining systems (Spasic et al., 2005; Ashburner et al., 2008). Up to recently, TM systems mainly use ontologies as terminologies to recognize biomedical terms, by mapping terms occurring in text to concepts in ontologies, or use ontologies to guide and constrain analysis of NLP results, by populating ontologies. In the latter case, ontologies are more actively used as a structured and semantic representation of domain knowledge.

The FrameNet project links Semantic Types (ST) of FrameNet to the Suggested Upper Merged Ontology (SUMO) classes (Scheffczyk et al., 2006). The main function of ST is to indicate the basic typing of fillers of semantic roles, e.g. "Sentient" defined for the semantic role "Cognizer" in the frame "Cogitation". The goal of the work is to combine frame semantics in FrameNet and the formal world knowledge from SUMO, for improving FrameNet capability for deductive reasoning.

BioFrameNet is a domain-specific FrameNet extension. Its *intracellular protein transport* frames are mapped to the Hunter Lab [1] knowledge base (HLKB) protein transport classes. The frame elements are taken from HLKB slots. BioFrameNet

[1] Website for Hunters Bioinformatics research lab: `http://compbio.uchsc.edu/`.

considered a collection of Gene References in Function (GRIF) texts that are annotated by the HLKB protein transport classes in the knowledge base. Predicates are extracted from this collection of GRIF texts.

PASBio and BioProp are the projects that aim to produce definitions of Predicate Argument Structure (PAS) frames. They do not offer a direct linking of the predicates or their arguments to domain or general ontologies. PASBio used a model for a hypothetical signal transduction pathway of an idealized cell, to motivate verb choices. BioProp annotated the arguments of 30 frequent biomedical verbs found in the GENIA corpus (Kim et al., 2003).

"This frame deals with the cellular process in which a protein or protein-complex, the *Transport_Entity*, moves from the *Transport_Origin* to a different location, the *Transport_Destination*. Sometimes the *Transport_Origin* and *Transport_Destination* are not specified or are the same location. The *Transport_Entity* undergoes directed movement into, out of or within a cell or between cells or within a multicellular organism. This activity could be aided or impeded by other substances, organelles or processes and could influence other cellular processes."

Table 1: The frame definition.

4 Case Study: *Protein Transport* Frame

In this section we present the frame *Protein Transport*. The frame is built completely based on the domain knowledge provided by the piece of GO describing the event. The core structure of the frame is the same as that of FrameNet. The description of the scenario evoked by the frame is provided, along with a list of the frame elements and their definitions. A list of lexical units (LUs) that evoke the frame is provided. In addition, example sentences that contain at least one of the LUs, are given annotations using definitions of the frame. The annotations follow FrameNet's guidelines for lexicographic annotation, described in (Ruppenhofer et al., 2005).

4.1 The Frame

Resources. The description of the frame uses the scenario defined in GO:0015031 protein transport from the GO biological process ontology. It is a subclass of GO:0006810 transport and GO:0045184 establishment of protein localization. The class has 177 descendant classes. A total of 581 class names and synonyms are collected for the study. In addition to that from GO concepts, synonyms are also gathered by querying the UMLS Metathesaurus (Schuyler et al., 1992).

Frame. The definition (see Table 1) follows the definition of GO:0015031 protein transport.

Frame Elements. By studying all the names and synonyms (we call them "term" in the paper), we defined all possible FEs for the frame (see Table 2). The first 4 FEs are considered as core FEs. Table 3 gives the number of the GO terms that indicate the FEs. For instance, in the term GO:003295 B cell receptor transport within lipid bilayer, `lipid bilayer` is the location `within` which protein transport happens. The term GO:0072322 protein transport across periplasmic space describes the path along which protein transport occurs. The term GO:0043953 protein transport by the Tat complex specifies a molecule that carries protein during the movement. GO:0030970 retrograde protein transport, ER to cytosol indicates the direction (`retrograde`) of the movement. An attribute (`SRP-independent`) of the event is described in the term GO:0006620 SRP-independent protein-membrane targeting ER.

Predicates. All lexical units in the frame are listed in Table 4. The first row gives the head of the GO terms (noun phrases). The number in the bracket indicates the number of GO terms with the head. If the verb derived from a head, can be used to describe the event that is expressed by the head, it is also included as a LU. GO terms, such as *related* and *broader* synonyms, may be not considered for collecting predicates. For example, fat body metabolism, a *broad* synonym of GO:0015032 storage protein import into fat body, is not considered.

Example Sentences. The example sentences are retrieved from the PubMed/MEDLINE database by using the GoPubMed (Doms and Schroeder, 2005), a knowledge-based search engine for biomedical text. The sentences to be annotated, are always the most relevant and from the latest publications. For

FEs	definition
Transport_ Entity (TE)	Protein or protein complex which is undergoing the motion event into, out of or within a cell, or between cells, or within a multicellular organism.
Transport_ Origin (TO)	The organelle, cell, tissue or gland from which the Transport_Entity moves to a different location.
Transport_ Destination (TDS)	The organelle, cell, tissue or gland to which the Transport_Entity moves from a different location.
Transport_ Condition (TC)	The event, substance, organelle or chemical environment which positively or negatively directly influences or is influenced by, the motion event. The substance organelle does not necessarily move with the Transport_Entity
Transport_ Location (TL)	The organelle, cell, tissue or gland where the motion event takes place when the origin and the destination are the same or when origin or destination is not specified.
Transport_ Path (TP)	The substance or organelle which helps the entity to move from the Transport_Origin to the Transport_Destination, sometimes by connecting the two locations, without itself undergoing translocation
Transport_ Transporter (TT)	The substance, organelle or cell crucial to the motion event, that moves along with the Transport_Entity, taking it from the Transport_Origin to the Transport_Destination.
Transport_ Direction (TDR)	The direction in which the motion event is taking place with respect to the Transport_Place, Transport_Origin, Transport_Destination or Transport_Location.
Transport_ Attribute (TA)	This describes the motion event in more detail by giving information on how (particular movement, speed etc.) the motion event occurs. It could also give information on characteristic or typical features of the motion event.

Table 2: The frame elements

#T	578	50	159	95	41	27	6	2	1
FEs	TE	TO	TDS	TC	TL	TP	TT	TDR	TA

Table 3: The number of the GO terms that describe the frame elements

the head of GO terms	delivery (1), egress (2), establishment of ... localization (19), exit (2), export (20), import (88), recycling (2), release (1), secretion (226), sorting (4), targeting (68), trafficking (1), translocation (76), transport (100), uptake (5)
LUs	delivery.n, deliver.v, egress.n, establishment of ... localization.n, exit.n, exit.v, export.n, export.v, import.n, import.v, recycling.n recycle.v, release.n, release.v, secretion.n, secrete.v, sort.v, sorting.n, target.v, targeting.n, translocation.n, translocate.v, transport.v, transport.n, trafficking.n, uptake.n

Table 4: The lexical units

[L.pneumophila$_{Transport_Origin\|\mathrm{NP.Ext}}$] [**translocate**$_{predicate}$] [more than 100 effector proteins$_{Transport_Entity\|\mathrm{NP.Obj}}$] [into host cytoplasm$_{Transport_Destination\|\mathrm{PP[into].Dep}}$] [using Dot/Icm T4BSS$_{Transport_Path\|\mathrm{VPing.Dep}}$], [modulating host cellular functions$_{Transport_Condition\|\mathrm{VPing.Dep}}$] to establish a replicative niche within host cells. (PMID: 20949065)

Table 5: An example sentence: the three layers of annotations are given as FE|PT.GF.

LUs derived from one head, we acquired sentences by using the GO terms with the head. The query starts from using the most general GO terms. In the case that the number of query results is huge, more specific terms are used instead. Minimally, 10 sentences are gathered for each LU, if applicable. In cases when only specific GO terms are available and the number of query results is too small, we generalize the query term. For example, the lexical units, `release.n` and `release.v`, are derived and only derived from GO:0002001 renin secretion into blood stream's synonym renin release into blood stream. No query result returns for the GO term. The general term "protein release" is used as the query term instead.

Table 5 shows an example sentence for the frame. For each sentence annotated, we mark the target LU, and collect and record syntactic and semantic information about the relevant frame's FEs. For each FE, three types of annotation are gathered. The first layer is the identity of the specific FE. In cases when the FE is explicitly realized, phrase type (PT, for example NP) and grammatical function (GF) of the realization are annotated. The GFs describe the ways in which the constituents satisfy abstract grammatical requirements of the target word. In cases when the FE is omitted, the type of its null instantiation is recorded. These three layers for all of the annotated sentences, along with complete frame and FE descriptions are used in summarizing valence patterns for each annotated LU.

4.2 Evaluation

4.2.1 Compared to BioFrameNet

We compared this frame to the frame *Protein_transport* in BioFrameNet [2]. The frame involves the phenomenon of intracellular protein transport. BioFrameNet considered a collection of GRIF texts that describe various types of intracellular protein transport phenomena. The GRIFs texts are annotated by HLKB protein transport classes. All the 5 HLKB protein transport classes are arranged in *is-a* hierarchy. The description of the top level class *protein transport* is taken from the definition of GO:0015031 protein transport which is a superclass of GO:0006886 intracellular protein transport in GO. For the frame, BioFrameNet provides definitions for 4 FEs, including *Transported_entity*, *Transport_origin*, *Transport_destination* and *Transport_locations*. The proposed FEs are taken from the slot definitions in the HLKB classes.

Table 6 illustrates the difference between the LUs in the 2 frames. The LUs that are not included in our corpus, can be classified into two groups. The first group include the LUs `enter.v`, `redistribution.n`, `return.v`, and `traffic.n`. They or their nominals are absent from GO biological process ontology terms. The second group includes those appear in GO, but in the terms that are not included in descendants of GO:0015031 protein transport.

The LUs, `endocytosis.n`, `internalization.n`, `recruitment.n`, do not appear in the descendants of GO:0015031 protein transport, but appear in GO terms that indeed describe protein transport event. `endocytosis` is the head of 9 GO terms, among which 2 concepts indeed describe an endocytotic process of protein (e.g.GO:0070086 ubiquitin-dependent endocytosis). 3 GO terms have `internalization` as the head. They all describe protein transport event (e.g. GO:0031623 receptor internalization). `recruitment.n` occurs in GO:0046799 recruitment of helicase-primase complex to DNA lesions and GO:0046799 recruitment of 3'-end processing factors to RNA polymerase II holoenzyme complex, which describe the movement of protein complex to another macro molecule.

The LUs, `efflux.n`, `entry.n`, `exocytosis.n`, `migrate.n`, `mobilization.n`, `move.v`, `movement.n`, `shuttle.n` and `shuttling.v`, appear in GO terms that are descendants of GO:0006810 transport. They are used to describe various kinds of transport events that protein is not involved in.

`shift.n` only occurs in GO:0003049 regulation of systemic arterial blood pressure by capillary fluid shift. `capillary fluid shift` describes a kind of transport event. `relocation.n` and `relocate.v` only appear in GO:0009902 chloroplast relocation which is considered as a kind of organelle organization.

Example Sentences. The number of example sentences for each lexical unit in BioFrameNet re-

[2] `http://dolbey.us/BioFN/BioFN.zip` (28-Mar-2009)

LUs only in Bio-FrameNet	efflux.n, endocytosis.n, enter.v, entry.n, exocytosis.n, internalization.n, migrate.v, mobilization.n, move.v, movement.n, recruitment.n, redistribution.n, relocate.v, relocation.n, return.v, shift.n, shuttle.v, shuttling.n, traffic.n
LUs in both corpus	delivery.n, exit.v, export.n, import.n, recycle.v, recycling.n, release.n, targeting.n, trafficking.n, translocate.v, translocation.n, transport.n, transport.v
LUs only in our corpus	deliver.v, egress.n, establishment of ... localization.n, exit.n, export.v, import.v, release.v, secretion.n, secrete.v, sort.v, sorting.n, target.v, uptake.n

Table 6: The comparison of LUs in the 2 frames

lies on the existing collection of GRIFs in HLKB. The number of annotated sentences for each LU ranges from 1 to over 200. 207 GRIFs use the LU `translocation.n`, and 10 GRIFs use `transport.v`.

In our corpus, minimally for each LU 10 annotated sentences are gathered, if applicable. Tables 7 and 8 show the realizations of the FEs for the LUs `translocation.n` and `translocate.v`. The second columns give the number of times that the FE is realized in the 10 sentences. The PT and GF layers and the number of times they occur are given in the last columns, in the format of PT GF (number of occurrences). There are differences between the valence patterns of two corpus. We notice that example sentences in BioFrameNet mainly describe about protein. Although protein transport is described, different topics may be covered in the sentences in our corpus.

4.2.2 Predicates in FrameNet and PropBank/VerbNet

We examined the gaps between the semantic classification of the LUs (or only verbs) in the frame, and in FrameNet and PropBank/VerbNet. Around half of the LUs from the frame are absent from FrameNet data. 5 LUs are used in describing protein transport event, with the same semantic sense as in FrameNet. We identified the FEs for *Protein Transport* frame based on the domain knowledge. The

FEs	#	Realizations
TE	10	PP[of] Dep (6); NP Dep (3); Poss Gen (1);
TO	1	PP[from] Dep (1);
TDS	7	A Dep (2); PP[into] Dep (2); PP[to] Dep (3);
TC	6	NP Ext (5); NP dep (1);
TL	2	PP[in] Dep (1); A Dep (1);
TP	1	PP[across] Dep (1);
TT	0	-
TD	0	-
TA	1	AJP Dep (1);

Table 7: FE realizations for annotations with `translocation.n`

FEs	#	Realizations
TE	10	PP[than].Dep (1); NP Ext (6); NP Obj (3);
TO	4	PP[from] Dep (2); PP[of] Dep (1); NP Ext (1);
TDS	9	PP[to] Dep (6); PP[into] Dep (3);
TC	6	NP Ext (1); PP[upon] Dep (2); PP[prior to] Dep (1); PP[during] Dep (1); VPing Dep (1); VPbrst Dep (1); VPfin Dep (1);
TL	0	-
TP	4	NP Ext(3); VPing Dep (1)
TT	0	-
TD	0	-
TA	2	PP[with] Dep (1); AVP Dep (1)

Table 8: FE realizations for annotations with `translocate.v`

LUs	FrameNet	SS
egress.n, establishment of ... localization, export.n, localization.n, localize.v, recycling.n, recycle.v, targeting.n, translocation.n, translocate.v, trafficking.n, uptake.n	-	-
delivery.n, deliver.v	Delivery	√
exit.v	Departing	√
export.v	Sending	√
	Exporting Import_export	
import.n	Importance	
import.v	Importing Import_export	
release.n, release.v	Releasing	
secrete.v	Emitting	√
sort.n	Type	
sort.v	Differentiation	
target.v	Aiming	
transport.n, transport.v	Bringing	√

Table 9: Predicates in FrameNet: If the predicate is used with the same semantic sense as in the FrameNet's frame, "semantic sense (SS)" is checked.

number of FEs and their definitions are very different from FrameNet data. Other LUs are used with different semantic senses.

Except `translocate`, all verbs are included in PropBank data. Half of the verb senses have been classified into VerbNet classes. Only 3 verbs are used with the same sense as in describing protein transport event.

5 Conclusion

In this paper we propose a method for building frame-based corpus for the domain of biomedicine. The corpus construction relies on domain knowledge provided by ontologies. We believe that ontological domain knowledge can instruct us and ease the tasks in building the corpora. We built a corpus for transport event completely on basis of the piece of domain knowledge provided by GO biological process ontology [3]. We compared the frame *Protein Transport* to the frame *Protein_transport* in BioFrameNet, and examined the gaps between the semantic classification of the target words in the domain-specific corpus and in FrameNet and PropBank/VerbNet.

In the future, we aim to extend the corpus to cover other biological events. GO ontologies will be the main resource to provide domain knowledge, but also other ontologies, such as pathway ontologies will be considered as important domain knowledge resources. The identification of frames and the relations between frames are needed to be investigated. In addition, we will study the definition of STs in the domain corpus and their mappings to classes in top domain ontologies, such as BioTop (Beißwanger et al., 2008).

verbs	VerbNet	PropBank
translocate	-	-
deliver, transport	send-11.1	with the same semantic sense
secrete	-	
exit	escape-51.1	with different semantic sense
release	free-80.1	
sort	classify-29.10	
target	confront-98	
export, import, localize, recycle	-	

Table 10: Verbs in PropBank/VerbNet

Acknowledgement

We acknowledge the financial support of the Center for Industrial Information Technology (CENIIT) and the foundation Stiftelsen Olle Engkvist Byggmästare.

References

Michael Ashburner, Catherine A. Ball, Judith A. Blake, David Botstein, Heather Butler, J. Michael Cherry, Allan P. Davis, Kara Dolinski, Selina S. Dwight, Janan

[3]The corpus is publicly available on `http://www.ida.liu.se/~hetan/bio-onto-frame-corpus`

T. Eppig, Midori A. Harris, David P. Hill, Laurie Issel-Tarver, Andrew Kasarskis, Suzanna Lewis, John C. Matese, Joel E. Richardson, Martin Ringwald, Gerald M. Rubin and Gavin Sherlock. 2000. Gene Ontology: tool for the unification of biology. *Nature Genetics*, 25:25-29.

Michael Ashburner, Ulf Leser and Dietrich Rebholz-Schuhmann (Eds.). 2008. *Ontologies and Text Mining for Life Sciences: Current Status and Future Perspectives*. 24.03-28.03, Dagstuhl Seminar Proceedings.

Steven Bethard, Zhiyong Lu, James H Martin and Lawrence Hunter. 2008. Semantic Role Labeling for Protein Transport Predicates. *BMC Bioinformatics*, 9:277.

Elena Beißwanger, Stefan Schulz, Holger Stenzhorn and Udo Hahn. 2008. BioTop: An upper domain ontology for the life sciences: A description of its current structure, contents and interfaces to OBO ontologies. *Applied Ontology*, 3(4):205-212.

Adress Doms and Michael Schroeder. 2005. GoPubMed: exploring PubMed with the Gene Ontology. *Nucleic Acids Research*, 33:W783-786.

Andrew Dolbey, Michael Ellsworth and Jan Scheffczyk. 2006. BioFrameNet: A Domain-Specific FrameNet Extension with Links to Biomedical Ontologies. *The proceedings of KR-MED*, 87-94.

Charles J. Fillmore. 1985. Frames and the semantics of understanding. *Quaderni di Semantica*, 6(2).

Charles J. Fillmore, Charles Wooters and Collin F. Baker. 2001. Building a Large Lexical Databank Which Provides Deep Semantics. *The Pacific Asian Conference on Language, Information and Computation*.

Halil Kilicoglu, Marcelo Fiszman, Graciela Rosemblat, Sean Marimpietri and Thomas C. Rindflesch. 2010. Arguments of nominals in semantic interpretation of biomedical text. *Proceedings of the 2010 Workshop on Biomedical Natural Language Processing (BioNLP'10)*.

Jin-Dong Kim, Tomoko Ohta, Yuka Teteisi and Jun'ichi Tsujii. 2003. GENIA corpus - a semantically annotated corpus for bio-textmining. *Bioinformatics*. 19(suppl. 1):180-182.

Paul Kingsbury and Karin Kipper. Deriving Verb-Meaning Clusters from Syntactic Structure. *Workshop on Text Meaning, held in conjunction with HLT/NAACL 2003*.

Karin Kipper, Hoa Trang Dang and Martha Palmer. 2000. Class-Based Construction of a Verb Lexicon. *AAAI-2000 Seventeenth National Conference on Artificial Intelligence*.

Mitchell Marcus, Grace Kim, Mary Ann Marcinkiewicz, Robert MacIntyre, Ann Bies, Mark Ferguson, Karen Katz and Britta Schasberger. 1994. The Penn Treebank: annotating predicate argument structure. *In Proceedings of the workshop on Human Language Technology (HLT '94)*.

Alexa T. McCray, Allen C. Browne and Olivier Bodenreider. 2002. The Lexical Properties of the Gene Ontology (GO). *Proceedings of AMIA Symposium* , 504-508.

Lluís Màrquez, Xavier Carreras, Kenneth C. Litkowski and Suzanne Stevenson. 2008. Semantic role labeling: an introduction to the special issue. *Computational Linguistics*, 34(2).

Philip V. Ogren, K. Bretonnel Cohen and Lawrence Hunter. 2005. Implications of compositionality in the gene ontology for its curation and usage. *Pacific Symposium on Biocomputing*, 10:174-185.

Martha Palmer, Daniel Gildea and Paul Kingsbury. 2005. The Proposition Bank: an annotated corpus of semantic roles. *Computational Linguistics*, 31:71-105.

Josef Ruppenhofer, Michael Ellsworth, Miriam R. L. Petruck, Christopher R. Johnson and Jan Scheffczyk. 2005. ICSI. *FrameNet II*: Extended Theory and Practice.

Irena Spasic, Sophia Ananiadou, John McNaught and Anand Kumar. 2005. Text mining and ontologies in biomedicine: making sense of raw text. *Briefings in Bioinformatics*, 6(3):239-251.

Jan Scheffczyk, Adam Pease and Michael Ellsworth. 2006. Linking FrameNet to the SUMO Ontology. *International Conference on Formal Ontology in Information Systems (FOIS 2006)*.

Peri L. Schuyler, William T. Hole, Mark S. Tuttle and David D. Sherertz. 1992. The UMLS Metathesaurus: representing different views of biomedical concepts. *Bulletin of the Medical Library Association*, 81(2):217-22.

He Tan. 2010. A study on the relation between linguistics-oriented and domain-specific semantics. *Proceedings of the 3rd International Workshop on Semantic Web Applications and Tools for the Life Sciences*.

Richard Tzong-Han Tsai, Wen-Chi Chou, Ying-Shan Su, Yu-Chun Lin, Cheng-Lung Sung, Hong-Jie Dai, Irene Tzu-Hsuan Yeh, Wei Ku, Ting-Yi Sung and Wen-Lian Hsu. 2006. BIOSMILE: adapting semantic role labeling for biomedical verbs: an exponential model coupled with automatically generated template features. *Proceedings of the 2005 Workshop on Biomedical Natural Language Processing (BioNLP'06)*.

Tuangthong Wattarujeekrit, Parantu K Shah and Nigel Collier. 2004. PASBio: predicate-argument structures for event extraction in molecular biology. *BMC Bioinformatics*, 5:155.

Building a Coreference-Annotated Corpus from the Domain of Biochemistry

Riza Theresa Batista-Navarro[1,2,3,†] and **Sophia Ananiadou**[1,2,††]
[1]National Centre for Text Mining, University of Manchester, United Kingdom
[2]School of Computer Science, University of Manchester, United Kingdom
[3]Department of Computer Science, University of the Philippines Diliman, Philippines
†batistar@cs.man.ac.uk, ††sophia.ananiadou@manchester.ac.uk

Abstract

One of the reasons for which the resolution of coreferences has remained a challenging information extraction task, especially in the biomedical domain, is the lack of training data in the form of annotated corpora. In order to address this issue, we developed the *HANAPIN* corpus. It consists of full-text articles from biochemistry literature, covering entities of several semantic types: chemical compounds, drug targets (e.g., proteins, enzymes, cell lines, pathogens), diseases, organisms and drug effects. All of the co-referring expressions pertaining to these semantic types were annotated based on the annotation scheme that we developed. We observed four general types of coreferences in the corpus: sortal, pronominal, abbreviation and numerical. Using the MASI distance metric, we obtained 84% in computing the inter-annotator agreement in terms of Krippendorff's alpha. Consisting of 20 full-text, open-access articles, the corpus will enable other researchers to use it as a resource for their own coreference resolution methodologies.

1 Introduction

Coreferences are linguistic expressions referring to the same real-world entity (Jurafsky and Martin, 2009). The process of grouping all co-referring expressions in text into respective coreference chains is known as *coreference resolution*. It was introduced as one of the tasks of the sixth Message Understanding Conference (MUC-6) in 1995 (Grishman and Sundheim, 1995) and is one of the information extraction tasks which have remained a challenge to this day. One of the reasons it is still considered an unresolved problem especially in the biomedical domain is the lack of coreference-annotated corpora which are needed for developing coreference resolution systems.

There exist only a handful of biomedical corpora which are annotated with coreference information. We have conducted a review of each of them, taking into consideration their sizes, document composition, domain, types of markable entities, types of coreference annotated, availability, and reliability in terms of inter-annotator agreement. Of these, only two corpora have been used in coreference resolution systems developed outside the research group that annotated them: MEDSTRACT (Castano et al., 2002), and the MEDCo[1] corpus of abstracts which was used by the different teams who participated in the Coreference Supporting Task of the BioNLP 2011 Shared Task[2]. These two corpora are widely used, despite the fact that they are composed only of abstracts.

Previous studies have shown the advantages of utilising full-text articles rather than abstracts in information extraction systems (Shah et al., 2003; Schumie et al., 2004; Cohen et al., 2010a). Furthermore, recent research on fact extraction (McIntosh and Curran, 2009) has demonstrated the need for processing full-text articles when identifying coreferent expressions pertaining to biomedical entities.

[1] http://nlp.i2r.a-star.edu.sg/medco.html

[2] http://sites.google.com/site/bionlpst/home/protein-gene-coreference-task

Proceedings of the 2011 Workshop on Biomedical Natural Language Processing, ACL-HLT 2011, pages 83–91,
Portland, Oregon, USA, June 23-24, 2011.

However, coreference-annotated corpora composed of full-text articles are not readily accessible. Currently, only the FlySlip corpus (Gasperin et al., 2007) is available for download. In this corpus, only gene-related entities were considered for coreference annotation. Thus, there is a need for developing full-text corpora with coreference annotations for more semantic types. This is currently being addressed by the CRAFT project (Cohen et al., 2010b) which seeks to develop a corpus of full-text articles with coreference annotations for more types of entities; it was not explicitly stated, however, exactly which types are being covered. Similarly, we are developing a corpus of full-text articles with coreference annotations, but to further the aim of covering as many semantic types as possible, we selected a domain that covers a variety of semantic concepts. Research literature from this biochemistry subdomain, *marine natural products chemistry*, contains references pertaining to chemical compounds, organisms, drug targets such as proteins, enzymes, nucleic acids, tissues, cells, cell components, cell lines and pathogens, drug effects, as well as diseases. We cover a number of entity types with the intention of providing more insight into how to disambiguate co-referring expressions of different semantic types.

An annotation scheme was developed, taking into consideration the coreference types which have been observed from the corpus, namely: sortal, pronominal, numerical and abbreviation. Three chemistry graduates were employed to annotate the corpus. To determine the reliability of the resulting annotations, we measured inter-annotator agreement in terms of Krippendorff's alpha.

2 Related Work

Coreference is often associated with the phenomenon of *anaphora* which is characterised by an expression (called an *anaphor*) that points back to an entity previously mentioned in the same discourse (called *antecedent*). Anaphora resolution is the process of determining the antecedent of an anaphor. While the output of anaphora resolution is a set of anaphor-antecedent pairs, that of coreference resolution is a set of coreference chains which can be treated as equivalence classes. Despite this difference, an overlap between them may be observed in several cases. Often, a number of anaphor-antecedent pairs from a discourse are coreferential or refer to the same entity in the same domain, and may be placed in the same coreference chain. For this reason, we also included in our review of biomedical corpora those which were annotated with anaphora information and refer to them henceforth as coreference-annotated corpora.

We determined the types of coreference annotated in each corpus we have reviewed, adapting Mitkov's classification of anaphora (Mitkov et al., 2000) which is also applicable to coreference. *Nominal coreference* is characterised by co-referring expressions pertaining to a noun. It is further divided into *pronominal coreference* and *sortal coreference* which use a pronoun and a lexical noun phrase, respectively, as co-referring expressions. Unlike nominal coreference, *verbal coreference* is characterised by co-referring expressions pertaining to verbs. Both nominal and verbal coreference can be broadly categorised according to the kind of relation as *direct* or *indirect*. In direct coreference, co-referring expressions are related by identity, synonymy or specialisation; in indirect coreference, they are related by associative relations such as meronymy or holonymy for nouns, and troponymy or entailment for verbs. Annotation of indirect coreference is usually more challenging as it requires more specialised domain knowledge.

Presently, there are five (5) different biomedical corpora which are annotated with coreference information: MEDSTRACT (Castano et al., 2002), MEDCo[3], FlySlip (Gasperin et al., 2007), the Colorado Richly Annotated Full Text (CRAFT) corpus (Cohen et al., 2010b) and DrugNerAr (Segura-Bedmar et al., 2009).

The MEDCo corpus has two subsets, one consisting of abstracts (which we shall refer to as MEDCo-A) and another consisting of full papers (MEDCo-B). The results of our review of all five corpora are presented in Table 1. Included in the last row (HANAPIN) are the attributes of the corpus that we have developed for comparison with existing corpora.

Three of them, MEDSTRACT, MEDCo and DrugNerAr, adapted an annotation scheme similar

[3]http://nlp.i2r.a-star.edu.sg/medco.html

Table 1: Comparison of Biomedical Corpora with Coreference Annotations

Corpus	Scheme Adapted	Document Composition	Domain/ Markables	Coreference Types	Availability	Format	Reliability
MEDSTRACT	MUCCS	100 abstracts	molecular biology/ UMLS types	direct nominal	publicly available	XML	unknown
MEDCo-A	MUCCS	1999 abstracts	human blood cell transcription factors/ GENIA Term Ontology types	direct nominal	publicly available	XML	Krippendorff's alpha: 83% on 15 abstracts
MEDCo-B	MUCCS	43 full papers	human blood cell transcription factors/ GENIA Term Ontology types	direct nominal	currently unavailable	XML	Krippendorff's alpha: 80.7% on 2 full papers
FlySlip	domain-specific	5 full papers	fruit fly genomics/ genetic entities	direct and indirect sortal	publicly available	XML	Kappa score: greater than 83% on each paper
CRAFT	OntoNotes	97 full papers	mouse genomics/ all encountered	direct nominal and verbal and	currently unavailable	SGML	Krippendorff's alpha: 61.9% on 10 full papers
DrugNerAr	MUCCS	49 DrugBank texts	drug-drug interactions/ drugs	direct nominal	publicly available	XML	unknown
HANAPIN	MEDCo	20 full papers	marine natural products chemistry/ chemical compounds, organisms, drug targets, drug effects, diseases	direct nominal, numerical & abbreviation	currently unavailable (to be released publicly)	XML	Krippendorff's alpha: 75% averaged over 20 papers; 84% using the MASI distance metric

to that of the Message Understanding Conference scheme or MUCCS (Hirschman, 1997). Using the Standard Generalized Markup Language (SGML) as annotation format, MUCCS creates a link between co-referring expressions by setting the value of an attribute of the referring element to the ID of the referent.

The same mechanism is used in the annotation of MEDSTRACT, MEDCo and DrugNerAr, but with respective extensions to account for more specific relations (e.g., appositive relation in the case of MEDCo). On the contrary, rather than linking the referring expression to its referent, an annotator explicitly places co-referring expressions in the same coreference chain with OntoNotes, the scheme adapted in annotating the CRAFT corpus. FlySlip can be considered unique in terms of its annotation scheme as it adapted a domain-specific scheme which was necessary since indirect coreferences were annotated. All corpora are available in the form of a mark-up language (SGML or XML).

The five corpora can be grouped into three according to general domain: molecular biology (MEDSTRACT and MEDCo), genomics (FlySlip and CRAFT), and pharmacology (DrugNerAr). MEDSTRACT and MEDCo both have coreference annotations for semantic types from the UMLS and the GENIA ontology, respectively, which can be broadly categorised into compound, organism, protein, gene and cell. Each of the FlySlip and DrugNerAr corpora, on the other hand, have annotations for only one general semantic type: gene-related entities and drugs, respectively. CRAFT is unique in this respect as its developers seek to annotate all co-referring expressions regardless of semantic type; the semantic types that have been encountered so far have not yet been reported, however.

In terms of coreference types for which annotations have been added, CRAFT is the only corpus with annotations for verbal coreference; all the rest have annotations only for pronominal and/or sortal coreference. With respect to coreference types according to relation, FlySlip is the only corpus with annotations for indirect coreference.

MEDCo-B, FlySlip and CRAFT are three existing corpora which are comprised of full-text articles. Among them, only FlySlip is currently publicly available.

The corpus that we have developed, which we call the HANAPIN corpus, is also intended for public release in the near future and covers five general semantic types. In the annotation scheme which was designed and used in HANAPIN, two additional coreference types were considered: abbreviations and numerical coreferences which are commonly used in chemistry research literature. These coreference types and the annotation scheme are further described in the succeeding section.

3 Methodology

3.1 Composition of Corpus Documents

Taking into consideration that the corpus should consist of full-text articles which can be distributed to the public, we gathered full-text articles from the journal *Marine Drugs*[4] which is under the PubMed Central Open Access subset[5]. The said journal covers subject areas such as marine natural products, medicine analysis, marine pharmacology, pharmaceutical biology, marine drugs development and marine biotechnology, among many others. From all of its articles from 2003 to 2009, we randomly selected twenty (20) which seemed to be a reasonable size considering that only five months were allocated for the annotation of the corpus, and that a previous study on biomedical corpora (Cohen et al., 2005) has shown that a corpus can possibly be widely used despite its small size. The experimental sections of the articles were not annotated as they contain very detailed descriptions of the methods carried out by the authors; according to a study (Shah et al., 2003), these usually contain technical data, instruments and measurements – types of information which are currently not of much interest to researchers doing biomedical information extraction, although they may be in the future. The corpus contains a total of 1,027 sentences or 27, 358 words.

3.2 Coreference Types

The coreferences observed in the corpus were categorised into four general nominal types: pronominal, sortal, numerical and abbreviation. Table 2 presents the subtypes of sortal and pronominal coreference, as well as examples for all types. We

[4]http://www.mdpi.com/journal/marinedrugs

[5]http://www.ncbi.nlm.nih.gov/pmc/about/openftlist.html

Table 2: Coreference Types with Examples

General Coreference Type	Subtype	Examples
pronominal	demonstrative	*this, that, these, those*
	personal	*it, they, its, their, theirs*
	indefinite	*another, few, other, some, all, any*
	distributive	*both, such, each, either, neither*
	relative	*which, that, whose*
sortal	definite	*the loihichelins*
	indefinite	*an alkaloid, a mycalamide*
	demonstrative	*this metabolite, these compounds*
	distributive	*both compounds*
	predicate nominative	*"Galactans are **polysaccharides**..."*
	appositive	*"Radiosumin, **an N-methyl dipeptide**..."*
numerical	N.A.	*"The structures of **1** and **2**..."* *"Compounds **1-3** inhibit..."*
abbreviation	N.A.	*"...as a membrane type 1 matrix metalloproteinase (MT1-MMP) inhibitor. Compound 1 inhibited **MT1-MMP** with..."*

have decided not to take into account verbal and indirect coreferences; only nominal and direct coreferences have been considered for the first release of the corpus.

3.2.1 Pronominal Coreference

This type of coreference is characterised by a pronoun referring to a noun phrase. The pronoun is used as a substitute to a noun. We have further identified the following subtypes of pronominal coreference: *demonstrative, personal, indefinite, distributive* and *relative*.

3.2.2 Sortal Coreference

Also referred to as lexical noun phrase coreference, sortal coreference is characterised by a noun phrase consisting of a head noun and its modifiers. The subtypes of sortal coreference which have been identified include: *definite, indefinite, demonstrative, distributive, predicate nominative* and *appositive*.

3.2.3 Numerical Coreference

In chemistry research literature, a number is conventionally used to refer to a chemical entity which was introduced using the same number. Oftentimes, a range of numbers is also used to refer to a number of compounds previously mentioned.

3.2.4 Abbreviation

In annotating the HANAPIN corpus, abbreviations were also considered as co-referring expressions. We distinguish them from the other coreference types to make the corpus of benefit to developers of abbreviation identification algorithms as well.

3.3 Annotation Scheme and Procedure

The annotation scheme used in MEDCo (which was based on MUCCS) was adapted and modified for the annotation of the HANAPIN corpus. We have selected the MEDCo scheme as it already differentiates between the pronominal and identity (equivalent to sortal) types, whereas MUCCS has only the identity type. There was a need, however, to extend the MEDCo scheme to further specialise the coreference types. The XML Concordancer (XConc) tool[6] was used in annotating the corpus. Configuring the said tool for our needs is straightforward as it only involved the customisation of a Document Type Definition (DTD) file.

3.3.1 Term Annotations

As a preliminary step, the scheme required that all terms which can be categorised into any of the

[6] http://www-tsujii.is.s.u-tokyo.ac.jp/GENIA/home/wiki.cgi?page=XConc+Suite

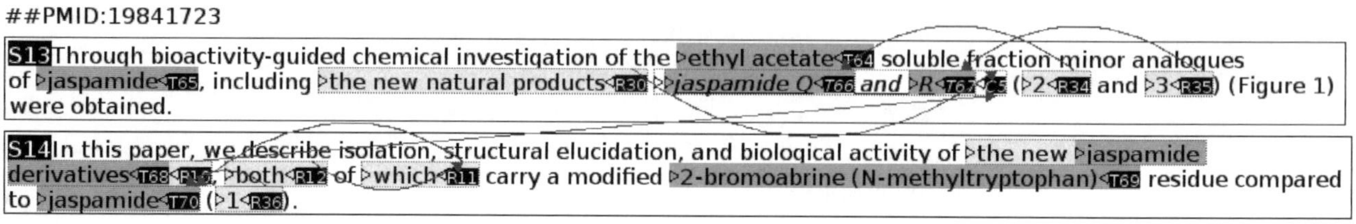

Figure 1: Sample annotations as shown in the XConc annotation tool. The sentences in this example come from one of the documents in the HANAPIN corpus, the *Marine Drugs* article with PubMed ID 19841723. For illustrative purposes, the first sentence in the example was slightly modified to demonstrate the use of the `cons` element.

following semantic types be annotated:

1. chemical compound
2. organism
3. drug effect
4. disease
5. drug target (further categorised into: protein, enzyme, nucleic acid, tissue, cell, cell component, cell line, pathogen)

For each markable, the annotator creates a `term` element which is assigned an ID and one of the semantic types above. The scheme supports the annotation of embedded terms, as well as terms in a discontinuous text region. The former entails placing a `term` element within another. The latter is done by dividing the discontinuous text into fragments and annotating each fragment in the same manner as an ordinary term element. The fragment elements are then grouped together as a constituent element (`cons`). Figure 1 presents a sample annotation of a discontinuous term (constituent `C5`) as viewed in XConc.

3.3.2 Co-referring Expressions

An annotator proceeds to the annotation of co-referring expressions after annotating all terms within a document. If an expression was found to be co-referring with another term, the annotator assigns the ID of the latter as the value of the `idref` attribute of the former. If the referring expression, however, is a noun phrase and not a term that was previously annotated during term annotation, it is marked as a `ref` element and then linked to its referent. Annotators delimit these expressions by including the necessary modifiers of the co-referring element (e.g., *the new jaspamide derivatives* instead of just *jaspamide derivatives*). A coreference type which could be any of pronominal, numerical, abbreviation, and sortal (further categorised into definite, indefinite, demonstrative, distributive, predicate nominative and appositive) is also assigned as the value of the `type` attribute of each link created. We decided not to further divide pronominal coreference into its subtypes as it became apparent during the annotation dry runs that there is only a handful of pronominal coreferences. Figure 1 shows co-referring expressions (connected by arrows) linked by the mechanism just described.

Listed below are some of the main points of the annotation guidelines:

1. A referring expression may be linked to multiple referents.
2. The more specific one between two co-referring expressions is considered as the referent. This means that there might be cases when the referent occurs later than the referring expression. For example, `R30:the new natural products` is the co-referring expression and `C5:jaspamide Q and R` is the referent in Figure 1.
3. In cases where there are multiple choices for the referent of a referring expression, the closest one may be chosen as long as it is (or will be) linked to the other choice expressions.
4. There are cases when more than one type of coreference applies. For example, in Figure 1, *the new natural products* is both an appositive and a definite noun phrase. In such cases, the appositive and predicate nominative types take precedence over the other sortal types.

```
<sentence id="S13">
    Through bioactivity-guided chemical investigation of the
    <term id="T64" sem="chem">ethyl acetate</term>
    soluble fraction minor analogues of
    <term id="T65" sem="chem">jaspamide</term>, including
    <ref id="R30" idref1="C5" type="appos">the new natural products</ref>
    <cons id="C5">
        <term id="T66" sem="chem">jaspamide Q</term> and
        <term id="T67" sem="chem">R</term>
    </cons> (
    <ref id="R34" idref1="T66" type="num">2</ref> and
    <ref id="R35" idref1="T67" type="num">3</ref>) (Figure 1) were obtained.
</sentence>
<sentence id="S14">In this paper, we describe isolation, structural elucidation,
    and biological activity of
    <ref id="R10" idref1="C5" type="definite">the new
        <term id="T68" sem="chem">jaspamide derivatives</term>
    </ref>,
    <ref id="R12" idref1="R11" type="pron">both</ref> of
    <ref id="R11" idref1="R10" type="pron">which</ref> carry a modified
    <term id="T69" sem="chem">2-bromoabrine (N-methyltryptophan)</term>
    residue compared to
    <term id="T70" sem="chem">jaspamide</term> (
    <ref id="R36" idref1="T70" type="num">1</ref>).
</sentence>
```

Figure 2: XML code generated by XConc for the sample annotations in Figure 1.

One could process the XML code (provided in Figure 2 for the reader's reference) to obtain the following coreference chains:

1. {`R30:the new natural products, C5:jaspamide Q and R, R10:the new jaspamide derivatives, R11:which, R12:both`}
2. {`T66:jaspamide Q, R34:2`}
3. {`T67:jaspamide R, R35:3`}
4. {`T70:jaspamide, R36:1`}

The complete annotation guidelines will be publicly released together with the annotated corpus.

4 Results

The three annotators were asked to complete the coreference annotations within five months. A biweekly meeting was held to address questions and issues which could not be addressed or resolved by means of the online project forum.

4.1 Statistics

As the HANAPIN corpus is the first of its kind from the biochemistry domain and aims to cover several semantic as well as coreference types, it is of interest to determine which of the types are most prevalent. To do this we computed statistics over the annotations (Figure 3). For each type, we obtained the average over the annotations from the three coders.

There is a total of 395 coreference chains (not including singleton chains or those with only one mention) in the entire corpus. The coreference chains are of the following semantic types: chemical compounds (70.89%), drug targets (12.66% that accounts for proteins, cell lines, pathogens, enzymes, cells, cell parts, nucleic acids and tissues), organisms (9.87%), drug effects (3.29%), and diseases (3.29%). Among the drug targets, the most prevalent are proteins, cell lines and pathogens.

A total of 760 coreference links have been found in the corpus. The most common among the types is the numerical one (46%), followed by the sortal type (33% that accounts for the definite, indefinite, demonstrative, appositive, predicate nominative and distributive types). Less common are the pronominal type (11%) and abbreviation (10%). Among the sortal coreferences, the most common are the definite and indefinite types, followed by the demonstrative type.

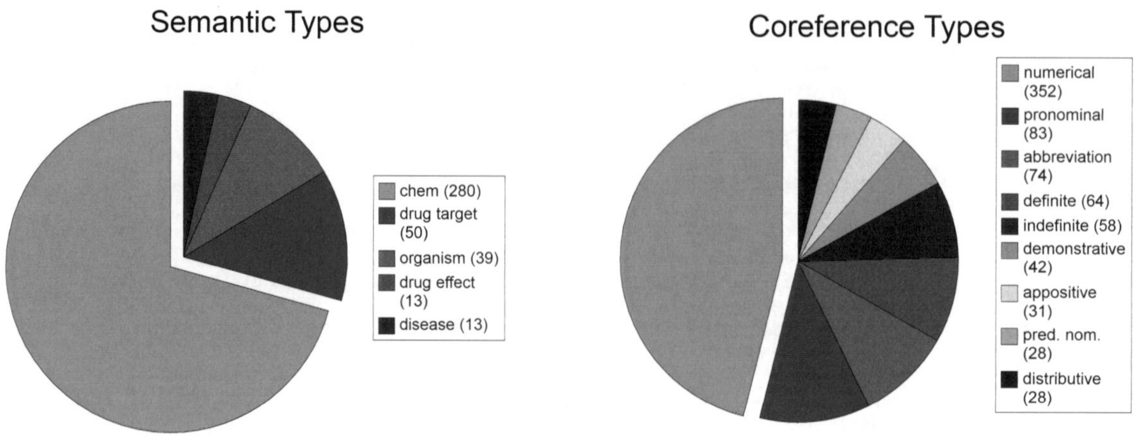

Figure 3: Distribution of semantic and coreference types in the HANAPIN corpus.

4.2 Corpus Reliability

Following Passoneau's proposed method for computing reliability for coreference annotation (Passoneau, 2004), we computed for the reliability of the corpus in terms of Krippendorff's alpha, a coefficient of agreement that allows for partial disagreement with the use of a distance metric based on the similarity between coreference chains. Passoneau's first proposed distance metric (d_P) assigns 0 for identity, 0.33 for subsumption, 0.67 for intersection and 1 for disjunction. There are, however, alternative distance metrics that consider the sizes of the coreference chains, such as Jaccard's coefficient of community (d_J) and Dice's coincidence index (d_D) which can be computed as follows (Artstein and Peosio, 2004):

$$d_J = 1 - \frac{|A \cap B|}{|A \cup B|}$$

$$d_D = 1 - \frac{2|A \cap B|}{|A| + |B|}$$

A new distance metric called Measuring Agreement on Set-valued Items (MASI) was then later proposed by Passoneau. It is obtained by getting the product of the original distance metric d_P and Jaccard's coefficient d_J.

Initially using Passoneau's first proposed distance metric d_P in computing for Krippendorff's alpha, we obtained an average of 75% over all documents in the HANAPIN corpus. Computing for alpha using the MASI distance metric gives 84%. Though there is no value of alpha that has been established to be an absolute indication of high agreement, previous works cited by Krippendorff have shown that values of alpha less than 67% indicate unreliability (Krippendorff, 1980). We can therefore regard the obtained values of alpha as satisfactory.

5 Conclusion and Future Work

A coreference-annotated corpus from the domain of biochemistry, consisting of full-text articles, has been developed. It was annotated following guidelines which covered coreference and semantic types that have not been covered in other biomedical corpora before. This was done to further the aim of providing researchers with more insight into the phenomenon of coreference in a cross-disciplinary domain. Results show that in this biochemistry domain, the most common types of coreference being used by authors are the numerical and sortal types. Verbal and indirect coreferences, however, have not been considered at this stage; the annotation of these types can be explored as part of future work on the corpus.

To measure reliability of the corpus, we determined inter-annotator agreement on all documents by computing for the value of Krippendorff's alpha. Using Passoneau's first proposed distance metric and the MASI distance metric, we obtained satisfactory values of 75% and 84%, respectively. The corpus and annotation guidelines will be released to the public to encourage and enable more researchers to develop improved biomedical coreference resolution methodologies.

Acknowledgements

The UK National Centre for Text Mining is funded by the UK Joint Information Systems Committee (JISC). The authors would also like to acknowledge the Office of the Chancellor, in collaboration with the Office of the Vice-Chancellor for Research and Development, of the University of the Philippines Diliman for funding support through the Outright Research Grant.

The authors also thank Paul Thompson for his feedback on the annotation guidelines, and the anonymous reviewers for their helpful comments.

References

Ron Artstein and Massimo Poesio. 2004. Inter-Coder Agreement for Computational Linguistics. *Computational Linguistics*, 34(4):555-596.

José Castaño, Jason Zhang and James Pustejovsky. 2002. Anaphora resolution in biomedical literature. *Proceedings of the International Symposium on Reference Resolution for NLP.*

K. Bretonnel Cohen, Philip V. Ogren, Lynne Fox and Lawrence E. Hunter. 2005. Empirical data on corpus design and usage in biomedical natural language processing. *AMIA Annual Symposium Proceedings*, pages 156-160.

K. Bretonnel Cohen, Helen L. Johnson, Karin Verspoor, Christophe Roeder, Lawrence E. Hunter. 2010. The structural and content aspects of abstracts versus bodies of full text journal articles are different. *BMC Bioinformatics*, 11(1):492.

K. Bretonnel Cohen, Arrick Lanfranchi, William Corvey, William A. Baumgartner Jr., Christophe Roeder, Philip V. Ogren, Martha Palmer and Lawrence E. Hunter. 2010. Annotation of all coreference in biomedical text: Guideline selection and adaptation. *Proceedings of the Second Workshop on Building and Evaluating Resources for Biomedical Text Mining (BioTxtM 2010), LREC 2010.*

Caroline Gasperin, Nikiforos Karamanis and Ruth Seal. 2007. Annotation of anaphoric relations in biomedical full-text articles using a domain-relevant scheme. *Proceedings of the 6th Discourse Anaphora and Anaphor Resolution Colloquium (DAARC 2007).*

Ralph Grishman and Beth Sundheim. 1995. Design of the MUC-6 Evaluation. *MUC '95: Proceedings of the 6th Message Understanding Conference*, pages 1-11.

Lynette Hirschman. 1997. MUC-7 Coreference Task Definition. *Message Understanding Conference 7 Proceedings.*

Daniel Jurafsky and James H. Martin. 2009. *Speech and Language Processing: An Introduction to Natural Language Processing, Computational Linguistics, and Speech Recognition.* Prentice-Hall, 2nd edition.

Klaus H. Krippendorff. 1980. Content Analysis: An Introduction to Its Methodology. Beverly Hills, CA: Sage Publications.

Tara McIntosh and James R.Curran. 2009. Challenges for automatically extracting molecular interactions from full-text articles. *BMC Bioinformatics*, 10(1):311.

Ruslan Mitkov, Richard Evans, Constantin Orasan, Catalina Barbu, Lisa Jones and Violeta Sotirova. 2005. Coreference and anaphora: developing annotating tools, annotated resources and annotation strategies. *Proceedings of the Discourse Anaphora and Anaphora Resolution Colloquium (DAARC 2000)*, pages 49-58.

Rebecca J. Passoneau. 2004. Computing reliability for coreference annotation. *Proceedings of the International Conference on Language Resouces (LREC).*

M. Schumie, M. Weeber, B. Schijvenaars, E. van Mulligen, C. van der Eijk, R. Jelier, B. Mons and J. Kors. 2004. Distribution of information in biomedical abstracts and full-text publications. *Bioinformatics*, 20(16):2597-2604.

Isabel Segura-Bedmar, Mario Crespo, César de Pablo-Sánchez and Paloma Martínez. 2009. Resolving anaphoras for the extraction of drug- drug interactions in pharmacological documents. *BMC Bioinformatics*, 11(Suppl 2):S1.

Parantu K. Shah, Carolina Perez-Iratxeta, Peer Bork and Miguel A. Andrade. 2003. Information extraction from full text scientific articles: Where are the keywords? *BMC Bioinformatics*, 4(1): 20.

Towards Morphologically Annotated Corpus of Hospital Discharge Reports in Polish

Małgorzata Marciniak
Institute of Computer Science PAS
ul. J.K. Ordona 21,
01-237 Warszawa, Poland
mm@ipipan.waw.pl

Agnieszka Mykowiecka
Institute of Computer Science PAS
ul. J.K. Ordona 21,
01-237 Warszawa, Poland
agn@ipipan.waw.pl

Abstract

The paper discuses problems in annotating a corpus containing Polish clinical data with low level linguistic information. We propose an approach to tokenization and automatic morphologic annotation of data that uses existing programs combined with a set of domain specific rules and vocabulary. Finally we present the results of manual verification of the annotation for a subset of data.

1 Introduction

Annotated corpora are knowledge resources indispensable to the design, testing and evaluation of language tools. Medical language differs significantly from the everyday language used in newspapers, magazines or fiction. Therefore, general language corpora are insufficient when creating tools for (bio)medical text processing.

There are several biomedical corpora available for English such as GENIA (Kim et al., 2010) — the best known and most used one, containing MEDLINE abstracts annotated on several levels; BioInfer (Pyysalo et al., 2007) targeted at protein, gene, and RNA relationships annotation; or CLEF (Roberts et al., 2009) containing 20,000 cancer patient records annotated with clinical relations. Medical corpora are also collected for lesser spoken languages, e.g. MEDLEX — Swedish medical corpus (Kokkinakis, 2006); IATROLEXI project for Greek (Tsalidis et al., 2007); or Norwegian corpus of patients' histories (Røst et al., 2008). The paper (Cohen et al., 2005) contains a survey of 6 biomedical corpora. The authors emphasize the importance of a standard format and give guidelines for careful annotation and evaluation of corpora.

The immediate goal of the paper is to establish and test a method of annotating Polish clinical data with low level linguistic information, i.e. token and morpheme descriptions. The research is done on a relatively small set of data (more than 450,000 tokens) but to gain the experience necessary to create a much larger annotated corpus of Polish medical texts. We would like to use our corpus to refine and test domain tools for: tagging, Named Entity Recognition or annotation of nominal phrases. We have already annotated the corpus with semantic information (Marciniak and Mykowiecka, 2011) using an existing rule based extraction system (Mykowiecka et al., 2009) and performed experiments with machine learning approaches to semantic labeling (Mykowiecka and Marciniak, 2011). Thus, to enable the realization of various scientific goals, a detailed and universal morphologic annotation of the corpus was introduced.

The division into tokens is the first level of text analysis. It is frequently performed without paying special attention to potential problems, just by dividing text on spaces, line breaks and punctuation marks. In many applications this is quite a satisfactory solution, but in case of texts that contain a lot of non-letter characters, using universal tokenization rules frequently causes problems. Some examples, in the case of using the Penn Treebank tokenization scheme in annotating the GENIA corpus were pointed out in (Teteisi and Tsujii, 2006). Jiang and Zhai (2007) show the importance of tokenization strategies in the biomedical domain, and the in-

Proceedings of the 2011 Workshop on Biomedical Natural Language Processing, ACL-HLT 2011, pages 92–100,
Portland, Oregon, USA, June 23-24, 2011.

fluence of this process on the results of information retrieval. Our approach consists of dividing text into simple tokens which can be grouped at subsequent levels of analysis using domain specific knowledge.

For languages with rich inflection, like Polish, morphological annotation is indispensable for further text analysis. As there are no Polish taggers which can analyze medical texts, nor medical lexicons containing inflected forms, we combine a general purpose tagger with a set of domain specific rules referring to a small data induced vocabulary. A portion of the automatically annotated data was checked by two linguists to assess data quality. The results obtained are given in 8. Currently, the entire dataset is undergoing manual verification.

2 Linguistic Characteristics of Texts

The corpus consists of 460 hospital discharge reports of diabetic patients, collected between the years 2001 and 2006 in one of Warsaw's hospitals. These documents are summaries of hospital treatment and are originally written in MS Word with spelling correction turned on, so the errors observed are mainly in words that are not included in the dictionary. The documents are converted into plain text files to facilitate their linguistic analysis and corpus construction. Clinical data include information serving identification purposes (names and addresses) which are substituted by symbolic codes before making the documents accessible for further analysis. The annonymization task was performed in order to make the data available for scientific purposes. We plan to inspect the data manually, to remove all indirect information enabling a patient's identification, and negotiate the terms for making the corpus publicly available.

Each document is 1.5 – 2.5 pages long, and begins with the identification information of the patient and his/her visit in hospital. Next, the following information is given in short form: significant past and current illnesses, diagnoses and patient's health at the beginning of the hospitalization. After these data, the document describes results of examinations such as height, weight, BMI and blood pressure, ophthalmology examinations, blood tests, lipid profile tests, radiology or ultrasound. This part of the document may also contain descriptions of attempts to select the best treatment for the patient. The summary of the document starts from the word *Epikryza* 'Discharge abstract'. Its length is about half a page of text. It contains: data about a patient's diabetes, a description of diabetic complications, and other illnesses, selected examination results and surgical interventions, information about education, diet observed, self monitoring, patient's reactions, and other remarks. Finally, all recommendations are mentioned, including information about prescribed diet, insulin treatment (type and doses) and oral medication.

Most information is given as free-form text, but the vocabulary of these documents is very specific, and significantly differs from texts included in corpora of general Polish like IPIPAN Corpus (Przepiórkowski, 2004) or NKJP (National Corpus of Polish, `http://nkjp.pl`). The texts contain many dates in different formats, and a lot of test results with numerical values, whose descriptions are omitted in NKJP. The texts contains also a lot of medication names, like *Cefepime* or *Acard* not present in any general Polish dictionary. Some of them are multi-word names like *Diaprel MR, Mono Mack Depot, Mixtard 10.* The same medication can be referred to in different ways depending on international or Polish spelling rules (e.g. *Amitriptylinum* and its Polish equivalent *Amitryptylina*). Polish names could be inflected by cases (e.g. *Amitryptyliny*$_{gen}$).

In documents, many diagnoses are written in Latin. In the following examples the whole phrases are in Latin: *Retinopathia diabetica simplex cum maculopathia oc. sin.* 'simple diabetic retinopathy with maculopathy of the left eye'; or *Laryngitis chronica. Otitis media purulenta chronica dex.* 'Chronic laryngitis. Chronic purulent inflammation of the middle right ear'. Sometimes foreign expressions are thrown into a Polish sentences: *Ascites duża ilość płynu w jamie brzusznej między pętlami jelit ...* 'Ascites a lot of fluid in abdominal cavity between intestinal loops ...' — only the first word is not in Polish.

3 Corpus description

The corpus is annotated with morphological and semantic information. The standard of annotation follows the TEI P5 guidelines advised for annotation of biomedical corpora, see (Erjavec et al., 2003). Our corpus format is based on the one accepted for the NKJP corpus (Przepiórkowski and Bański, 2009). According to this scheme, every annotation is described in a separate file. Each discharge document is represented by a catalog containing the following five files:

- *xxx.txt* – plain text of the original annonymized document;
- *xxx.xml* – text of the document (in the form as in *xxx.txt* file) divided into numbered sections which are in turn divided into paragraphs;
- *xxx_segm.xml* – token limits and types (29 classes);
- *xxx_morph.xml* – morphological information (lemmas and morphological feature values);
- *xxx_sem.xml* – semantic labels and limits.

4 Tokenization

The first level of text analysis is its segmentation into tokens. In general, most tokens in texts are lowercase words, words beginning with a capital letter and punctuation marks. The most common (thus the most important) tokenization problem is then to decide whether a particular dot ends a sentence or belongs to the preceding abbreviation (or both). In some texts there are also many numbers representing dates, time points, time intervals or various numerical values. For texts in which uniform standards of expressing these notions are obeyed, recognizing such complex tokens is much easier and simplifies further text analysis.

In medical texts the problem of non-word tokens is harder than in the case of newspapers or novel content as they constitute a much larger portion of the text itself. Apart from descriptions of time (dates, hours, periods of time) there are numbers that refer to values of different medical tests or medicine doses and sizes. There are also many specific names which sometimes contain non-letter characters (e.g. *Na+*) as well as locally used abbreviations and acronyms. An additional difficulty is caused by the lack of will to obey writing standards. Physicians use different ways of describing dates (e.g. *02.09.2004, 30.09/1.10.2003, 06/01/2004, 14.05.05, 28 .04. 05, 12.05.2005r.*) or time (*8:00 vs 8.00*). They also do not pay enough attention to punctuation rules and mix Polish and English standards of writing decimal numbers. In Polish we use a comma not a dot, but the influence of English results in common usage of the decimal point. Sometimes both notations can be found in the same line of text. Further, the sequence *'2,3'* may mean either *'2.3'* or two separate values: *'2'* and *'3'*.

Two tools used in the process of constructing the corpus have embedded tokenizers. The first one is a part of the information extraction system SProUT (Drożdżyński et al., 2004) which was used to write grammars identifying semantically important pieces of text. The general assumption adopted while building its tokenizer was "not to interpret too much", which means that tokens are relatively simple and do not rely on any semantic interpretation. Their self explanatory names, together with token examples and their frequencies in the entire input data set, are listed in table 1.

Two other tokenization modules are embedded in the TaKIPI tagger used to disambiguate the morphological descriptions of word forms (Piasecki, 2007). The first one divides all character sequences into words and non-words which are assigned the *ign* label. The second tokenizer interprets these non-word sequences and assigns them *ttime, tdate, turi* (for sequences with dots inside) and *tsym* labels. It also applies a different identification strategy for token limits – for all non-word tokens only a space or a line break ends a token. Although treating a date (*15.10.2004r*) or a range (*1500-2000*) as one token is appropriate, in the case of sequences where spaces are omitted by mistake, the resulting tokens are often too long (e.g. *'dnia13/14.07.04'*, *'iVS-1,5'*).

After analyzing the results given by three different tokenizers we decided to use the token classes identified by the SProUT tokenizer and align its results with the results of the 'simple' TaKIPI tokenizer. SProUT tokens which were longer than TaKIPI tokens, e.g. *'1x2mg'*, *'100mg'*, *'50x16x18'*, were divided into smaller onces. The changes introduced to token limits concern those tokens of the *other_symbol* type which contain punctuation marks. The *other_symbol* class comprises sequences which do not fit into any other class, i.e.

symbols for which separate classes are not defined (e.g. '=') and mixed sequences of letters and digits. In this latter case a token ends only when a space or a line break is encountered. The most typical case when this strategy fails in our data is the sequence *'HbA1c:'* as the name of the test according to the tokenizer rules is classified as an 'other_symbol' the following colon is not separated. There are also other similar sequences: *'HbA1c=9,1%:'* or *'(HbA1C'*. To make the results more uniform we divided these tokens on punctuation characters. This process resulted in replacing 1226 complex tokens by 4627 simple ones. Among these newly created tokens the most numerous class was *lowercase_word* and numbers which were formed after separating numbers and unit names, e.g. *10g, 100cm* and sequences describing repetitions or sizes, like *2x3, 2mmx5mm*. The longest sequence of this kind was *'ml/min.,GFR/C-G/-37,5ml/min/1,73m2'*. This string was divided into 18 tokens by TAKIPI but finally represented as 23 tokens in the corpus. Finally, in the entire data set 465004 tokens (1802864 characters) were identified. The most numerous class represents numbers – 18.8% (9% of characters), all punctuation characters constitute 25% of the total number of tokens (6.5% characters).

Table 1: Token types and number of occurrences

token class name & examples	numbers initial	numbers final
all_capital_word: ALT, B, HDL, HM	18369	18416
any_natural_number	85766	87246
apostrophe	14	14
back_slash	7	7
closing_bracket	2661	2663
colon	12426	12427
comma	28799	28831
dot	47261	47269
exclamation_sign	49	49
first_capital_word: Al, Amikacin, Wysokie	43136	43269
hyphen	4720	4725
lowercase_word: antygen, aorta	192305	193368
mixed_word_first_capital: AgHBs, Ilo, NovoRapid	513	514
mixed_word_first_lower: antyHBS, dlAST	989	1003
number_word_first_capital: 200Hz, 14HN	48	0
number_word_first_lower: 100ml, 200r 1kaps	650	0
opening_bracket	3344	3355
other_symbol: (132x60mm), 1,34x3,25, HbA1c=10,3%,	3161	2868
percentage_tok	4461	4478
question_mark	207	209
quotation	1	1
semicolon	455	455
slash	10340	10353
word_number_first_capital: AST34, B6	1195	1195
word_number_first_lower: mm3, pH6	1865	1854
word_with_hyphen_first_capital: B-hCG, Anty-HBs	163	163
word_with_hyphen_first_lower: m-ce, p-ciał	402	402
all tokens	463307	465004

5 Morphological analysies

Morphological annotation was based on the results obtained by the publicly available Polish POS tagger TaKIPI that cooperates with *Morfeusz* SIAT (Woliński, 2006) — a general-purpose morphological analyzer of Polish. For each word, it assigns all possible interpretations containing: its base form, part of speech, and complete morphological characterization (e.g. case, gender, number, aspect if relevant). The description is exhaustive and aimed at further syntactic analyses of texts.

The annotation is done in three steps. In the first one the documents are analyzed and disambiguated by TaKIPI. TaKIPI can be combined with the *Guesser* module (Piasecki and Radziszewski, 2007) which suggests tags for words which are not in the dictionary. We decided to use this module because otherwise 70600 tokens representing words and acronyms that occur in the documents would be assigned an unknown description. The gain from its usage is however not so evident, as tags and base forms suggested by *Guesser* are quite often incorrect – in one test set, only 272 forms out of 1345 were analyzed correctly.

The analyses of TaKIPI results shows that there are many systematic errors. They can be corrected globally. An example of such an error is the description of medication names produced by *Guesser*. Their morphologic tags are often correct, but the problem is with gender assignment in case of masculine forms. In Polish there are three subtypes of masculine gender: personal, animate and inanimate, and *Guesser* quite often uses personal masculine gender instead of the inanimate one while analyzing medication names. The second most common problem concerns base forms, because all base forms created by the module are written with a small letter. So in the case of proper names, all base forms have to be corrected. Moreover, TaKIPI do not disambiguate all tags – certain forms still have more than one possible description.

Thus, to limit the number of manual changes needed in the final version of the corpus, we post-process the results with a set of rules (see section 7) created on the basis of a list of all different token descriptions. The rules mainly correct the annotations of domain related tokens like acronyms and units: *BMI, HbA1c, RR, USG, Hz* or *kcal*; medication names e.g. *Diaprel*, its *diaprel* base form is changed into *Diaprel*; and other domain terms like *dekarboksylazie* ('decarboxylase$_{loc}$') for which the masculine base form was suggested *dekarboksylaz* instead of feminine *dekarboksylaza*. Moreover, tags of misspelled tokens and foreign words are assigned to tokens during this stage and if there is more than one description attached to a token, then the more probable in the domain is chosen.

Finally, the morphology analyses are manually corrected. This is done by two linguists. The results are compared and corrected by a third annotator. The first results are described in section 8.

6 Tags

For each token, TaKIPI assigns its base form, POS, and full morphological description. For example, the token *badania* that has the base form *badanie* 'examination' is classified in all 579 occurrences as a neutral noun. In 566 cases it is classified as a singular form in genitive and is assigned the tag **subst:sg:gen:n** (substantive:singular:genitive:neutral); in 13 cases as a plural noun including 8 nominative forms, 4 accusative and even one vocative (unreliable in medical texts). TaKIPI assigns the unknown tag (ign) to numbers, so we introduced the **number** tag to represent numerical values in the corpus. It is assigned to 18.8% of tokens.

The set of potential morphological tags consists of more than 4000 elements. In our corpus only 450 different tags are represented, in comparison to over 1000 tags used in the general Polish IPIPAN corpus (Przepiórkowski, 2005).

In the rest of this section we describe tags used for the classification of strings that are not properly classified by TaKIPI. If no tag described in the section suits a token, the tag **tsym** is assigned to it. In particular, all patient codes (like *d2005_006*) have the **tsym** tag.

6.1 Errors

Spelling errors in the corpus are left as they are. Misspelled tokens are assigned the base form equal to the token, and one of the following tags depending on the type of error:

- **err_spell** describes misspelled tokens like *bia3ko* instead of *białko* ('protein'). In the corpus we provide additional information with the corrected input token, its base form and morphological tag.

- **err_conj** describes concatenations like *cukrzycowej2000* ('diabetic2000'). In this case we add the correct form *cukrzycowej 2000* to the corpus but do not add its description.

- **err_disj_f** describes the first part of an incorrectly disjointed word. For example the word *ciśnienie* ('pressure') was divided into two parts *ci* and *śnienie*, (by chance, both are valid Polish words).

- **err_disj_r** describes the second part of the incorrectly disjointed word.

The last three categories can be supplemented with **spell** description if necessary. For example the token *Bylaw* is a concatenation of the misspelled word *Była* ('was') with the preposition *w* ('in'). This token has the tag **err_conj_spell**, and the *Była w* correction is added.

6.2 Abbreviations

There are many abbreviations in the documents. Some of them are used in general Polish like *prof* ('professor') or *dr* ('doctor), but there are many abbreviations that are specific to the medical domain. For example in the descriptions of USG examinations the letter *t* denotes *tętnica* ('artery'), while *tt* refers to the same word in plural, although usually there is no number related difference e.g. *wit* ('vitamin') can be used in plural and singular context. Sometimes it is not a single word but the whole phrase which is abbreviated, e.g. *NLPZ* is the acronym of the noun phrase *Niesterydowe Leki PrzeciwZapalne* 'Non-Steroidal Anti-Inflammatory Drugs', and *wpw* is an abbreviation of the prepositional phrase *w polu widzenia* 'in field of view'.

Abbreviations and acronyms obtain the tag **acron**. Moreover, it is possible to insert the full form corresponding to them.

Acronyms denoting units obtain the tag **unit**. Units in common usage are not explained: *mm, kg, h*, but if a unit is typical to the medical domain, its full form is given (e.g. *HBD* means *tydzień ciaży* 'week of pregnancy').

We also distinguish two tags describing prefixes and suffixes. The token *makro* ('macro') in the phrase *makro i mikroangiopatia* ('macro and microangiopathy') has the tag **prefix**, while the **suffix** tag describes, for example, the part *ma* of the string *10-ma* which indicates instrumental case of number 10, like in: *cukrzyca rozpoznana przed 10-ma laty* ('diabetes diagnosed 10 years ago').

6.3 Foreign Words

Foreign words receive the **foreign** tag. This tag can be elaborated with information on the part of speech, so for example, *Acne* has the tag **foreign_subst**. It is possible to attach a Polish translation to foreign words.

7 Correction Rules

Correction rules are created on the basis of a list of different tokens, their base form, and tags that occurred in the corpus. Each rule is applied to all matching form descriptions of tokens in the already tagged data.

We use the method of global changes because we want to decrease the number of manual corrections in the corpus on the final, manual stage. It should be noted that without context it is impossible to correct all morphological tags. We can only eliminate evident errors but we cannot decide, for example, if a particular description of a token *badanie* 'examination' (see section 6) is correct or not. All these tags can be verified only if we know the context where they occurred. However, quite a lot of changes can be made correctly in any context, e.g. changes of gender of a medication name ($Lorinden_f$ into $Lorinden_{m3}$), or in the prevailing number of cases, e.g. assigning to *zwolnienie* the *gerund* tag 'slowing' (11 occurrences) instead of less frequent in the texts *noun* 'sick leave' only one occurrence (TaKIPI leaves both descriptions).

There are two main types of correction rules of which syntax is given in (1–2). '#' is a separator; the character '>' indicates the new token description that is applied to the corpus; after || additional information can be noted. In case of rule (1) it could be a text that explains the meaning of acronyms, abbreviations or foreign words, while for rule (2), a corrected token, base form and tag can be given. This additional information might be used for creating a corpus without spelling errors, dictionaries of abbreviations or foreign words used in the medical domain.

(1) token#base form#tag#>
token#new base form#new tag#
|| 'string' (optionally)

(2) token#base form#tag#>
token#token#error_spell# ||
corr. token#corr. base form#new tag#

The first scheme is useful for changing the base form or the tag of a token. See example (3) where the first letter of the base form is capitalized and personal masculine gender *m1* is changed into inanimate masculine gender *m3*.

(3) Insulatard#insulatard#subst:sg:nom:m1#>
Insulatard#Insulatard#subst:sg:nom:m3#

The second scheme is applied to a token *graniach* 'ridges' (in mountain) that represents the existing but unreliable word in the medical domain. For all of its occurrences in our data (3 cases) it is substituted by *granicach* 'limits' by the following correction rule:

(4) graniach#grań#subst:pl:loc:f#>
granicach#granicach#err_spell# ||
granicach#granica#subst:pl:loc:f#

If there is more than one interpretation left by TaKIPI, all are mentioned before the character '>'. See example (5) where two different base forms are possible for the token *barku* and both have the same tag assigned. The first base form *bark* ('shoulder') is definitely more probable in the medical domain than the second one *barek* ('small bar' or 'cocktail cabinet'), so the rule chooses the first description.

(5) barku#bark#subst:sg:gen:m3##barek#
subst:sg:gen:m3#>barku#bark#subst:sg:gen:m3#

Table 2 presents the frequencies of top level morphological classes: directly after running the tagger, after changing the token limits and after applying automatic changes. In the last column the number of different forms in every POS class is presented.

Most part of speech names are self explanatory, the full list and description of all morphological tags can be found in (Przepiórkowski, 2004), the newly introduced tags are marked with *. Of all words (all tags apart form *interp*unction, *number* and *tsym*) the most numerous groups are nouns (*subs*tantive) – 54% and *adj*ectives – 15% of wordform occurrences.

Table 2: Morpheme types and numbers of occurrences

POS tag	tagger results	after tok. change	final corpus	different forms
	number of tag occurences			
adj	35305	35041	36848	3576
adv	2323	2323	2437	245
conj	5852	5852	5680	36
prep	29400	29400	26120	71
pron	302	302	142	21
subst	82215	82215	105311	5093
verb forms:	24743	24741	19912	2001
fin	2173	2173	1900	190
ger	9778	9778	4677	423
ppas	5593	5593	6170	551
other	7199	7197	7165	837
qub	4244	4242	2452	67
num	703	703	703	34
ign	160951	163629	0	0
acron*	0	0	30003	678
unit*	0	0	28290	82
prefix*	0	0	13	5
suffix*	0	0	36	6
tsym*	0	0	534	462
interp	115323	116556	116556	21
number*	0	0	87898	1386
err_disj*	0	0	179	129
err_spell*	0	0	560	440
foreign*	0	0	1330	184
total	461361	465004	465004	14537

If we don't take into account *number*, *tsym* and the punctuation tokens, we have a corpus of 348461 tokens (TW) out of which 78854 (29.81%) were changed. The most frequent changes concerned introducing domain related *unit* and *acronym* classes (nearly 72% of changes). Quite a number of changes were responsible for the capitalization of proper name lemmata. In table 3 the numbers of some other types of changes are presented.

Table 3: Morphological tag changes

type of change	number	% of changes	% of TW
base form			
capitalization only	6164	13.8	4.12
other	25503	32.34	9.64
POS			
to acron & unit	56697	71.90	21.43
to other	10547	13.37	3.99
grammatical features (without acron and unit)			
only case	109	0.13	0.04
only gender	1663	2.11	0.62
other	13215	16.75	4.99

Table 4: Manual correction

	basic tags	all tags
all tokens	8919	8919
without numbers and interp	4972	4972
unchanged	4497	4451
changed	475	521
same changes accepted	226	228
same changes not accepted	1	1
different changes none accepted	4	5
different changes. accepted 1	3	4
different changes. accepted 2	40	42
only 1st annot. changes - accepted	15	48
only 2nd annot. changes - accepted	128	124
only 1nd annot. changes - not accepted	47	47
only 2nd annot. changes - not accepted	0	0

8 Manual Correction

The process of manual correction of the corpus is now in progress. It is performed using an editor specially prepared for visualization and facilitation of the task of correcting the corpus annotation at all levels. In this section we present conclusions on the bases of 8 documents corrected by two annotators (highly experienced linguists). In the case of inconsistent corrections the opinion of a third annotator was taken into account. The process of annotation checking took about 2x20 hours.

From a total number of 8919 tokens in the dataset, the verification of 4972 (words, acronyms, units) was essential, the remaining 3947 tokens represent numbers, punctuation and **tsym** tokens. The correction rules changed the descriptions of 1717 (34%) tokens, only 87 cases were limited to the change of a lowercase letter into a capital letter of the base form. Manual verification left 4497 token descriptions unchanged, while 10.6% of descriptions were modified (evaluation of TaKIPI by Karwińska and Przepiórkowski (2009) reports 91.3% accuracy). Kappa coefficient was equal to 0.983 for part of speech and 0.982 for case assignment (when it is applicable). The results of manual correction are given in table 4. The 'basic tags' column gives the number of changes of the base form and tag, while the 'all tags' column takes into account all changes, including descriptions of the correct word form in case of spelling errors, explanations of acronyms or units.

More detailed analysis of annotation inconsistencies shows two main sources of errors:

- lack of precision in guidelines resulted in choosing different base forms in case of spelling errors and different labeling of cases with the lack of diacritics which resulted in correct but not the desired forms;

- some errors were unnoticed by one of the annotators (just cost of manual work), e.g. in the data there are many strings 'W' and 'w' which may be either acronyms or prepositions.

There are only a few cases that represent real morphological difficulties, e.g. differentiating adjectives and participles (5 cases among the annotators). Some examples of different case and gender assignments were also observed. They are mostly errors consisting in correcting only one feature instead of two, or a wrong choice of a case for long phrases.

9 Conclusions and Further Work

The problems described in the paper are twofold, some of them are language independent like tokenization, description of: abbreviations, acronyms, foreign expressions and spelling errors; while the others are specific for rich-morphology languages. Our experiment showed that analyzing specialized texts written in highly inflected language with a general purpose morphologic analyzer can give satisfactory results if it is combined with manually created global domain dependent rules. Our rules were created on the basis of a sorted list of all token descriptions. That allowed us to analyze a group of tokens with the same base form e.g. an inflected noun. Additional information concerning the frequency of each description, indicated which token corrections would be important.

Unfortunately, the process of rule creation is time-consuming (it took about 90 hours to create them). To speed up the process we postulate to prepare three sets of tokens for which rules will be created separately. The first one shall contain tokens which are not recognized by a morphological analyzer, and hence requiring transformation rules to be created for them. The second set shall contain tokens with more than one interpretation, for which a decision is necessary. Finally we propose to take into account the set of frequent descriptions. Infrequent tokens can be left to the manual correction stage as it is easier to correct them knowing the context.

At the moment our corpus contains three annotation levels – segmentation into tokens, morphological tags and semantic annotation. After the first phase of corpus creation we decided to introduce an additional level of annotation — extended tokenization, see (Marcus Hassler, 2006). Current tokenization divides text into simple unstructured fragments. This solution makes it easy to address any important fragment of a text, but leaves the interpretation of all complex strings to the next levels of analysis. A new extended tokenization is planned to create higher level tokens, semantically motivated. It will allow the annotation of complex strings like: dates (*02.12.2004, 02/12/2004*); decimal numbers; ranges (*10 - 15, 10-15*); sizes and frequencies (*10 x 15, 10x15*); complex units (mm/h);abbreviations with full stops (*r. – rok* 'year'); acronyms containing non-letter characters (*K+*); complex medication names (*Mono Mack Depot*).

Extended tokens can be recognized by rules taking into account two aspects: specificity of the domain and problems resulting from careless typing. In the case of abbreviations and acronyms, the best method is to use dictionaries, but some heuristics can be useful too. Electronic dictionaries of acronyms and abbreviations are not available for Polish, but on the basis of annotated data, a domain specific lexicon can be created. Moreover, we want to test ideas from (Kokkinakis, 2008), the author presents a method for the application of the MeSH lexicon (that contains English and Latin data) to Swedish medical corpus annotation. We will use a similar approach for acronyms and complex medication name recognition.

References

K. Bretonnel Cohen, Lynne Fox, Philip V. Ogren, and Lawrence Hunter. 2005. Corpus design for biomedical natural language processing. In *Proceedings of the ACL-ISMB Workshop on Linking Biological Literature, Ontologies and Databases: Mining Biological Semantics*, pages 38–45, Detroit, June. Association for Computational Linguistics.

Witold Drożdżyński, Hans-Ulrich Krieger, Jakub Piskorski, Ulrich Schäfer, and Feiyu Xu. 2004. Shallow Processing with Unification and Typed Feature Structures — Foundations and Applications. *German AI Journal KI-Zeitschrift*, 01/04.

Toma Erjavec, Yuka Tateisi, Jin dong Kim, Tomoko Ohta, and Jun ichi Tsujii. 2003. Encoding Biomedical Resources in TEI: the Case of the GENIA Corpus. In *Proceedings of the ACL 2003, Workshop on Natural Language Processing in Biomedicine*, pages 97–104.

Jing Jiang and Chengxiang Zhai. 2007. An Empirical Study of Tokenization Strategies for Biomedical Information Retrieval. *Information Retrieval*, 10(4–5):341–363.

Danuta Karwańska and Adam Przepiórkowski. 2009. On the evaluation of two Polish taggers. In *The proceedings of Practical Applications in Language and Computers PALC 2009*.

Jin-Dong Kim, Tomoko Ohtai, and Jun'ichi Tsujii. 2010. Multilevel Annotation for Information Extraction Introduction to the GENIA Annotation. In *Linguistic Modeling of Information and Markup Languages*, pages 125–142. Springer.

Dimitrios Kokkinakis. 2006. Collection, Encoding and Linguistic Processing of a Swedish Medical Corpus - The MEDLEX Experience. In *Proceedings of the Fifth International Language Resources and Evaluation (LREC'06)*, pages 1200–1205.

Dimitrios Kokkinakis. 2008. A Semantically Annotated Swedish Medical Corpus. In *Proceedings of the Sixth International Language Resources and Evaluation (LREC'08)*, pages 32–38.

Małgorzata Marciniak and Agnieszka Mykowiecka. 2011. Construction of a medical corpus based on information extraction results. *Control & Cybernetics*. in preparation.

Günther Fliedl Marcus Hassler. 2006. Text preparation throughextended tokenization. *Data Mining VII: Data, Text and Web Mining and their Business Applications*, 37.

Agnieszka Mykowiecka and Małgorzata Marciniak. 2011. Automatic semantic labeling of medical texts with feature structures. In *The Text Speech and Dialogue Conference 2011 (submitted)*.

Agnieszka Mykowiecka, Małgorzata Marciniak, and Anna Kupść. 2009. Rule-based information extraction from patients' clinical data. *Journal of Biomedical Informatics*, 42:923–936.

Maciej Piasecki and Adam Radziszewski. 2007. Polish Morphological Guesser Based on a Statistical A Tergo Index. In *2nd International Symposium Advances in Artificial Intelligence and Applications (AAIA'07), wisła, Poland*, pages 247–256.

Maciej Piasecki. 2007. Polish tagger TaKIPI: Rule based construction and optimisation. *Task Quarterly*, 11(1–2):151–167.

Adam Przepiórkowski and Piotr Bański. 2009. XML text intechange format in the National Corpus of Polish. In *The proceedings of Practical Applications in Language and Computers PALC 2009*, pages 245–250.

Adam Przepiórkowski. 2004. *Korpus IPI PAN. Wersja wstępna / The IPI PAN Corpus: Preliminary version*. IPI PAN.

Adam Przepiórkowski. 2005. The IPI PAN Corpus in numbers. In Zygmunt Vetulani, editor, *Proc. of the* 2nd Language & Technology Conference, Poznań, Poland.

Sampo Pyysalo, Filip Ginter, Juho Heimonen, Jari Björne, Jorma Boberg, Jouni Jörvinen, and Tapio Salakoski. 2007. BioInfer: a corpus for information extraction in the biomedical domain. *BMC Bioinformatics*, 8.

Angus Roberts, Robert Gaizauskas, Mark Hepple, George Demetriou, Yikun Guo, Ian Roberts, and Andrea Setzer. 2009. Building a semantically annotated corpus of clinical texts. *Journal of Biomedical Informatics*, 42(5):950–966.

Thomas Brox Røst, Ola Huseth, Øystein Nytrø, and Anders Grimsmo. 2008. Lessons from developing an annotated corpus of patient histories. *Journal of Computing Science and Engineering*, 2(2):162–179.

Yuka Teteisi and Jun'ichi Tsujii. 2006. GENIA Annotation Guidelines for Tokenization and POS Tagging. Technical report, Tsujii Laboratory, University of Tokyo.

Christos Tsalidis, Giorgos Orphanos, Elena Mantzari, Mavina Pantazara, Christos Diolis, and Aristides Vagelatos. 2007. Developing a Greek biomedical corpus towards text mining. In *Proceedings of the Corpus Linguistics Conference (CL2007)*.

Marcin Woliński. 2006. Morfeusz — a Practical Tool for the Morphological Analysis of Polish. In Mieczysław Kłopotek, Sławomir Wierzchoń, and Krzysztof Trojanowski, editors, *IIS:IIPWM'06 Proceedings, Ustron, Poland*, pages 503–512. Springer.

In Search of Protein Locations

Catherine Blake[1,2]
clblake@illinois.edu

Wu Zheng[1]
wuzheng2@illinois.edu

[1] Graduate School of Library and Information Science
[2] Computer Science and Medical Information Science
University of Illinois, Urbana Champaign, IL, USA

Abstract

We present a bootstrapping approach to infer new proteins, locations and protein-location pairs by combining UniProt seed protein-location pairs with dependency paths from a large collection of text. Of the top 20 system proposed protein-location pairs, 18 were in UniProt or supported by online evidence. Interestingly, 3 of the top 20 locations identified by the system were in the UniProt description, but missing from the formal ontology.

1 Introduction

Identifying subcellular protein locations is an important problem because the protein location can shed light on the protein function. Our goal is to identify new proteins, new locations and new protein-location relationships directly from full-text scientific articles. As with many ontological relations, location relations can be described as a binary predicate comprising two arguments, Location(X, Y) indicates that X is located in Y, such as Location (ClC-5, luminal membrane) from the sentence: ClC-5 *specific signal also appeared to be localized close to the* luminal membrane *of the intestinal crypt.*

Identifying protein subcellular locations has been framed as a classification task, where features include sequences, motifs and amino acid composition (Höglund, et al, 2006) and protein networks (Lee et al., 2008). The SherLoc system (Shatkay et al., 2007) includes text features the EpiLoc system (Brady & Shatkay, 2008) represents text from Medline abstracts as a vector of terms and uses a support vector machine to predict the most likely location for a new protein. Classification accuracy varies between species, locations, and datasets.

We take an alternative strategy in this paper and propose a bootstrapping algorithm similar to (Gildea & Jurafsky, 2001). The proposed system builds on earlier work (Zheng & Blake, 2010) by considering a larger set of seed terms and by removing syntactic path constraints.

2 Approach

The proposed bootstrapping algorithm is depicted in Figure 1. The system identifies lexico-syntactic patterns from sentences that include a given set of seed terms. Those the patterns are then used to infer new proteins, new locations, and new protein-location relationships. The system thus requires (a) an existing collection of known entity pairs that participate in a location relationship (called the seed terms) (b) a corpora of texts that report location relationships and (c) a syntactic path representation.

Our experiments use seed protein-location relationships from the UniProt knowledge base (www.uniprot.org). The complete knowledge base comprises more than 80,000 protein names for a range of species. The system uses the location and the location synonyms from the UniProt controlled vocabulary of subcellular locations and membrane topologies and orientations (www.uniprot.org/docs/subcell release 2011_2). The system also used a list of protein terms that were created by identifying words that immediately precede the word *protein* or *proteins* in the TREC collection. Two-thirds of the top 100 proteins in the TREC collection were used as seed terms and the remaining 1/3 were used to evaluate system performance.

The system was developed and evaluated using different subsets of the Genomics Text Retrieval (TREC) collection (Hersh, & Voorhees, 2009). Specifically 5533 articles in JBC 2002 were used for development and ~11,000 articles in JBC 2004 and 2005 were used in the evaluation.

The syntactic paths used the dependency tree representation produced by the Stanford Parser (Klein & Manning., 2003) (version 1.6.4).

Proceedings of the 2011 Workshop on Biomedical Natural Language Processing, ACL-HLT 2011, pages 101–102,
Portland, Oregon, USA, June 23-24, 2011.

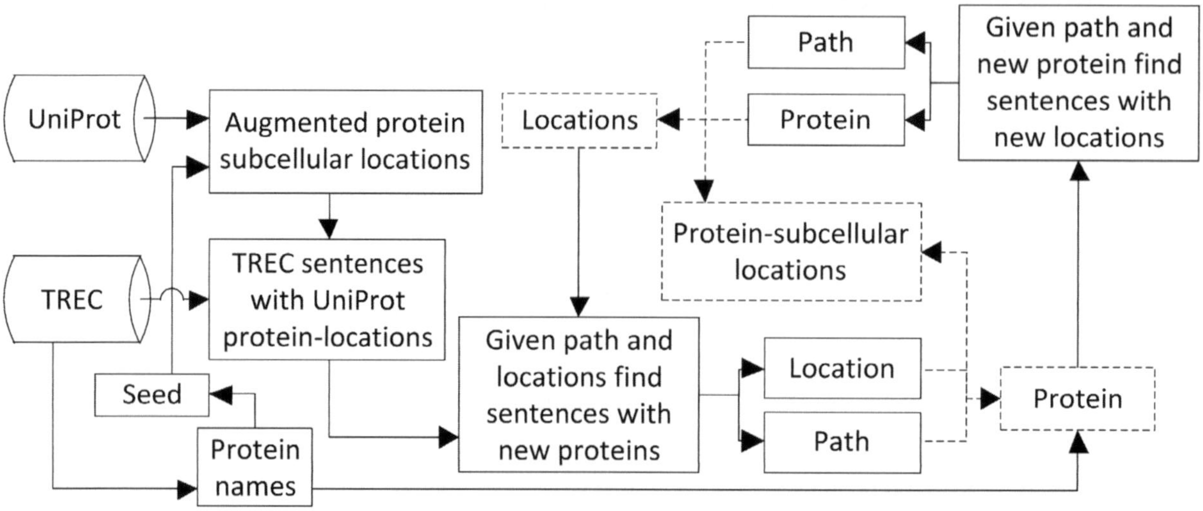

Figure 1 – The Bootstrapping approach used to generate new proteins, subcellular locations and protein location pairs. Inferred proteins and locations are depicted with a dashed line.

3 Results

The system identified 792 new proteins in the first iteration. All but 3 of the most frequent 20 proteins were in UniProt. All proteins in the test set were identified, but only 10 were in the top 100 proteins.

The system identified just over 1,200 new protein-location pairs after the first bootstrapping step. We evaluated the twenty most frequent pairs. Two erroneous proteins in the previous step caused two protein-location pair errors. UniProt reported 13 of the remaining 18 protein-location pairs. The five remaining pairs, were supported by online sources and in sentences within the collection.

The system identified 493 new locations after the second bootstrapping step and we evaluated the top 20. Sentences in the collection suggest that 9 of the new locations are in fact locations, but that they may not be subcellular locations and that 8 proposed locations are too general. Interestingly, 3 of the top 20 locations identified by the system are mentioned in the UniProt definitions, but are not included in the control vocabulary as a synonym, which suggests the need for automated approaches such as this to supplement manual efforts.

Acknowledgments

This material is based in part upon work supported by the National Science Foundation under Grant IIS-0812522. Any opinions, findings, and conclusions or recommendations expressed in this material are those of the author(s) and do not necessarily reflect the views of the National Science Foundation.

References

Brady, S., & Shatkay, H. 2008. EpiLoc: a (working) text-based system for predicting protein subcellular location., Pac Symp Biocomput (pp. 604-615).

Gildea, D., & Jurafsky, D. 2001. Automatic labeling of semantic roles. Computational Linguistics, 99(9): 1-43.

Hersh, W., & Voorhees, E. (2009). TREC genomics special issue overview. Information Retrieval, 12(1), 1-15.

Höglund, A., Dönnes, P., Blum, T., Adolph, H.W., & Kohlbacher, O. 2006. MultiLoc: prediction of protein subcellular localization using N-terminal targeting sequences, sequence motifs and amino acid composition. Bioinformatics, 22(10):1158-1165.

Klein, D., & Manning., C.D. 2003. In Accurate Unlexicalized Parsing (pp. 423-430). Paper presented at the In Proceedings of the 41st Annual Meeting of the Association for Computational Linguistics (ACL-2003).

Lee, K., Chuang, H.-Y., Beyer, A., Sung, M.-K., Huh, W.-K., Lee, B., et al. 2008. Protein networks markedly improve prediction of subcellular localization in multiple eukaryotic species. Nucleic Acids Research, 36(20), e136.

Shatkay, H., Höglund, A., Brady, S., Blum, T., Dönnes, P., & Kohlbacher, O. 2007. SherLoc: high-accuracy prediction of protein subcellular localization by integrating text and protein sequence data Bioinformatics, 23(11), 1410-1417.

Zheng, W., & Blake, C. 2010. Bootstrapping Location Relations from Text. American Society for Information Science and Technology, Pittsburgh, PA.

Automatic extraction of data deposition sentences: where do the research results go?

Aurélie Névéol, W. John Wilbur, Zhiyong Lu

National Center for Biotechnology Information
U.S. National Library of Medicine
Bethesda, MD 20894, USA
{Aurelie.Neveol,John.Wilbur,zhiyong.lu}@nih.gov

Abstract

Research in the biomedical domain can have a major impact through open sharing of data produced. In this study, we use machine learning for the automatic identification of data deposition sentences in research articles. Articles containing deposition sentences are correctly identified with 73% f-measure. These results show the potential impact of our method for literature curation.

1 Background

Research in the biomedical domain aims at furthering the knowledge of biological processes and improving human health. Major contributions towards this goal can be achieved by sharing the results of research efforts with the community, including datasets produced in the course of the research work. While such sharing behavior is encouraged by funding agencies and scientific journals, recent work has shown that the ratio of data sharing is still modest compared to actual data production. For instance, Ochsner et al. (2008) found the deposition rate of microarray data to be less than 50% for work published in 2007.

Information about the declaration of data deposition in research papers can be used both for data curation and for the analysis of emerging research trends. Our long-term research interest is in assessing the value of deposition sentences for predicting future trends of data production. The initial step of automatically identifying deposition sentences would then lead to an assessment of the need for storage space of incoming data in public repositories.

2 Objective

In this study, we aim at automatically performing a fine-grained identification of biological data deposition sentences in biomedical text. That is, we aim at identifying articles containing deposition sentences, extracting the specific sentences and characterizing the information contained in the sentences in terms of data type and deposition location (e.g. database, accession numbers).

3 Material and Methods

Data deposition sentences. A collection of sentences reporting the deposition of biological data (such as microarray data, protein structure, gene sequences) in public repositories was compiled based on previous work that we extended. We take these sentences as a primary method of identifying articles reporting on research that produced the kind of data deposited in public repositories. (1) and (2) show examples of such sentences. In contrast, (3) and (4) contain elements related to data deposition while focusing on other topics.

(1) The sequences reported in this paper have been deposited in the GenBank database (accession numbers AF034483 for susceptible strain RC688s and AF034484 for resistant strain HD198r).

(2) The microarray data were submitted to MIAMExpress at the EMBL-EBI.

(3) Histone TAG Arrays are a repurposing of a microarray design originally created to represent the TAG sequences in the Yeast Knockout collection (Yuan et al 2005 NCBI GEO Accession Number GPL1444).

(4) The primary sequence of native Acinetobacter CMO is identical to the gene sequence for chnB deposited under accession number AB006902.

Proceedings of the 2011 Workshop on Biomedical Natural Language Processing, ACL-HLT 2011, pages 103–104,
Portland, Oregon, USA, June 23-24, 2011.

Sentence classification. A Support Vector Machine (SVM) classifier was built using a corpus of 583 positive data deposition sentences and 578 other negative sentences. Several sets of features were tested, including the following: sentence tokens, associated part-of-speech tags obtained using MEDPOST[1], relative position of the sentence in the article, identification of elements related to data deposition (data, deposition action, database, accession number) obtained using a CRF model[2].

Article classification. The automatic classification of articles relied on sentence analysis. The full text of articles was segmented into sentences, which were then scored by the sentence-level SVM classifier described above. An article is classified as positive if its top-scored sentence is scored higher than a threshold, which is predetermined as the 25th percentile score for positive sentences in the training set.

Evaluation corpus. A corpus composed of 670 PubMed Central articles was used to evaluate article classification. 200 articles were considered as "positive" for data deposition based on MEDLINE gold standard annotations in the [si] field used to curate newly reported accession numbers.

4 Results

Table 1 shows the performance of selected SVM models for article classification on the test set. While differences were very small for cross-validation on the training set, they are emphasized on the test set.

Features	P	R	F
Tokens, position, part-of-speech tags	52%	56%	54%
Token, position, CRF+, part-of-speech tags	65%	58%	62%
Tokens, position, CRF+/-, part-of-speech tags	**69%**	**78%**	**73%**

Table 1: Precision, Recall and F-measure of SVM models for article classification on test set.

5 Discussion and Conclusion

Portability of the method. Although trained mainly on microarray data deposition sentences, the method adapts well to the identification of other data deposition sentences, e.g. gene sequences, protein coordinates.

Comparison to other work. Our approach is not directly comparable to any of the previous studies. At the article level, we perform an automatic classification of articles containing data deposition sentences, in contrast with Oshner et al. who performed a one-time manual classification. Piwowar et al used machine learning and rule-based algorithms for article classification. However, they relied on identifying the names of five predetermined databases in the full text of articles. Our approach is generic and aiming at the automatic identification of any biological data deposition in any public repository. Furthermore, our approach also retrieves specific data deposition sentences where data and deposition location are identified. At the sentence level, this is also different from the classification of databank accession number sentences performed by Kim et al. (2010) in two ways: first, we focus on retrieving sentences containing accession numbers if they are deposition sentences (vs. data re-use, etc.) and second, we are also interested in retrieving data deposition sentences that do not contain accession numbers.

Error analysis. Almost half of the articles classified as containing a deposition sentence by our method but not by the gold standard were found to indeed contain a deposition sentence.

Conclusion. These results show the potential impact of our method for literature curation. In addition, it provides a robust tool for future work assessing the need for storage space of incoming data in public repositories.

Acknowledgments

This research was supported by the Intramural Research Program of the NIH, NLM.

References

Jongwoo Kim, Daniel Le, Georges R. Thoma. Naïve bayes and SVM classifiers for classifying databank accession number sentences from online biomedical articles. Proc. SPIE 2010 (7534): 7534OU-OU8

Scott A. Ochsner, Davd L Steffen, Christian J Stoeckert Jr, Neil J. McKenna. Much room for improvement in deposition rates of expression microarray datasets. Nat Methods. 2008 Dec;5(12):991.

Heather A. Piwowar, Wendy W. Chapman. Identifying data sharing in biomedical literature.AMIA Annu Symp Proc. 2008 Nov 6:596-600.

[1] http://www.ncbi.nlm.nih.gov/staff/lsmith/MedPost.html
[2] http://mallet.cs.umass.edu/

From Pathways to Biomolecular Events: Opportunities and Challenges

Tomoko Ohta[*] **Sampo Pyysalo**[*] **Jun'ichi Tsujii**[†]
[*]Department of Computer Science, University of Tokyo, Tokyo, Japan
[†]Microsoft Research Asia, Beijing, China
{okap,smp}@is.s.u-tokyo.ac.jp, jtsujii@microsoft.com

Abstract

The construction of pathways is a major focus of present-day biology. Typical pathways involve large numbers of entities of various types whose associations are represented as reactions involving arbitrary numbers of reactants, outputs and modifiers. Until recently, few information extraction approaches were capable of resolving the level of detail in text required to support the annotation of such pathway representations. We argue that event representations of the type popularized by the BioNLP Shared Task are potentially applicable for pathway annotation support. As a step toward realizing this possibility, we study the mapping from a formal pathway representation to the event representation in order to identify remaining challenges in event extraction for pathway annotation support. Following initial analysis, we present a detailed study of protein association and dissociation reactions, proposing a new event class and representation for the latter and, as a step toward its automatic extraction, introduce a manually annotated resource incorporating the type among a total of nearly 1300 annotated event instances. As a further practical contribution, we introduce the first pathway-to-event conversion software for SBML/CellDesigner pathways and discuss the opportunities arising from the ability to convert the substantial existing pathway resources to events.

1 Introduction

For most of the previous decade of biomedical information extraction (IE), efforts have focused on foundational tasks such as named entity detection and their database normalization (Krallinger et al., 2008) and simple IE targets, most commonly binary entity relations representing associations such as protein-protein interactions (Pyysalo et al., 2008; Tikk et al., 2010). In recent years, an increasing number of resources and methods pursuing more detailed representations of extracted information are becoming available (Pyysalo et al., 2007; Kim et al., 2008; Thompson et al., 2009; Björne et al., 2010). The main thrust of this move toward structured, fine-grained information extraction falls under the heading of *event extraction* (Ananiadou et al., 2010), an approach popularized and represented in particular by the BioNLP Shared Task (BioNLP ST) (Kim et al., 2009a; Kim et al., 2011).

While a detailed representation of extracted information on biomolecular events has several potential applications ranging from semantic search to database curation support (Ananiadou et al., 2010), the number of practical applications making use of this technology has arguably so far been rather limited. In this study, we pursue in particular the opportunities that event extraction holds for pathway annotation support,[1] arguing that the match between

[1]Throughout this paper, we call the projected task *pathway annotation support*. There is no established task with this label, and we do not envision this to be a specific single task. Rather, we intend the term to refer to a set of tasks where information extraction/text mining methods are applied in some role to contribute directly to pathway curation, including, for example, the identification of specific texts in the literature relevant to annotated reactions, the automatic suggestion of further entities or reactions to add to a pathway, or even the fully automatic generation of entire pathways from scratch.

Proceedings of the 2011 Workshop on Biomedical Natural Language Processing, ACL-HLT 2011, pages 105–113,
Portland, Oregon, USA, June 23-24, 2011.

representations that biologists employ to capture reactions between biomolecules in pathways and the event representation of the BioNLP ST task makes pathway-oriented applications a potential "killer application" for event extraction technology.

The fit between these representations is not accidental – the design of the BioNLP ST event representation has been informed by that of popular pathway models – nor is it novel to suggest to support pathway extraction through information methods in general (see e.g. (Rzhetsky et al., 2004)) or through event extraction specifically (Oda et al., 2008). However, our study differs from previous efforts in two key aspects. First, instead of being driven by information extraction and defining a representation fitting its results, we specifically adopt the perspective and model of a widely applied standard database representation and proceed from the pathway to events in text. Second, while previous work on event extraction for pathway annotation has been exploratory in nature or has otherwise had limited practical impact, we introduce and release a first software implementation of a conversion from a standard pathway format to the event format, thus making a large amount of pathway data available for use in event extraction and taking a concrete step toward reliable, routine mappings between the two representations.

2 Representations and Resources

Before proceeding to consider the mapping between the two, we first briefly introduce the pathway and event representations in focus in this study and the applied pathway resources.

2.1 Pathways

The biomolecular curation community has created and made available an enormous amount of pathway resources: for example, as of April 2011, the Pathguide pathway resource list[2] includes references to 325 pathway-related resources – many of which are themselves pathway databases containing hundreds of individual models. These resources involve a formidable variety of different, largely independently developed formats and representations of which only few pairs have tools supporting mutual conversion. To address the challenges of interoperability that this diversity implies, a number of standardization efforts for pathway representations have been introduced.

In this work, we consider two widely adopted pathway representation formats: Systems Biology Markup Language (SBML)[3] (Hucka et al., 2003) and Biological Pathway Exchange (BioPAX)[4] (Demir et al., 2010). SBML is an XML-based machine-readable data exchange format that supports a formal mathematical representation of chemical reactions (including e.g. kinetic parameters), allowing biochemical simulation. BioPAX is an RDF/OWL-based standard language to represent bio-molecular and cellular networks designed to enable data integration, exchange, visualization and analysis. Despite significantly different choices in storage format, the represented information content of the two is broadly compatible. In the following, we refer to established correspondences and mappings when relating the two (see e.g. (Mi and Thomas, 2009)).

As an interchange format aimed to support a large variety of specific representations, the SBML standard itself does not define a fixed set of types of physical entities or biochemical reactions. However, the standard defines an extension mechanism allowing additional information, including such types, to be defined. As specific, fixed types with established semantics are a requirement for practical conversion between the different representations, we thus rely in this work not only on SBML core, but also a minimal set of the extensions introduced by the popular CellDesigner pathway modeling tool (Funahashi et al., 2008). In the following, we assume throughout the availability of CellDesigner extensions when discussing SBML features.

For pathway data, in this study we use the full set of pathways contained in the Panther and Payao pathway repositories in SBML form. Panther (Protein ANalysis THrough Evolutionary Relationships) is a gene function-based classification system that hosts a large collection of pathways. The Panther repository consists of 165 pathways, including 153 signaling and 12 metabolic pathways. All pathways

[2]http://www.pathguide.org/

[3]http://sbml.org

[4]http://www.biopax.org

Figure 1: Illustration of the event representation.

were drawn on CellDesigner by manual curation and thus include CellDesigner SBML extensions (Mi and Thomas, 2009). Payao is a community-based SBML model tagging platform (Matsuoka et al., 2010) that allows a community to share models, tag and add comments, and search relevant literature (Kemper et al., 2010). Currently, 28 models are registered in Payao. As in Panther, all Payao pathways include CellDesigner extensions.

2.2 Event Representation

The application of *event representations* in biomedical IE is a relatively recent development, following the introduction of corpus resources annotating structured, n-ary associations of entities with detailed types (Pyysalo et al., 2007; Kim et al., 2008; Thompson et al., 2009)) and popularized in particular by the BioNLP Shared Task (BioNLP ST) events (Kim et al., 2009b; Kim et al., 2011). In this paper, we use *event* in the BioNLP ST sense, to refer specifically to the representation where each event is assigned a type from a fixed ontology, bound to a specific expression in text stating its occurrence (the *trigger* or *text binding*), and associated with an arbitrary number of participants (similarly text-bound entities or other events), for which the roles in which they are involved in the event are defined from a fixed small inventory of event argument types (e.g. *Theme*, *Cause*, *Site*). These concepts are illustrated in Figure 1.

3 Analysis of Pathway-Event Mapping

We next present an analysis of the relationship between the two representations, considering features required from IE systems for efficient support of pathway annotation support.

We assume throughout that the target on the pathway side is restricted to the broad, central biological content of pathways, excluding information only related to e.g. simulation support or pathway visualization/layout.

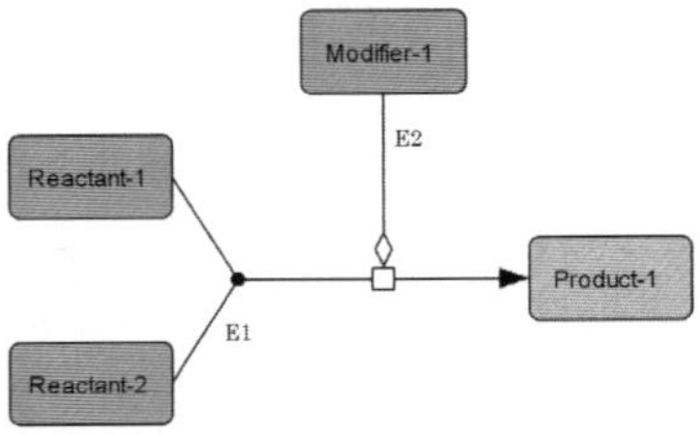

Figure 2: Illustration of a generalized pathway reaction.

3.1 Top-level concepts

Both SBML and BioPAX involve two (largely comparable) top-level concepts that form the core of the representation: entity (species/physical entity) and reaction (interaction). In the following we focus primarily on entities and reactions, deferring consideration of detailed concepts such as modification state and compartment localization to Section 3.3.

The concept of a reaction in the considered pathway representations centrally involves three sets of entities: reactants, products, and modifiers. As the names suggest, the reaction produces the set of product entities from the reactant entities and is affected by the modifiers. Figure 2 shows an illustration of a generalized reaction. Pathway reactions find a reasonably good analogy in events in the event representation. While the event representation does not differentiate "reactants" from "products" in these terms, the roles assigned to event participants allow comparable interpretation. There is no single concept in the event representation directly comparable to reaction modifiers. However, the semantics of specific modification types (see Section 3.3) correspond broadly to those of regulation in the event representation, suggesting that modification be represented using a separate event of the appropriate type with the modifying entities participating in the *Cause* role (Kim et al., 2008). Figure 3 illustrates the event structure proposed to correspond to the reaction of Figure 2, with the added assumptions that the reaction and modification types (unspecified in Figure 2) are Association (BioPAX:ComplexAssembly) and Modulation (BioPAX:Control).

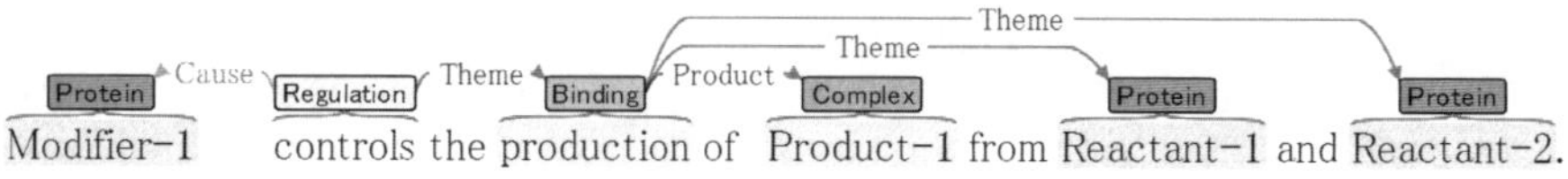

Figure 3: Illustration of a generalized event structure with four entities and two events (REGULATION and BINDING). Note that the text is only present as filler to satisfy the requirement that events are bound to specific expressions in text. The *Product* role is not a standard role in event representation but newly proposed in this study.

Pathway		Event		
CellDesigner	BioPAX	ST'09	ST'11	GENIA
Protein	Protein	Protein	Protein	Protein
RNA	RNA	Protein	Protein	RNA
AntiSenseRNA	RNA	Protein	Protein	RNA
Gene	DNA	Protein	Protein	DNA
Simple molecule	Small molecule	-	Chemical	Inorganic compound
Ion	Small molecule	-	Chemical	Inorganic compound
Drug	PhysicalEntity	-	Chemical	Inorganic compound
Hetero/homodimer	Complex	-	-	Protein complex

Table 1: Entity type comparison between pathways and events.

The mapping of top-level concepts that we consider thus unifies physical entities in pathways with the entities of the BioNLP ST representation, and pathway *reaction* with *event*.[5]

To be able to efficiently support (some aspect of) pathway annotation through IE, the applied extraction model should be able, for both entities and reactions, to 1) recognize mentions of all relevant types of entity/reaction and 2) differentiate between entity/reaction types at the same or finer granularity as the pathway representation. For example, an IE system that does not detect mentions of protein complexes cannot efficiently support aspects of pathway annotation that involve this type; a system that detects proteins and complexes with no distinction between the two will be similarly limited. In the following, we consider entity and reaction types separately to determine to what extent these requirements are filled by presently available resources for event extraction, in particular the GENIA corpus (Kim et al., 2008) and the BioNLP ST 2009 (Kim et al., 2009b) and 2011 corpora.

3.2 Entities

Table 1 shows a comparison of the primary entity types between SBML/CellDesigner, BioPAX, and the event representations. There is significant difference in the resolution of gene and gene product types between the pathway representations and that applied in ST'09 and ST'11: while both pathway representations and the GENIA corpus differentiate the DNA, RNA and protein forms, the STs fold the three types into a single one, PROTEIN.[6] The CHEMICAL type defined in ST'11 (ID task) overlaps largely with BioPAX SMALL MOLECULE, a type that SBML/CellDesigner further splits into two specific types, and further partly covers the definition of the SBML/CellDesigner type Drug. The same holds (with somewhat less specificity) for GENIA INORGANIC COMPOUND. Finally, although annotated in GENIA, the category of protein complexes has no correspondence among the entities considered in the BioNLP ST representation.

Thus, information extraction systems applying the core BioNLP ST entity types will entirely lack coverage for protein complexes and will not be able

[5]Pathways and IE/text mining use many of the same terms with (sometimes subtly) different meanings. We use largely IE terminology, using e.g. *entity* instead of *species* (SBML) and *entity type* instead of *physical entity class* (BioPAX) / *species type* (SBML) For the pathway associations, we have adopted *reaction* (SBML term) in favor of *interaction* (BioPAX). With *event*, we refer to the BioNLP ST sense of the word; we make no use of the SBML "event" concept.

[6]While the term PROTEIN appears to suggest that the class consists only of protein forms, these entities are in fact annotated in the BioNLP ST data according to the GENIA gene/gene product guidelines (Ohta et al., 2009) and thus include also DNA and RNA forms. The type could arguably more accurately be named GENE OR GENE PRODUCT.

Pathway		Event		
CellDesigner	BioPAX	ST'09	ST'11	GENIA
State transition	BiochemicalReaction		(see Table 3)	
Truncation	BiochemicalReaction	Catabolism	Catabolism	Catabolism
Transcription	BiochemicalReaction	Transcription	Transcription	Transcription
Translation	BiochemicalReaction	-	-	Translation
Association	ComplexAssembly	Binding	Binding	Binding
Dissociation	ComplexAssembly	-	-	-
Transport	Transport w/reaction	Localization	Localization	Localization
Degradation	Degradation	Catabolism	Catabolism	Catabolism
Catalysis	Catalysis	Positive regulation	Positive regulation	Positive regulation
Physical stimulation	Control	Positive regulation	Positive regulation	Positive regulation
Modulation	Control	Regulation	Regulation	Regulation
Trigger	Control	Positive regulation	Positive regulation	Positive regulation
Inhibition	Control	Negative regulation	Negative regulation	Negative regulation

Table 2: Reaction type comparison between pathways and events.

to fully resolve the detailed type of gene and gene product types applied in the pathway representations. While these distinctions exist in the full GENIA corpus, it has not been frequently applied in event extraction in its complete form and is unlikely to be adopted over the widely applied ST resources. Finally, none of the event representations differentiate the pathway small molecule/drug types. We discuss the implications of these ambiguities in detail below. By contrast, we note that both SBML/CellDesigner and BioPAX entity types cover the scope of the major BioNLP ST types and have comparable or finer granularity in each case.

3.3 Reactions

Table 2 shows a comparison between the reaction types of the two considered pathway representations and those of the BioNLP ST event representation. The full semantics of the generic reaction type State transition (BioPAX: BiochemicalReaction) cannot be resolved from the type alone; we defer discussion of this type.

Contrary to the event types, we find that for reaction types even the least comprehensive BioNLP ST'09 event representation has high coverage of the pathway reaction types as well as a largely comparable level of granularity in its types. While neither of the BioNLP ST models defines a TRANSLATION type, the adoption of the GENIA representation – matching that for TRANSCRIPTION – for this simple and relatively rare event type would likely be relatively straightforward. A more substantial omission in all of the event representations is the lack of a Dissociation event type. As dissociation is the "reverse" reaction of (protein) BINDING and central to many pathways, its omission from the event model is both surprising as well as potentially limiting for applications of event extraction to pathway annotation support.

The detailed resolution of pathway reactions provided by the event types has implications on the impact of the ambiguity noted between the single type covering genes and gene products in the event representation as opposed to the distinct DNA/RNA/protein types applied in the pathways. Arguably, for many practical cases the specific type of an entity of the broad gene/gene product type is unambiguously resolved by the events it participates in: for example, any gene/gene product that is modified through phosphorylation (or similar reaction) is necessarily a protein.[7] Similarly, only proteins will be involved in e.g. localization between nucleus and cytoplasm. On a more detailed level, BINDING events resolves their arguments in part through their *Site* argument: binding to a promoter implies DNA, while binding to a C-terminus implies protein. Thus, we can (with some reservation) forward the argument that it is not necessary to disambiguate all gene/gene product mentions on the entity level for pathway annotation support, and that successful event extraction will provide disambiguation in cases where the distinction matters.

[7]DNA methylation notwithstanding; the BioNLP ST'11 EPI task demonstrated that protein and DNA methylation can be disambiguated on the event type level without entity type distinctions.

Pathway			Event		
SBML/CellDesigner			ST'09	ST'11	GENIA
in:Compartment$_1$	→	in:Compartment$_2$	Localization	Localization	Localization
residue:state:∅	→	residue:state:Phosphorylated	Phosphorylation	Phosphorylation	Phosphorylation
residue:state:Phosphorylated	→	residue:state:∅	-	Dephosphorylation	Dephosphorylation
residue:state:∅	→	residue:state:Methylated	-	Methylation	Methylation
residue:state:Methylated	→	residue:state:∅	-	Demethylation	Demethylation
residue:state:∅	→	residue:state:Ubiquitinated	-	Ubiquitination	Ubiquitination
residue:state:Ubiquitinated	→	residue:state:∅	-	Deubiquitination	Deubiquitination
species:state:inactive	→	species:state:active	Positive regulation	Positive regulation	Positive regulation
species:state:active	→	species:state:inactive	Negative regulation	Negative regulation	Negative regulation

Table 3: Interpretation and comparison of state transitions.

Finally, the pathway representations define generic reaction types (State transition/BiochemicalReaction) that do not alone have specific interpretations. To resolve the event involved in these reactions it is necessary to compare the state of the reactants against that of the matching products. Table 3 shows how specific state transitions map to event types (this detailed comparison was performed only for SBML/CellDesigner pathways). We find here a good correspondence for transitions affecting a single aspect of entity state. While generic pathway transitions can change any number of such aspects, we suggest that decomposition into events where one event corresponds to one point change in state is a reasonable approximation of the biological interpretation: for example, a reaction changing one residue state into Methylated and another into Phosphorylated would map into two events, METHYLATION and PHOSPHORYLATION.

In summary of the preceding comparison of the core pathway and event representations, we found that in addition to additional ambiguity in e.g. gene and gene product types, the popular BioNLP ST representations lack a protein complex type and further that none of the considered event models define a (protein) dissociation event. To address these latter omissions, we present in the following section a case study of dissociation reactions as a step toward their automatic extraction. We further noted that pathway types cover the event types well and have similar or higher granularity in nearly all instances. This suggests to us that mapping from the pathway representation to events is more straightforward than vice versa. To follow up on these opportunities, we introduce such a mapping in Section 5, in following the correspondences outlined above.

4 Protein Association and Dissociation

In the analysis presented above, we noted a major reaction type defined in both considered pathway representations that had no equivalent in the event representation: dissociation. In this section, we present a study of this reaction type and its expression as statements in text through the creation of event-style annotation for dissociation statements.

4.1 Target data

Among the large set of pathways available, we chose to focus on the Payao mTOR pathway (Caron et al., 2010) because it is a large, recently introduced pathway with high-quality annotations that involves numerous dissociation reactions. The Payao pathways are further annotated with detailed literature references, providing a PubMed citation for nearly each entity and reaction in the pathway. To acquire texts for event annotation, we followed the references in the pathway annotation and retrieved the full set of PubMed abstracts associated with the pathway, over 400 in total. We then annotated 60 of these abstracts that were either marked as relevant to dissociation events in the pathway or were found to include dissociation statements in manual analysis. These abstracts were not included in any previously annotated domain corpus. Further, as we aimed specifically to be able to identify event structures for which no previous annotations exist, we could not rely on (initial) automatic annotation.

4.2 Annotation guidelines

We performed exhaustive manual entity and event annotation in the event representation for the selected 60 abstracts. For entity annotation, we initially considered adopting the gene/gene product annotation guidelines (Ohta et al., 2009) applied in the BioNLP ST 2009 as well as in the majority of the 2011 tasks. However, the requirement of these guidelines to mark only specific gene/protein names would exclude a substantial number of the entities marked in the pathway, as many refer to gene/protein families or groups instead of specific individual genes or proteins. We thus chose to adopt the pathway annotation itself for defining the scope of our entity annotation: we generated a listing of all the names appearing in the target pathway and annotated their mentions, extrapolating from this rich set of examples to guide us in decisions on how to annotate references to entities not appearing in the pathway. For event annotation, we adapted the GENIA event corpus annotation guidelines (Kim et al., 2008), further developing a specific representation and guidelines for annotating dissociation events based on an early iteration of exploratory annotation.

Annotation was performed by a single biology PhD with extensive experience in event annotation (TO). While we could thus not directly assess inter-annotator consistency, we note that our recent comparable efforts have been evaluated by comparing independently created annotations at approximately 90% F-score for entity annotations and approximately 80% F-score for event annotations (BioNLP Shared Task primary evaluation criteria) (Pyysalo et al., 2011; Ohta et al., 2011).

4.3 Representing Association and Dissociation

Based on our analysis of 107 protein dissociation statements annotated in the corpus and a corresponding study of the "reverse", statements of protein association in the corpus, we propose the following extensions for the BioNLP ST event representation. First, the introduction of the event type DISSOCIATION, taking as its primary argument a single *Theme* identifying a participating entity of the type COMPLEX. Second, we propose the new role type *Product*, in the annotation of DISSOCIATION events an optional (secondary) argument identifying the PROTEIN entities that are released in the dissociation event. This argument should be annotated (or extracted) only when explicitly stated in text. Third, for symmetry in the representation, more detail in extracted information, and to have a representation

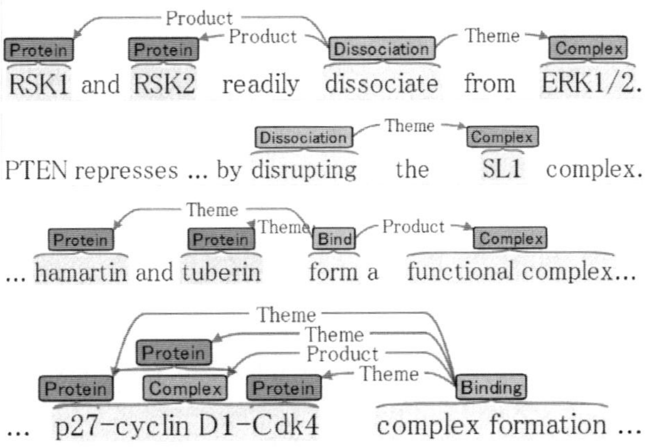

Figure 4: Examples annotated with the proposed event representation for DISSOCIATION and BINDING events with the proposed *Product* role marking formed complex.

Item	Count
Abstract	60
Word	11960
Protein	1483
Complex	201
Event	1284

Table 4: Annotation statistics.

more compatible with the pathway representation for protein associations, we propose to extend the representation for BINDING, adding *Product* as an optional argument identifying a COMPLEX participant in BINDING events marking statements of complex formation stating the complex. The extended event representations are illustrated in Figure 4.

4.4 Annotation statistics

Table 4 presents the statistics of the created annotation. While covering a relatively modest number of abstracts, the annotation density is very high, relating perhaps in part to the fact that many of the referenced documents are reviews condensing a wealth of information into the abstract.

5 Pathway-to-event conversion

As an additional practical contribution and outcome of our analysis of the mapping from the pathway representation to the event representation, we created software implementing this mapping from SBML with CellDesigner extensions to the event representation. This conversion otherwise follows the conventions of the event model, but lacks specific text bindings for the mentioned entities and event expressions (triggers). To maximize the applicability of the conversion, we chose to forgo e.g. the CellDesigner plugin architecture and to instead create an entirely standalone software based on python and libxml2. We tested this conversion on the 165 Panther and 28 Payao pathways to assure its robustness.

Conversion from pathways into the event representation opens up a number of opportunities, such as the ability to directly query large-scale event repositories (e.g. (Björne et al., 2010)) for specific pathway reactions. For pathways where reactions are marked with literature references, conversion further allows event annotations relevant to specific documents to be created automatically, sparing manual annotation costs. While such event annotations will not be bound to specific text expressions, they could be used through the application of techniques such as distant supervision (Mintz et al., 2009). As a first attempt, the conversion introduced in this work is limited in a number of ways, but we hope it can serve as a starting point for both wider adoption of pathway resources for event extraction and further research into accurate conversions between the two. The conversion software, `SBML-to-event`, is freely available for research purposes.

6 Discussion and Conclusions

Over the last decade, the bio-community has invested enormous efforts in the construction of detailed models of the function of a large variety of biological systems in the form of pathways. These efforts toward building systemic understanding of the functioning of organisms remain a central focus of present-day biology, and their support through information extraction and text mining perhaps the greatest potential contribution that the biomedical natural language processing community could make toward the broader bio-community.

We have argued that while recent developments in BioNLP are highly promising for approaching practical support of pathway annotation through information extraction, the BioNLP community has not yet made the most of the substantial resources in the form of existing pathways and that pursuing mapping from pathways to the event representation might be both more realistic and more fruitful than the other way around. As a first step in what we hope will lead to broadened understanding of the different perspectives, communication between the communities, and better uses resources, we have introduced a fully automatic mapping from SBML/CellDesigner pathways into the BioNLP ST-style event representation. As a first effort this mapping has many limitations and imperfections that we hope the BioNLP community will take as a challenge to do better.

Noting in analysis that dissociation reactions are not covered in previously proposed event representations, we also presented a detailed case study focusing on statements describing protein association and dissociation reactions in PubMed abstracts relevant to the mTOR pathway. Based on exploratory annotation, we proposed a novel event class DISSOCIATION, thus taking a step toward covering this arguably most significant omission in the event representation.

The pathway-bound event annotations created in this study, exhaustive annotation of all relevant entities and events in 60 abstracts, consist in total of annotation identifying nearly 1500 protein and 200 complex mentions and over 1200 events involving these entities in text. These annotations are freely available for use in research at `http://www-tsujii.is.s.u-tokyo.ac.jp/GENIA`.

Acknowledgments

We would like to thank Hiroaki Kitano, Yukiko Matsuoka and Samik Ghosh of the Systems Biology Institute for their generosity in providing their time and expertise in helping us understand the CellDesigner and SBML pathway representations. This work was partially supported by Grant-in-Aid for Specially Promoted Research (MEXT, Japan).

References

Sophia Ananiadou, Sampo Pyysalo, Jun'ichi Tsujii, and Douglas B. Kell. 2010. Event extraction for systems biology by text mining the literature. *Trends in Biotechnology*, 28(7):381–390.

Jari Björne, Filip Ginter, Sampo Pyysalo, Jun'ichi Tsujii, and Tapio Salakoski. 2010. Complex event extraction at PubMed scale. *Bioinformatics*, 26(12):i382–390.

E. Caron, S. Ghosh, Y. Matsuoka, D. Ashton-Beaucage, M. Therrien, S. Lemieux, C. Perreault, P.P. Roux, and H. Kitano. 2010. A comprehensive map of the mTOR signaling network. *Molecular Systems Biology*, 6(1).

E. Demir, M.P. Cary, S. Paley, K. Fukuda, C. Lemer, I. Vastrik, G. Wu, P. D'Eustachio, C. Schaefer, J. Luciano, et al. 2010. The BioPAX community standard for pathway data sharing. *Nature biotechnology*, 28(9):935–942.

A. Funahashi, Y. Matsuoka, A. Jouraku, M. Morohashi, N. Kikuchi, and H. Kitano. 2008. CellDesigner 3.5: a versatile modeling tool for biochemical networks. *Proceedings of the IEEE*, 96(8):1254–1265.

M. Hucka, A. Finney, H. M. Sauro, H. Bolouri, J. C. Doyle, and H Kitano et al. 2003. The systems biology markup language (SBML): a medium for representation and exchange of biochemical network models. *Bioinformatics*, 19(4):524–531.

B. Kemper, T. Matsuzaki, Y. Matsuoka, Y. Tsuruoka, H. Kitano, S. Ananiadou, and J. Tsujii. 2010. PathText: a text mining integrator for biological pathway visualizations. *Bioinformatics*, 26(12):i374.

Jin-Dong Kim, Tomoko Ohta, and Jun'ichi Tsujii. 2008. Corpus annotation for mining biomedical events from literature. *BMC Bioinformatics*, 9(10).

Jin-Dong Kim, Tomoko Ohta, Sampo Pyysalo, Yoshinobu Kano, and Jun'ichi Tsujii. 2009a. Overview of BioNLP'09 shared task on event extraction. In *Proceedings of BioNLP 2009 Shared Task*.

Jin-Dong Kim, Tomoko Ohta, Sampo Pyysalo, Yoshinobu Kano, and Jun'ichi Tsujii. 2009b. Overview of BioNLP'09 Shared Task on Event Extraction. In *Proceedings of Natural Language Processing in Biomedicine (BioNLP) NAACL 2009 Workshop*, pages 1–9.

Jin-Dong Kim, Sampo Pyysalo, Tomoko Ohta, Robert Bossy, Ngan Nguyen, and Jun'ichi Tsujii. 2011. Overview of BioNLP Shared Task 2011. In *Proceedings of BioNLP 2011*.

M. Krallinger, A. Morgan, L. Smith, F. Leitner, L. Tanabe, J. Wilbur, L. Hirschman, and A. Valencia. 2008. Evaluation of text-mining systems for biology: overview of the Second BioCreative community challenge. *Genome biology*, 9(Suppl 2):S1.

Y. Matsuoka, S. Ghosh, N. Kikuchi, and H. Kitano. 2010. Payao: a community platform for SBML pathway model curation. *Bioinformatics*, 26(10):1381.

H. Mi and P. Thomas. 2009. PANTHER pathway: an ontology-based pathway database coupled with data analysis tools. *Methods Mol. Biol*, 563:123–140.

Mike Mintz, Steven Bills, Rion Snow, and Daniel Jurafsky. 2009. Distant supervision for relation extraction without labeled data. In *Proceedings of ACL'09*, pages 1003–1011.

Kanae Oda, Jin-Dong Kim, Tomoko Ohta, Daisuke Okanohara, Takuya Matsuzaki, Yuka Tateisi, and Jun'ichi Tsujii. 2008. New challenges for text mining: Mapping between text and manually curated pathways. *BMC Bioinformatics*, 9(Suppl 3):S5.

Tomoko Ohta, Jin-Dong Kim, Sampo Pyysalo, Yue Wang, and Jun'ichi Tsujii. 2009. Incorporating GENETAG-style annotation to GENIA corpus. In *Proceedings of BioNLP'09*, pages 106–107.

Tomoko Ohta, Sampo Pyysalo, and Jun'ichi Tsujii. 2011. Overview of the Epigenetics and Post-translational Modifications (EPI) task of BioNLP Shared Task 2011. In *Proceedings of BioNLP 2011*.

Sampo Pyysalo, Filip Ginter, Juho Heimonen, Jari Björne, Jorma Boberg, Jouni Järvinen, and Tapio Salakoski. 2007. BioInfer: A corpus for information extraction in the biomedical domain. *BMC Bioinformatics*, 8(50).

Sampo Pyysalo, Antti Airola, Juho Heimonen, and Jari Björne. 2008. Comparative analysis of five protein-protein interaction corpora. *BMC Bioinformatics*, 9(Suppl. 3):S6.

Sampo Pyysalo, Tomoko Ohta, Rafal Rak, Dan Sullivan, Chunhong Mao, Chunxia Wang, Bruno Sobral, Jun'ichi Tsujii, and Sophia Ananiadou. 2011. Overview of the Infectious Diseases (ID) task of BioNLP Shared Task 2011. In *Proceedings of BioNLP 2011*.

Andrey Rzhetsky, Ivan Iossifov, Tomohiro Koike, Michael Krauthammer, Pauline Kra, Mitzi Morris, Hong Yu, Pablo Ariel Duboué, Wubin Weng, W. John Wilbur, Vasileios Hatzivassiloglou, and Carol Friedman. 2004. GeneWays: A system for extracting, analyzing, visualizing, and integrating molecular pathway data. *Journal of Biomedical Informatics*, 37(1):43–53.

Paul Thompson, Syed Iqbal, John McNaught, and Sophia Ananiadou. 2009. Construction of an annotated corpus to support biomedical information extraction. *BMC Bioinformatics*, 10(1):349.

Domonkos Tikk, Philippe Thomas, Peter Palaga, Jörg Hakenberg, and Ulf Leser. 2010. A comprehensive benchmark of kernel methods to extract protein-protein interactions from literature. *PLoS Comput Biol*, 6(7):e1000837, 07.

Towards Exhaustive Protein Modification Event Extraction

Sampo Pyysalo* **Tomoko Ohta*** **Makoto Miwa*** **Jun'ichi Tsujii†**
*Department of Computer Science, University of Tokyo, Tokyo, Japan
†Microsoft Research Asia, Beijing, China
{smp,okap,mmiwa}@is.s.u-tokyo.ac.jp, jtsujii@microsoft.com

Abstract

Protein modifications, in particular post-translational modifications, have a central role in bringing about the full repertoire of protein functions, and the identification of specific protein modifications is important for understanding biological systems. This task presents a number of opportunities for the automatic support of manual curation efforts. However, the sheer number of different types of protein modifications is a daunting challenge for automatic extraction that has so far not been met in full, with most studies focusing on single modifications or a few prominent ones. In this work, aim to meet this challenge: we analyse protein modification types through ontologies, databases, and literature and introduce a corpus of 360 abstracts manually annotated in the BioNLP Shared Task event representation for over 4500 mentions of proteins and 1000 statements of modification events of nearly 40 different types. We argue that together with existing resources, this corpus provides sufficient coverage of modification types to make effectively exhaustive extraction of protein modifications from text feasible.

1 Introduction

In the decade following the sequencing of the human genome, the critical role of protein modifications in establishing the full set of protein functions from forms transcribed from the fixed DNA is increasingly appreciated, reflected in the rise of proteomics as an extension and complement to genetics in efforts to understand gene and protein functions. The mapping of the space of modifications of specific proteins is a formidable undertaking: the number of known *types* of post-translational modifications (PTMs) is as high as 300 (Witze et al., 2007) with new types identified regularly (e.g. (Brennan and Barford, 2009)), and the number of specific molecular variants of proteins in cells may be several orders of magnitude larger than that encoded in the genome; up to millions for humans (Walsh, 2006). Automatic extraction of protein modifications from the massive literature on the topic could contribute significantly to addressing these challenges.

Biomedical information extraction (IE) has advanced substantially in recent years, shifting from the detection of simple binary associations such as protein-protein interactions toward resources and methods for the extraction of multiple types of structured associations of varying numbers participants in specific roles. These IE approaches are frequently termed *event extraction* (Ananiadou et al., 2010). While protein modifications have been considered in numerous IE studies in the domain (e.g. (Friedman et al., 2001; Rzhetsky et al., 2004; Hu et al., 2005; Narayanaswamy et al., 2005; Saric et al., 2006; Yuan et al., 2006; Lee et al., 2008; Ohta et al., 2010), event extraction efforts have brought increased focus also on the extraction of protein modifications: in the BioNLP Shared Task series that has popularized event extraction, the 2009 shared task (Kim et al., 2009) involved the extraction of nine event types including one PTM, and in the 2011 follow-up event (Kim et al., 2011) the Epigenetics and Post-translational modifications (EPI) task (Ohta et al., 2011) targeted six PTM types, their re-

Proceedings of the 2011 Workshop on Biomedical Natural Language Processing, ACL-HLT 2011, pages 114–123,
Portland, Oregon, USA, June 23-24, 2011. ©2011 Association for Computational Linguistics

verse reactions, and statements regarding their catalysis. The results of these tasks were promising, suggesting that the single PTM type could be extracted at over 80% F-score (Buyko et al., 2009) and the core arguments of the larger set at nearly 70% F-score (Björne and Salakoski, 2011).

The increasing availability of systems capable of detailed IE for protein modifications, their high performance also for multiple modifications types, and demonstrations of the scalability of the technology to the full scale of the literature (Björne et al., 2010) are highly encouraging for automatic extraction of protein modifications. However, previous efforts have been restricted by the relatively narrow scope of targeted modification types. In the present study, we seek to address the task in full by identifying all modifications of substantial biological significance and creating an annotated resource with effectively complete type-level coverage. We additionally present preliminary extraction results to assess the difficulty of exhaustive modification extraction.

2 Event representation

To be able to benefit from the substantial number of existing resources and systems for event extraction, we apply the event representation of the BioNLP Shared Task (ST) for annotating protein modifications. Specifically, we directly extend the approach of the BioNLP ST 2011 EPI task (Ohta et al., 2011). In brief, in the applied representation, each event is marked as being expressed by a specific span of text (the *event trigger*) and assigned a type from a fixed ontology defining event types. Events can take a conceptually open-ended number of participants, each of which is similarly bound to a specific textual expression and marked as participating in the event in a specific role. In this work, we apply three roles: *Theme* identifies the entity or event that is affected by the event (e.g. the protein that is modified), *Cause* its cause, and *Site* specifies a specific part on the *Theme* participant that is affected, i.e. the modification site or region. Further, events are primary objects of annotation and can thus in turn be participants in other events as well as being marked as e.g. explicitly negated ("is not phosphorylated") or stated speculatively ("may be phosphorylated"). An event annotation example is shown in Figure 1.

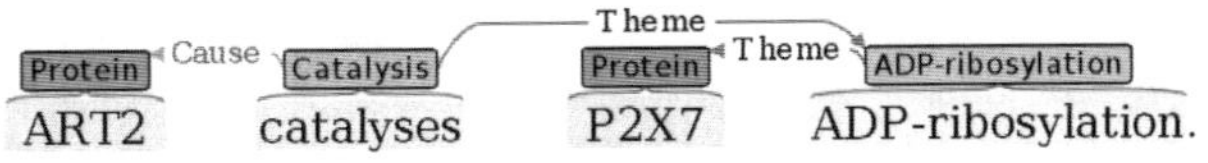

Figure 1: Illustration of the event representation. An event of type ADP-RIBOSYLATION (expressed through the text "ADP-ribosylation") with a PROTEIN ("P2X7") participant in the *Theme* role is in turn the *Theme* of a CATALYSIS event with another PROTEIN ("ART2") as its *Cause*.

3 Protein Modifications

We next present our selection of protein modification types relevant to event annotation and an extended analysis of their relative prominence.

3.1 Protein Modifications in Ontologies

For mapping and structuring the space of protein modification concepts, we primarily build on the community-standard Gene Ontology (GO) (Ashburner et al., 2000). GO has substantial representation of protein modifications: the sub-ontology rooted at `protein modification process` (*GO:0006464*) in the GO biological process ontology contains 805 terms[1] (including both leaf and internal nodes). This set of terms is the starting point for our selection of modifications types to target.

First, many specific GO terms can be excluded due to the different approach to semantic representation taken in event annotation: while GO terms represent detailed concepts without explicit structure (see e.g. (Ogren et al., 2004)), the event representation is structured, allowing more general terms to be applied while capturing the same information. For example, many GO modification terms have child nodes that identify the target (substrate) of modification, e.g. `protein phosphorylation` has the child `actin phosphorylation`. In the event representation, the target of modification is captured through the *Theme* argument. Similarly, GO terms may identify the site or region of modification, which becomes a *Site* argument in the event representation (see Figure 2). To avoid redundancy, we exclude GO terms that differ from a more general included term only in specifying a substrate or modification site. We similarly exclude terms that specify a catalyst or refer to regulation of modifi-

[1]GO structure and statistics from data retrieved Dec. 2010.

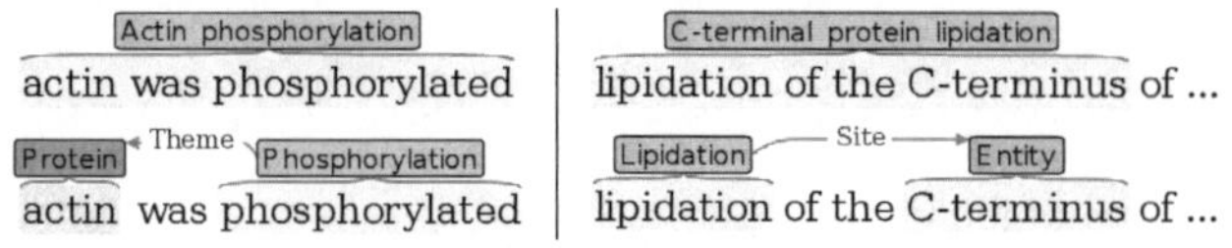

Figure 2: Comparison of hypothetical text-bound GO annotation with specific terms (top) and event annotation with general GO terms (bottom).

cation, as these are captured using separate events in the applied representation, as illustrated in Figure 1. For an analogous reason, we do not separately include type-level distinctions for "magnitude" variants of terms (e.g. `monoubiquitination`, `polyubiquitination`) as these can be systematically modeled as aspects that can mark any event (cf. the low/neutral/high *Manner* of Nawaz et al. (2010)).

Second, a number of the GO terms identify reactions that are in scope of previously defined (non-modification) event types in existing resources. To avoid introducing redundant or conflicting annotation with e.g. the GENIA Event corpus (Kim et al., 2008) or BioNLP ST resources, we excluded terms that involve predominantly (or exclusively) non-covalent binding (included in the scope of the event type BINDING) and terms involving the removal of or binding between the amino acids of a protein, including protein maturation by peptide bond cleavage (annotated – arguably somewhat inaccurately – as PROTEIN CATABOLISM in GENIA/BioNLP ST data). By contrast, we do differentiate between reactions involving the addition of chemical groups or small proteins and those involving their removal, including e.g. PALMITOYLATION and DEPALMITOYLATION as distinct types. To preserve the ontology structure, we further include also internal nodes appearing in GO for the purposes of structuring the ontology (e.g. `small protein conjugation or removal`), although we only apply more specific leaf nodes in event annotation.

This selection, aiming to identify the maximal subset of the protein modification branch of the GO ontology relevant to event annotation, resulted in the inclusion of 74 terms, approximately 9% of the branch total. Table 1 shows the relevant part of the GO `protein modification` subontology term structure, showing each term only once[2] and excluding very rare terms for space. (A detailed description of other information in the table is given in the following sections.)

In addition to GO, we consider protein modifications in the MeSH ontology,[3] used to index PubMed citations with concepts relevant to them. Further, for resolving cases not appearing in GO, we refer to the Uniprot controlled vocabulary of posttranslational modifications[4] and the Proteomics Standards Initiative Protein Modification Ontology[5] (PSI-MOD) (Montecchi-Palazzi et al., 2008).

3.2 Protein Modifications in Databases

A substantial number of databases tracking protein modifications from a variety of perspectives exist, and new ones are introduced regularly. The databases range from the specific (e.g. (Gupta et al., 1999; Diella et al., 2004; Zhang et al., 2010)) to the broad in scope (Lee et al., 2005; Li et al., 2009). Information on protein modifications is also found in general protein knowledge resources such as Swiss-Prot (Boeckmann et al., 2003) and PIR (Wu et al., 2003). The relative number of entries relevant to each protein modification in such resources is one possible proxy for the biological significance of the various modifications. We apply two such estimates in this work.

One of the primary applications of GO is the use of the ontology terms to annotate gene products, identifying their functions. These annotations, provided by a variety of groups in different efforts (e.g. (Camon et al., 2004)), are readily available in GO and used in various GO tools as a reflection of the prominence of each of the ontology concepts. As GO is a community standard with wide participation and a primary source in this work, we give these annotation numbers priority in introducing an additional filter: we exclude from detailed analysis any term that has no gene product association annotations, taking this as an indication that the modifica-

[2]GO allows multiple inheritance, and e.g. `protein palmitoylation` occurs under both `protein lipidation` and `protein acylation` reflecting the biological definition.

[3]`http://www.nlm.nih.gov/mesh/meshhome.html`

[4]`http://www.uniprot.org/docs/ptmlist`

[5]`http://www.psidev.info/MOD`

Term	GO ID	GPA	SysPTM	PubMed	GENIA	Ohta'10	EPI	This study
phosphorylation	GO:0006468	8246	24705	93584	546	3	130	85
small protein conj./removal	GO:0070647							
small protein conjugation	GO:0032446							
ubiquitination	GO:0016567	1724	439	4842	6	-	340	52
sumoylation	GO:0016925	121	260	886	-	-	-	101
neddylation	GO:0045116	66	2	100	-	-	-	52
ufmylation	GO:0071569	33	-	1	-	-	-	-
urmylation	GO:0032447	16	-	7	-	-	-	-
pupylation	GO:0070490	11	-	15	-	-	-	-
small protein removal	GO:0070646							
deubiquitination	GO:0016579	360	-	206	0	-	17	2
deneddylation	GO:0000338	45	-	39	-	-	-	8
desumoylation	GO:0016926	20	-	45	-	-	-	3
dephosphorylation	GO:0006470	1479	121	8339	28	-	3	1
glycosylation	GO:0006486	1145	2982	12619	-	122	347	62
acylation	GO:0043543	1	-	1728	-	-	-	71
acetylation	GO:0006473	522	2000	4423	7	90	337	17
palmitoylation	GO:0018345	49	198	1009	-	-	-	187
myristoylation	GO:0018377	27	150	895	-	-	-	34
octanoylation	GO:0018190	4	-	11	-	-	-	-
palmitoleylation	GO:0045234	3	-	0	-	-	-	-
alkylation	GO:0008213	0						
methylation	GO:0006479	552	499	9749	-	90	374	18
lipidation	GO:0006497	34	51	258	-	-	-	16
prenylation	GO:0018342	64	111	822	-	-	-	71
farnesylation	GO:0018343	19	-	118	-	-	-	48
geranylgeranylation	GO:0018344	26	-	79	-	-	-	30
deacylation	GO:0035601	1	-	331	-	-	-	1
deacetylation	GO:0006476	320	6	1056	1	-	50	4
depalmitoylation	GO:0002084	9	-	81	-	-	-	9
ADP-ribosylation	GO:0006471	261	9	3113	-	-	-	52
cofactor linkage	GO:0018065							
lipoylation	GO:0009249	53	-	49	-	-	-	14
FAD linkage	GO:0018293	46	-	6	-	-	-	-
pyridoxal-5-phosphate linkage	GO:0018352	6	-	0	-	-	-	-
dealkylation	GO:0008214	0						
demethylation	GO:0006482	116	-	1465	-	-	13	1
deglycosylation	GO:0006517	22	1	1204	-	-	27	0
ISG15-protein conjugation	GO:0032020	20	-	3	-	-	-	-
arginylation	GO:0016598	20	-	46	-	-	-	-
hydroxylation	GO:0018126	20	226	2948	-	103	139	3
sulfation	GO:0006477	18	132	960	-	-	-	37
carboxylation	GO:0018214	17	7	595	-	-	-	34
nucleotidylation	GO:0018175	0						
adenylylation	GO:0018117	16	-	116	-	-	-	-
uridylylation	GO:0018177	1	-	105	-	-	-	-
polyglycylation	GO:0018094	17	-	14	-	-	-	-
de-ADP-ribosylation	GO:0051725	16	-	7	-	-	-	5
nitrosylation	GO:0017014	14	-	670	-	-	-	-
glutathionylation	GO:0010731	11	-	279	-	-	-	-
biotinylation	GO:0009305	8	-	1247	-	-	-	4
deglutathionylation	GO:0080058	3	-	42	-	-	-	-
delipidation	GO:0051697	3	-	303	-	-	-	-
oxidation	GO:0018158	3	475	23413	-	-	-	21
phosphopantetheinylation	GO:0018215	3	-	26	-	-	-	-
tyrosinylation	GO:0018322	2	-	2	-	-	-	-
deamination	GO:0018277	1	-	840	-	-	-	-
esterification	GO:0018350	1	-	1180	-	-	-	-
glucuronidation	GO:0018411	1	-	705	-	-	-	-
polyamination	GO:0018184	1	-	13	-	-	-	-

Table 1: Protein modifications and protein modification resources. GO terms shown abbreviated, mostly by removing "protein" (e.g. "acylation" instead of "protein acylation"). Terms with 0 GPA not shown except when required for structure. Columns: GPA: number of Gene Product Associations for each term in GO (not including counts of more specific child nodes), SysPTM: number of SysPTM modification entries (excluding sites), PubMed: PubMed query matches (see Section 3.3), GENIA: GENIA corpus (Kim et al., 2008), Ohta'10: corpus introduced in Ohta et al. (2010), EPI: BioNLP ST'11 EPI task corpus (Ohta et al., 2011) (excluding test set).

tion is not presently established as having high biological significance.[6]

In addition to the GO associations, we include an estimate based on dedicated protein modification databases. We chose to use the integrated SysPTM resource (Li et al., 2009), which incorporates data from five databases, four webservers, and manual extraction from the literature. In its initial release, SysPTM included information on "nearly 50 modification types" on over 30,000 proteins. The columns labeled *GPA* and *SysPTM* in Table 1 show the number of gene product associations for each selected type in GO and entries per type in SysPTM, respectively.

3.3 Protein Modifications in domain literature

As a final estimate of the relative prominence of the various protein modification types, we estimated the relative frequency with which they are discussed in the literature through simple PubMed search, querying the Entrez system for each modification in its basic nominalized form (e.g. *phosphorylation*) in a protein-related article. Specifically, for each modification string MOD we searched Entrez for

"MOD"[TIAB] AND "protein"[TIAB]

The modifier *[TIAB]* specifies to search the title and abstract. The literal string "protein" is included to improve the estimate by removing references that involve the modification of non-proteins or related concepts that happen to share the term.[7] While this query is far from a perfect estimate of the actual number of protein modifications, we expect it to be a useful as a rough indicator of their relative frequencies and more straightforward to assess than more involved statistical analyses (e.g. (Pyysalo et al., 2010)). The results for these queries are given in the *PubMed* column of Table 1.

[6]We are also aware that GO coverage of protein modifications is not perfect: for example, citrullination, eliminylation, sialylation, as well as a number of reverse reactions for addition reactions in the ontology (e.g. demyristoylation) are not included at the time of this writing. As for terms with no gene product associations, we accept these omissions as indicating that these modifications are not biologically prominent.

[7]For example, search for only *dehydration* – a modification with zero GPA in GO – matches nearly 10 times as many documents as search including *protein*, implying that most of the hits for the former query likely do not concern protein modification by dehydration. By contrast, the majority of hits for *phosphorylation* match also *phosphorylation AND protein*.

3.4 Protein Modifications in Event Resources

The rightmost four columns of Table 1 present the number of annotations for each modification type in previously introduced event-annotated resources following the BioNLP ST representation as well as those annotated in the present study. While modification annotations are found also in other corpora (e.g. (Wu et al., 2003; Pyysalo et al., 2007)), we only include here resources readily compatible with the BioNLP ST representation.

Separating for the moment from consideration the question of what level of practical extraction performance can be supported by these event annotations, we can now provide an estimate of the upper bound on the coverage of relevant modification statements for each of the three proxies (GO GPA, SysPTM DB entries, PubMed query hits) simply by dividing the sum of instances of modifications for which annotations exist by the total. Thus, for example, there are 8246 GPA annotations for `Phosphorylation` and a total of 15597 GPA annotations, so the BioNLP ST'09 data (containing only PHOSPHORYLATION events) could by the GPA estimate cover 8246/15597, or approximately 53% of individual modifications.[8]

For the total coverage of the set of types for which event annotation is available given the corpus introduced in this study, the coverage estimates are: GO GPA: 98.2%, SysPTM 99.6%, PubMed 97.5%. Thus, we estimate that correct extraction of the included types would, depending on whether one takes a gene association, database entry, or literature mention point of view, cover between 97.5% to 99.6% of protein modification instances – a level of coverage we suggest is effectively exhaustive for most practical purposes. We next briefly describe our annotation effort before discarding the assumption that correct extraction is possible and measuring actual extraction performance.

4 Annotation

This section presents the entity and event annotation approach, document selection, and the statistics of the created annotation.

[8]The remarkably high coverage for a single type reflects the Zipfian distribution of the modification types; see e.g. Ohta et al. (2010).

4.1 Entity and Event Annotation

To maximize compatibility with existing event-annotated resources, we chose to follow the general representation and annotation guidelines applied in the annotation of GENIA/BioNLP ST resources, specifically the BioNLP ST 2011 EPI task corpus. Correspondingly, we followed the GENIA gene/gene product (Ohta et al., 2009) annotation guidelines for marking protein mentions, extended the GENIA event corpus guidelines (Kim et al., 2008) for the annotation of protein modification events, and marked CATALYSIS events following the EPI task representation. For compatibility, we also marked event negation and speculation as in these resources. We followed the GO definitions for individual modification types, and in the rare cases where a modification discussed in text had no existing GO definition, we extrapolated from the way in which protein modifications are generally defined in GO, consulting other domain ontologies and resources (Section 3.1) as necessary.

4.2 Document Selection

As the distribution of protein modifications in PubMed is extremely skewed, random sampling would recover almost solely instances of major types such as phosphorylation. As we are interested also in the extraction of very rare modifications, we applied a document selection strategy targeted at individual modification types. We applied one of two primary strategies depending on whether each targeted modification type had a corresponding MeSH term or not. If a MeSH term specific to the modification exists, we queried PubMed for the MeSH term, thus avoiding searches for specific forms of expression that might bias the search. In cases where no specific MeSH term existed, we searched the text of documents marked with the generic MeSH term `protein processing, post-translational` for mentions of likely forms of expression for the modification.[9] Finally, in a few isolated instances we applied custom text-based PubMed searches with broader coverage. Then, as many of the modifications are not limited to protein substrates, to select documents relating specifically to *protein* modification we proceeded to tagged a large random sample of selected documents with the BANNER named entity tagger (Leaman and Gonzalez, 2008) trained on the GENETAG corpus (Tanabe et al., 2005) and removed documents with fewer than five automatically tagged gene/protein-related entities. The remaining documents were then randomly sampled for annotation.[10]

[9]Specifically, we applied a regular expression incorporating the basic form of modification expression and allowing variance through relevant affixes and inflections derived from an initial set of annotations for documents for which MeSH terms were defined.

Item	Count
Abstract	360
Word	76806
Protein	4698
Event type	37
Event instance	1142

Table 2: Annotation statistics.

4.3 Corpus Statistics

We initially aimed to annotate balanced numbers of modification types in order of their estimated prominence, with particular focus on previously untargeted reaction types involving the addition of chemical groups or small proteins. However, it became apparent in the annotation process that the extreme rarity of some of the modifications as well as the tendency for more frequent modifications to be discussed in texts mentioning rare ones made this impossible. Thus, while preserving the goal of establishing broadly balanced numbers of major new modifications, we allowed the number of rare reactions to remain modest.

Table 2 summarizes the statistics of the final corpus, and the rightmost column of Table 1 shows per-type counts. We note that as reactions involving the removal of chemical groups or small proteins were not separately targeted, only few events of such types were annotated. We did not separately measure inter-annotator agreement for this effort, but note that this work is an extension of the EPI corpus annotation, for which comparison of independently created event annotations indicated an F-score of 82% for the full task and 89% for the core targets (see Section 5.1) (Ohta et al., 2011).

[10]This strategy, including MeSH-based search, was applied also in the BioNLP Shared Task 2011 EPI task document selection.

5 Experiments

To assess actual extraction performance, we performed experiments using a state-of-the art event extraction system.

5.1 Experimental Setup

We first split the corpus into a training/development portion and a held out set for testing, placing half of the abstracts into each set. The split was stratified by event type to assure that relatively even numbers of each event type were present in both sets. All development was performed using cross-validation on the visible portion of the data, and a single final experiment was performed on the test dataset.

To assure that our results are comparable with those published in recent event extraction studies, we adopted the standard evaluation criteria of the BioNLP Shared Task. The evaluation is event instance-based and uses the standard precision/recall/F_1-score metrics. We modified the shared task evaluation software to support the newly defined event types and ran experiments with the standard *approximate span matching* and *partial recursive matching* criteria (see (Kim et al., 2009)). We further follow the EPI task evaluation in reporting results separately for the extraction of only *Theme* and *Cause* arguments (*core* task) and for the *full* argument set.

5.2 Event extraction method

We applied the EventMine event extraction system (Miwa et al., 2010a; Miwa et al., 2010b), an SVM-based pipeline system using an architecture similar to that of the best-performing system in the BioNLP ST'09 (Björne et al., 2009); we refer to the studies of Miwa et al. for detailed description of the base system. For analysing sentence structure, we applied the mogura 2.4.1 (Matsuzaki and Miyao, 2007) and GDep beta2 (Sagae and Tsujii, 2007) parsers.

For the present study, we modified the base EventMine system as follows. First, to improve efficiency and generalizability, instead of using all words as trigger candidates as in the base system, we filtered candidates using a dictionary extracted from training data and expanded by using the UMLS specialist lexicon (Bodenreider, 2004) and the "hypernyms" and "similar to" relations in WordNet (Fellbaum, 1998). Second, to allow generalization across argument types, we added support for solving a single classification problem for event argument detection instead of solving multiple classification problems separated by argument types. Finally, to facilitate the use of other event resources for extraction, we added functionality to incorporate models trained by other corpora as reference models, using predictions from these models as features in classification.

5.3 Experimental results

We first performed a set of experiments to determine whether models can beneficially generalize across different modification event types. The EventMine pipeline has separate classification stages for event trigger detection, event-argument detection, and the extraction of complete event structures. Each of these stages involves a separate set of features and output labels, some of which derive directly from the involved event types: for example, in determining whether a specific entity is the *Theme* of an event triggered by the string "phosphorylation", the system by default uses the predicted event type (PHOSPHORYLATION) among its features. It is possible to force the model to generalize across event types by replacing specific types with placeholders, for example replacing PHOSPHORYLATION, METHYLATION, etc. with MODIFICATION.

In preliminary experiments on the development set, we experimented with a number of such generalizations. Results indicated that while some generalization was essential for achieving good extraction performance, most implementation variants produced broadly comparable results. We chose the following generalizations for the final test: in the trigger detection model, no generalization was performed (allowing specific types to be extracted), for argument detection, all instances of event types were replaced with a generic type (EVENT), and for event structure prediction, all instances of specific modification event types (but not CATALYSIS) were replaced with a generic type (MODIFICATION). Results comparing the initial, ungeneralized model to the generalized one are shown in the top two rows of Table 3. The results indicate that generalization is clearly beneficial: attempting to learn each of the event types in isolation leaves F-score results approximately 4-5% points lower than when general-

	Core	Full
Initial	39.40 / 46.36 / 42.60	31.39 / 38.88 / 34.74
Generalized	39.02 / 61.18 / 47.65	31.07 / 51.89 / 38.87
+Model	41.28 / 61.28 / 49.33	33.66 / 53.06 / 41.19
+Ann	38.46 / 66.99 / 48.87	32.36 / 59.17 / 41.84
+Model +Ann	41.84 / 66.17 / 51.26	33.98 / 56.00 / 42.30
Test data	45.69 / 62.35 / 52.74	38.03 / 54.57 / 44.82

Table 3: Experimental results.

izing across types. A learning curve for the generalized model is shown in Figure 3. While there is some indication of decreasing slope toward use of the full dataset, the curve suggests performance could be further improved through additional annotation efforts.

In a second set of experiments, we investigated the compatibility of the newly introduced annotations with existing event resources by incorporating their annotations either directly as training data (+Ann) or indirectly through features from predictions from a model trained on existing resources (+Model), as well as their combination. We performed experiments with the BioNLP Shared Task 2011 EPI task corpus[11] and the generalized setting. The results of these experiments are given in the middle rows of Table 3. We find substantial benefit from either form of existing resource integration alone, and, interestingly, an indication that the benefits of the two approaches can be combined. This result indicates that the newly introduced corpus is compatible with the EPI corpus, a major previously introduced resource for protein modification event extraction. Evaluation on the test data (bottom row of Table 3) confirmed that development data results were not overfit and generalized well to previously unseen data.

6 Discussion and Conclusions

We have presented an effort to directly address the challenges involved in the exhaustive extraction of protein modifications in text. We analysed the Gene Ontology `protein modification process` subontology from the perspective of event extraction for information extraction, arguing that due largely to the structured nature of the event representation,

[11]When combining EPI annotations directly as additional training abstracts, we filtered out abstracts including possible "missing" annotations for modification types not annotated in EPI data using a simple regular expression.

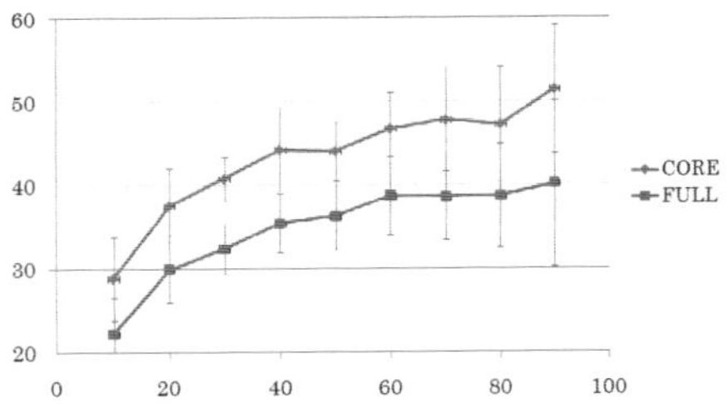

Figure 3: Learning curve.

74 of the 805 ontology terms suffice to capture the general modification types included. Through an analysis of the relative prominence of protein modifications in ontology annotations, domain databases, and literature, we then filtered and prioritized these types, estimating that correct extraction of the most prominent half of these types would give 97.5%-99.6% coverage of protein modifications, a level that is effectively exhaustive for practical purposes.

To support modification event extraction and to estimate actual extraction performance, we then proceeded to manually annotate a corpus of 360 PubMed abstracts selected for relevance to the selected modification types. The resulting corpus annotation marks over 4500 proteins and over 1000 instances of modification events and more than triples the number of specific protein modification types for which text-bound event annotations are available. Experiments using a state-of-the-art event extraction system showed that a machine learning method can beneficially generalize features across different protein modification event types and that incorporation of BioNLP Shared Task EPI corpus annotations can improve performance, demonstrating the compatibility of the created resource with existing event corpora. Using the best settings on the test data, we found that the core extraction task can be performed at 53% F-score.

The corpus created in this study is freely available for use in research from `http://www-tsujii.is.s.u-tokyo.ac.jp/GENIA`.

Acknowledgments

We would like to thank Yo Shidahara and Yoshihiro Okuda of NalaPro Technologies for their efforts in creating the corpus annotation. This work was supported by Grant-in-Aid for Specially Promoted Research (MEXT, Japan).

References

Sophia Ananiadou, Sampo Pyysalo, Jun'ichi Tsujii, and Douglas B. Kell. 2010. Event extraction for systems biology by text mining the literature. *Trends in Biotechnology*, 28(7):381–390.

M Ashburner, CA Ball, JA Blake, D Botstein, H Butler, JM Cherry, AP Davis, K Dolinski, SS Dwight, JT Eppig, MA Harris, DP Hill, L Issel-Tarver, A Kasarskis, S Lewis, JC Matese, JE Richardson, M Ringwald, GM Rubin, and G Sherlock. 2000. Gene ontology: tool for the unification of biology. *Nature genetics*, 25:25–29.

Jari Björne and Tapio Salakoski. 2011. Generalizing biomedical event extraction. In *Proceedings of the BioNLP 2011 Workshop Companion Volume for Shared Task*.

Jari Björne, Juho Heimonen, Filip Ginter, Antti Airola, Tapio Pahikkala, and Tapio Salakoski. 2009. Extracting complex biological events with rich graph-based feature sets. In *Proceedings of BioNLP'09 Shared Task*, pages 10–18.

Jari Björne, Filip Ginter, Sampo Pyysalo, Jun'ichi Tsujii, and Tapio Salakoski. 2010. Scaling up biomedical event extraction to the entire pubmed. In *Proceedings of the 2010 Workshop on Biomedical Natural Language Processing*, pages 28–36.

Olivier Bodenreider. 2004. The Unified Medical Language System (UMLS): integrating biomedical terminology. *Nucleic Acids Research*, 32(Database issue):D267–70.

B. Boeckmann, A. Bairoch, R. Apweiler, M.C. Blatter, A. Estreicher, E. Gasteiger, M.J. Martin, K. Michoud, C. O'Donovan, I. Phan, et al. 2003. The SWISS-PROT protein knowledgebase and its supplement TrEMBL in 2003. *Nucleic acids research*, 31(1):365.

D.F. Brennan and D. Barford. 2009. Eliminylation: a post-translational modification catalyzed by phosphothreonine lyases. *Trends in biochemical sciences*, 34(3):108–114.

Ekaterina Buyko, Erik Faessler, Joachim Wermter, and Udo Hahn. 2009. Event extraction from trimmed dependency graphs. In *Proceedings of the BioNLP'09 Shared Task*, pages 19–27, Boulder, Colorado, June. Association for Computational Linguistics.

Evelyn Camon, Michele Magrane, Daniel Barrell, Vivian Lee, Emily Dimmer, John Maslen, David Binns, Nicola Harte, Rodrigo Lopez, and Rolf Apweiler. 2004. The Gene Ontology Annotation (GOA) Database: sharing knowledge in Uniprot with Gene Ontology. *Nucl. Acids Res.*, 32(suppl 1):D262–266.

Francesca Diella, Scott Cameron, Christine Gemund, Rune Linding, Allegra Via, Bernhard Kuster, Thomas Sicheritz-Ponten, Nikolaj Blom, and Toby Gibson. 2004. Phospho.elm: A database of experimentally verified phosphorylation sites in eukaryotic proteins. *BMC Bioinformatics*, 5(1):79.

C. Fellbaum. 1998. Wordnet: an electronic lexical database. In *International Conference on Computational Linguistics*.

Carol Friedman, Pauline Kra, Hong Yu, Michael Krauthammer, and Andrey Rzhetsky. 2001. GENIES: A natural-language processing system for the extraction of molecular pathways from journal articles. *Bioinformatics*, 17(Suppl. 1):S74–S82.

Ramneek Gupta, Hanne Birch, Kristoffer Rapacki, Sren Brunak, and Jan E. Hansen. 1999. O-glycbase version 4.0: a revised database of o-glycosylated proteins. *Nucleic Acids Research*, 27(1):370–372.

Z. Z. Hu, M. Narayanaswamy, K. E. Ravikumar, K. Vijay-Shanker, and C. H. Wu. 2005. Literature mining and database annotation of protein phosphorylation using a rule-based system. *Bioinformatics*, 21(11):2759–2765.

Jin-Dong Kim, Tomoko Ohta, and Jun'ichi Tsujii. 2008. Corpus annotation for mining biomedical events from lterature. *BMC Bioinformatics*, 9(1):10.

Jin-Dong Kim, Tomoko Ohta, Sampo Pyysalo, Yoshinobu Kano, and Jun'ichi Tsujii. 2009. Overview of BioNLP'09 Shared Task on Event Extraction. In *Proceedings of Natural Language Processing in Biomedicine (BioNLP) NAACL 2009 Workshop*, pages 1–9.

Jin-Dong Kim, Sampo Pyysalo, Tomoko Ohta, Robert Bossy, and Jun'ichi Tsujii. 2011. Overview of bionlp shared task 2011. In *Proceedings of the BioNLP 2011 Workshop Companion Volume for Shared Task*.

Robert Leaman and Graciela Gonzalez. 2008. BANNER: An executable survey of advances in biomedical named entity recognition. In *Proceedings of PSB'08*, pages 652–663.

Tzong-Yi Lee, Hsien-Da Huang, Jui-Hung Hung, Hsi-Yuan Huang, Yuh-Shyong Yang, and Tzu-Hao Wang. 2005. dbPTM: an information repository of protein post-translational modification. *Nucleic Acids Research*, 34(suppl 1):D622–D627.

Hodong Lee, Gwan-Su Yi, and Jong C. Park. 2008. E3Miner: a text mining tool for ubiquitin-protein ligases. *Nucl. Acids Res.*, 36(suppl.2):W416–422.

Hong Li, Xiaobin Xing, Guohui Ding, Qingrun Li, Chuan Wang, Lu Xie, Rong Zeng, and Yixue Li. 2009. Sysptm: A systematic resource for proteomic research on post-translational modifications. *Molecular & Cellular Proteomics*, 8(8):1839–1849.

Takuya Matsuzaki and Yusuke Miyao. 2007. Efficient HPSG parsing with supertagging and CFG-filtering. In *In Proceedings of IJCAI-07*, pages 1671–1676.

Makoto Miwa, Sampo Pyysalo, Tadayoshi Hara, and Jun'ichi Tsujii. 2010a. Evaluating dependency representations for event extraction. In *Proceedings of Coling'10*, pages 779–787.

Makoto Miwa, Rune Sætre, Jin-Dong Kim, and Jun'ichi Tsujii. 2010b. Event extraction with complex event classification using rich features. *Journal of Bioinformatics and Computational Biology (JBCB)*, 8(1):131–146.

Luisa Montecchi-Palazzi, Ron Beavis, Pierre-Alain Binz, Robert Chalkley, John Cottrell, David Creasy, Jim Shofstahl, Sean Seymour, and John Garavelli. 2008. The PSI-MOD community standard for representation of protein modification data. *Nature Biotechnology*, 26:864–866.

M. Narayanaswamy, K. E. Ravikumar, and K. Vijay-Shanker. 2005. Beyond the clause: extraction of phosphorylation information from medline abstracts. *Bioinformatics*, 21(suppl.1):i319–327.

R. Nawaz, P. Thompson, J. McNaught, and S. Ananiadou. 2010. Meta-Knowledge Annotation of Bio-Events. *Proceedings of LREC 2010*, pages 2498–2507.

P.V. Ogren, K.B. Cohen, G.K. Acquaah-Mensah, J. Eberlein, and L. Hunter. 2004. The compositional structure of Gene Ontology terms. In *Pacific Symposium on Biocomputing*, page 214.

Tomoko Ohta, Jin-Dong Kim, Sampo Pyysalo, Yue Wang, and Jun'ichi Tsujii. 2009. Incorporating GENETAG-style annotation to GENIA corpus. In *Proceedings of BioNLP'09*, pages 106–107.

Tomoko Ohta, Sampo Pyysalo, Makoto Miwa, Jin-Dong Kim, and Jun'ichi Tsujii. 2010. Event extraction for post-translational modifications. In *Proceedings of BioNLP'10*, pages 19–27.

Tomoko Ohta, Sampo Pyysalo, and Jun'ichi Tsujii. 2011. Overview of the Epigenetics and Post-translational Modifications (EPI) task of BioNLP Shared Task 2011. In *Proceedings of the BioNLP 2011 Workshop Companion Volume for Shared Task*.

Sampo Pyysalo, Filip Ginter, Juho Heimonen, Jari Björne, Jorma Boberg, Jouni Järvinen, and Tapio Salakoski. 2007. BioInfer: A corpus for information extraction in the biomedical domain. *BMC Bioinformatics*, 8(50).

Sampo Pyysalo, Tomoko Ohta, and Jun'ichi Tsujii. 2010. An analysis of gene/protein associations at pubmed scale. In *Proceedings of the fourth International Symposium for Semantic Mining in Biomedicine (SMBM 2010)*.

Andrey Rzhetsky, Ivan Iossifov, Tomohiro Koike, Michael Krauthammer, Pauline Kra, Mitzi Morris, Hong Yu, Pablo Ariel Duboué, Wubin Weng, W. John Wilbur, Vasileios Hatzivassiloglou, and Carol Friedman. 2004. GeneWays: A system for extracting, analyzing, visualizing, and integrating molecular pathway data. *Journal of Biomedical Informatics*, 37(1):43–53.

Kenji Sagae and Jun'ichi Tsujii. 2007. Dependency parsing and domain adaptation with LR models and parser ensembles. In *Proceedings of EMNLP-CoNLL'07*, pages 1044–1050.

Jasmin Saric, Lars Juhl Jensen, Rossitza Ouzounova, Isabel Rojas, and Peer Bork. 2006. Extraction of regulatory gene/protein networks from Medline. *Bioinformatics*, 22(6):645–650.

Lorraine Tanabe, Natalie Xie, Lynne Thom, Wayne Matten, and John Wilbur. 2005. Genetag: a tagged corpus for gene/protein named entity recognition. *BMC Bioinformatics*, 6(Suppl 1):S3.

Christopher Walsh. 2006. *Posttranslational modification of proteins: expanding nature's inventory*. Roberts & Company Publishers.

Eric S Witze, William M Old, Katheryn A Resing, and Natalie G Ahn. 2007. Mapping protein post-translational modifications with mass spectrometry. *Nature Methods*, 4:798–806.

Cathy H. Wu, Lai-Su L. Yeh, Hongzhan Huang, Leslie Arminski, Jorge Castro-Alvear, Yongxing Chen, Zhangzhi Hu, Panagiotis Kourtesis, Robert S. Ledley, Baris E. Suzek, C.R. Vinayaka, Jian Zhang, and Winona C. Barker. 2003. The Protein Information Resource. *Nucl. Acids Res.*, 31(1):345–347.

X. Yuan, ZZ Hu, HT Wu, M. Torii, M. Narayanaswamy, KE Ravikumar, K. Vijay-Shanker, and CH Wu. 2006. An online literature mining tool for protein phosphorylation. *Bioinformatics*, 22(13):1668.

Yan Zhang, Jie Lv, Hongbo Liu, Jiang Zhu, Jianzhong Su, Qiong Wu, Yunfeng Qi, Fang Wang, and Xia Li. 2010. Hhmd: the human histone modification database. *Nucleic Acids Research*, 38(suppl 1):D149–D154.

A Study on Dependency Tree Kernels for Automatic Extraction of Protein-Protein Interaction

Md. Faisal Mahbub Chowdhury [†‡] and **Alberto Lavelli** [‡] and **Alessandro Moschitti** [†]
[†] Department of Information Engineering and Computer Science, University of Trento, Italy
[‡] Human Language Technology Research Unit, Fondazione Bruno Kessler, Trento, Italy
{chowdhury,lavelli}@fbk.eu, moschitti@disi.unitn.it

Abstract

Kernel methods are considered the most effective techniques for various relation extraction (RE) tasks as they provide higher accuracy than other approaches. In this paper, we introduce new dependency tree (DT) kernels for RE by improving on previously proposed dependency tree structures. These are further enhanced to design more effective approaches that we call mildly extended dependency tree (MEDT) kernels. The empirical results on the protein-protein interaction (PPI) extraction task on the AIMed corpus show that tree kernels based on our proposed DT structures achieve higher accuracy than previously proposed DT and phrase structure tree (PST) kernels.

1 Introduction

Relation extraction (RE) aims at identifying instances of pre-defined relation types in text as for example the extraction of protein-protein interaction (PPI) from the following sentence:

> "Native *C8* also formed a heterodimer with *C5*, and low concentrations of polyionic ligands such as protamine and suramin inhibited the interaction."

After identification of the relevant named entities (NE, in this case *proteins*) *C8* and *C5*, the RE task determines whether there is a PPI relationship between the entities above (which is *true* in the example).

Kernel based approaches for RE have drawn a lot of interest in recent years since they can exploit a huge amount of features without an explicit representation. Some of these approaches are structure kernels (e.g. tree kernels), which carry out structural similarities between instances of relations, represented as phrase structures or dependency trees, in terms of common substructures. Other kernels simply use techniques such as bag-of-words, subsequences, etc. to map the syntactic and contextual information to flat features, and later compute similarity.

One variation of tree kernels is the dependency tree (DT) kernel (Culotta and Sorensen, 2004; Nguyen et al., 2009). A DT kernel (DTK) is a tree kernel that is computed on a dependency tree (or subtree). A dependency tree encodes grammatical relations between words in a sentence where the words are nodes, and dependency types (i.e. grammatical functions of children nodes with respect to their parents) are edges. The main advantage of a DT in comparison with phrase structure tree (PST) is that the former allows for relating two words directly (and in more compact substructures than PST) even if they are far apart in the corresponding sentence according to their lexical word order.

Several kernel approaches exploit syntactic dependencies among words for PPI extraction from biomedical text in the form of dependency graphs or dependency paths (e.g. Kim et al. (2010) or Airola et al. (2008)). However, to the best of our knowledge, there are only few works on the use of DT kernels for this task. Therefore, exploring the potential of DTKs applied to different structures is a worthwhile research direction. A DTK, pioneered by Culotta and Sorensen (2004), is typically applied to the minimal or smallest common subtree that includes a target pair of entities. Such subtree reduces

Proceedings of the 2011 Workshop on Biomedical Natural Language Processing, ACL-HLT 2011, pages 124–133,
Portland, Oregon, USA, June 23-24, 2011.

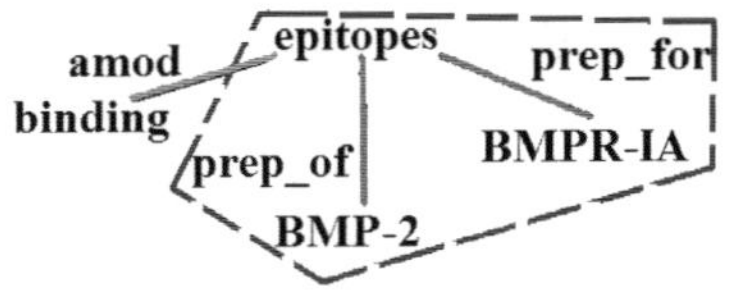

Figure 1: Part of the DT for the sentence "The binding epitopes of *BMP-2* for *BMPR-IA* was characterized using BMP-2 mutant proteins". The dotted area indicates the minimal subtree.

unnecessary information by placing word(s) closer to its dependent(s) inside the tree and emphasizes local features of relations. Nevertheless, there are cases where a minimal subtree might not contain important cue words or predicates. For example, consider the following sentence where a PPI relation holds between *BMP-2* and *BMPR-IA*, but the minimal subtree does not contain the cue word "binding" as shown in Figure 1:

> The binding epitopes of **BMP-2** for **BMPR-IA** was characterized using BMP-2 mutant proteins.

In this paper we investigate two assumptions. The first is that a DTK based on a mild extension of minimal subtrees would produce better results than the DTK on minimal subtrees. The second is that previously proposed DT structures can be further improved by introducing simplified representation of the entities as well as augmenting nodes in the DT tree structure with relevant features. This paper presents an evaluation of the above assumptions.

More specifically, the contributions of this paper are the following:

- We propose the use of new DT structures, which are improvement on the structures defined in Nguyen et al. (2009) with the most general (in terms of substructures) DTK, i.e. Partial Tree Kernel (PTK) (Moschitti, 2006).

- We firstly propose the use of the Unlexicalized PTK (Severyn and Moschitti, 2010) with our dependency structures, which significantly improves PTK.

- We compare the performance of the proposed DTKs on PPI with the one of PST kernels and show that, on biomedical text, DT kernels perform better.

- Finally, we introduce a novel approach (called mildly extended dependency tree (MEDT) kernel[1], which achieves the best performance among various (both DT and PST) tree kernels.

The remainder of the paper is organized as follows. In Section 2, we introduce tree kernels and relation extraction and we also review previous work. Section 3 describes the unlexicalized PTK (uPTK). Then, in Section 4, we define our proposed DT structures including MEDT. Section 5 describes the experimental results on the AIMed corpus (Bunescu et al., 2005) and discusses their outcomes. Finally, we conclude with a summary of our study as well as plans for future work.

2 Background and Related Work

The main stream work for Relation Extraction uses kernel methods. In particular, as the syntactic structure is very important to derive the relationships between entities in text, several tree kernels have been designed and experimented. In this section, we introduce such kernels, the problem of relation extraction and we also focus on the biomedical domain.

2.1 Tree Kernel types

The objective behind the use of tree kernels is to compute the similarity between two instances through counting similarities of their sub-structures. Among the different proposed methods, two of the most effective approaches are Subset Tree (SST) kernel (Collins and Duffy, 2001) and Partial Tree Kernel (PTK) (Moschitti, 2006).

The SST kernel generalizes the subtree kernel (Vishwanathan and Smola, 2002), which considers all common subtrees in the tree representation of two compared sentences. In other words, two subtrees are identical if the node labels and order of children are identical for all nodes. The SST kernel relaxes the constraint that requires leaves to be always included in the sub-structures. In SST, for a given node, either none or all of its children have to be included in the resulting subset tree. An extension of

[1]We defined new structures, which as it is well known it corresponds to define a new kernel.

the SST kernel is the SST+bow (bag-of-words) kernel (Zhang and Lee, 2003; Moschitti, 2006a), which considers individual leaves as sub-structures as well.

The PT kernel (Moschitti, 2006) is more flexible than SST by virtually allowing any tree substructure; the only constraint is that the order of child nodes must be identical. Both SST and PT kernels are convolution tree kernels[2].

The PT kernel is the most complete in terms of structures. However, the massive presence of child node subsequences and single child nodes, which in a DT often correspond to words, may cause overfitting. Thus we propose the use of the unlexicalized (i.e. PT kernel without leaves) tree kernel (uPTK) (Severyn and Moschitti, 2010), in which structures composed by only one lexical element, i.e. single nodes, are removed from the feature space (see Section 3).

2.2 Relation Extraction using Tree Kernels

A first version of dependency tree kernels (DTKs) was proposed by Culotta and Sorensen (2004). In their approach, they find the smallest common subtree in the DT that includes a given pair of entities. Then, each node of the subtree is represented as a feature vector. Finally, these vectors are used to compute similarity. However, the tree kernel they defined is not a convolution kernel, and hence it generates a much lower number of sub-structures resulting in lower performance.

For any two entities *e1* and *e2* in a DT, Nguyen et al. (2009) defined the following three dependency structures to be exploited by convolution tree kernels:

- Dependency Words (DW) tree: a DW tree is the minimal subtree of a DT, which includes *e1* and *e2*. An extra node is inserted as parent of the corresponding NE, labeled with the NE category. Only words are considered in this tree.

- Grammatical Relation (GR) tree: a GR tree is similar to a DW tree except that words are replaced by their grammatical functions, e.g. `prep`, `nsubj`, etc.

- Grammatical Relation and Words (GRW) tree: a GRW tree is the minimal subtree that uses both words and grammatical functions, where the latter are inserted as parent nodes of the former.

Using PTK for the above dependency tree structures, the authors achieved an F-measure of 56.3 (for DW), 60.2 (for GR) and 58.5 (for GRW) on the ACE 2004 corpus[3].

Moschitti (2004) proposed the so called path-enclosed tree (PET)[4] of a PST for Semantic Role Labeling. This was later adapted by Zhang et al. (2005) for relation extraction. A PET is the smallest common subtree of a PST, which includes the two entities involved in a relation.

Zhou et al. (2007) proposed the so called context-sensitive tree kernel approach based on PST, which expands PET to include necessary contextual information. The expansion is carried out by some heuristics tuned on the target RE task.

Nguyen et al. (2009) improved the PET representation by inserting extra nodes for denoting the NE category of the entities inside the subtree. They also used sequence kernels from tree paths, which provided higher accuracy.

2.3 Relation Extraction in the biomedical domain

There are several benchmarks for the PPI task, which adopt different PPI annotations. Consequently the experimental results obtained by different approaches are often difficult to compare. Pyysalo et al. (2008) put together these corpora (including the AIMed corpus used in this paper) in a common format for comparative evaluation. Each of these corpora is known as *converted corpus* of the corresponding original corpus.

Several kernel-based RE approaches have been reported to date for the PPI task. These are based on various methods such as subsequence kernel (Lodhi et al., 2002; Bunescu and Mooney, 2006), dependency graph kernel (Bunescu and Mooney, 2005), etc. Different work exploited dependency analyses with different kernel approaches such as bag-of-

[2]Convolution kernels aim to capture structural information in term of sub-structures, providing a viable alternative to flat features (Moschitti, 2004).

[3]http://projects.ldc.upenn.edu/ace/

[4]Also known as shortest path-enclosed tree or SPT (Zhou et al., 2007).

words kernel (e.g. Miwa et al. (2009)), graph based kernel (e.g. Kim et al. (2010)), etc. However, there are only few researches that attempted the exploitation of tree kernels on dependency tree structures.

Sætre et al. (2007) used DT kernels on AIMed corpus and achieved an F-score of 37.1. The results were far better when they combined the output of the dependency parser with that of a Head-driven Phrase Structure Grammar (HPSG) parser, and applied tree kernel on it. Miwa et al. (2009) also proposed a hybrid kernel [5], which is a composition of all-dependency-paths kernel (Airola et al., 2008), bag-of-words kernel and SST kernel. They used multiple parser inputs. Their system is the current state-of-the-art for PPI extraction on several benchmarks. Interestingly, they applied SST kernel on the shortest dependency paths between pairs of proteins and achieved a relatively high F-score of 55.1. However, the trees they constructed from the shortest dependency paths are actually not dependency trees. In a dependency tree, there is only one node for each individual word whereas in their constructed trees (please refer to Fig. 6 of Miwa et al. (2009)), a word (that belongs to the shortest path) has as many node representations as the number of dependency relations with other words (those belonging to the shortest path). Perhaps, this redundancy of information might be the reason their approach achieved higher result. In addition to work on PPI pair extraction, there has been some approaches that exploited dependency parse analyses along with kernel methods for *identifying sentences* that might contain PPI pairs (e.g. Erkan et al. (2007)).

In this paper, we focus on finding the best representation based on a single structure. We speculate that this can be helpful to improve the state-of-the-art using several combinations of structures and features. As a first step, we decided to use uPTK, which is more robust to overfitting as the description in the next section unveil.

[5]The term "hybrid kernel" is identical to "combined kernel". It refers to those kernels that combine multiple types of kernels (e.g., tree kernels, graph kernels, etc)

3 Unlexicalized Partial Tree Kernel (uPTK)

The uPTK was firstly proposed in (Severyn and Moschitti, 2010) and experimented with semantic role labeling (SRL). The results showed no improvement for such task but it is well known that in SRL lexical information is essential (so in that case it could have been inappropriate). The uPTK definition follows the general setting of tree kernels.

A tree kernel function over two trees, T_1 and T_2, is defined as

$$TK(T_1, T_2) = \sum_{n_1 \in N_{T_1}} \sum_{n_2 \in N_{T_2}} \Delta(n_1, n_2),$$

where N_{T_1} and N_{T_2} are the sets of nodes in T_1 and T_2, respectively, and

$$\Delta(n_1, n_2) = \sum_{i=1}^{|\mathcal{F}|} \chi_i(n_1)\chi_i(n_2).$$

The Δ function is equal to the number of common fragments rooted in nodes n_1 and n_2 and thus depends on the fragment type.

The algorithm for the uPTK computation straightforwardly follows from the definition of the Δ function of PTK provided in (Moschitti, 2006). Given two nodes n_1 and n_2 in the corresponding two trees T_1 and T_2, Δ is evaluated as follows:

1. if the node labels of n_1 and n_2 are different then $\Delta(n_1, n_2) = 0$;
2. else $\Delta(n_1, n_2) = \mu\Big(\lambda^2 + \sum_{\vec{I}_1, \vec{I}_2, l(\vec{I}_1)=l(\vec{I}_2)} \lambda^{d(\vec{I}_1)+d(\vec{I}_2)} \prod_{j=1}^{l(\vec{I}_1)} \Delta(c_{n_1}(\vec{I}_{1j}), c_{n_2}(\vec{I}_{2j}))\Big)$,

where:

1. $\vec{I}_1 = \langle h_1, h_2, h_3, ..\rangle$ and $\vec{I}_2 = \langle k_1, k_2, k_3, ..\rangle$ are index sequences associated with the ordered child sequences c_{n_1} of n_1 and c_{n_2} of n_2, respectively;
2. $\vec{I}_{1j}$ and $\vec{I}_{2j}$ point to the j-th child in the corresponding sequence;
3. $l(\cdot)$ returns the sequence length, i.e. the number of children;

4. $d(\vec{I_1}) = \vec{I}_{1l(\vec{I_1})} - \vec{I}_{11} + 1$ and $d(\vec{I_2}) = \vec{I}_{2l(\vec{I_2})} - \vec{I}_{21} + 1$; and

5. μ and λ are two decay factors for the size of the tree and for the length of the child subsequences with respect to the original sequence, i.e. we account for gaps.

The uPTK can be obtained by removing λ^2 from the equation in step 2. An efficient algorithm for the computation of PTK is given in (Moschitti, 2006). This evaluates Δ by summing the contribution of tree structures coming from different types of sequences, e.g. those composed by p children such as:

$$\Delta(n_1, n_2) = \mu\big(\lambda^2 + \textstyle\sum_{p=1}^{lm} \Delta_p(c_{n_1}, c_{n_2})\big), \quad (1)$$

where Δ_p evaluates the number of common subtrees rooted in subsequences of exactly p children (of n_1 and n_2) and $lm = min\{l(c_{n1}), l(c_{n2})\}$. It is easy to verify that we can use the recursive computation of Δ_p by simply removing λ^2 from Eq. 1.

4 Proposed dependency structures and MEDT kernel

Our objective is twofold: (a) the definition of improved DT structures and (b) the design of new DT kernels to include important words residing outside of the shortest dependency tree, which are neglected in current approaches. For achieving point (a), we modify the DW, GR and GRW structures, previously proposed by Nguyen et al. (2009). The new proposed structures are the following:

- Grammatical Relation and lemma (GRL) tree: A GRL tree is similar to a GRW tree except that words are replaced by their corresponding lemmas.

- Grammatical Relation, PoS and lemma (GRPL) tree: A GRPL tree is an extension of a GRL tree, where the part-of-speech (PoS) tag of each of the corresponding words is inserted as a new node between its grammatical function and its lemma, i.e. the new node becomes the parent node of the node containing the lemma.

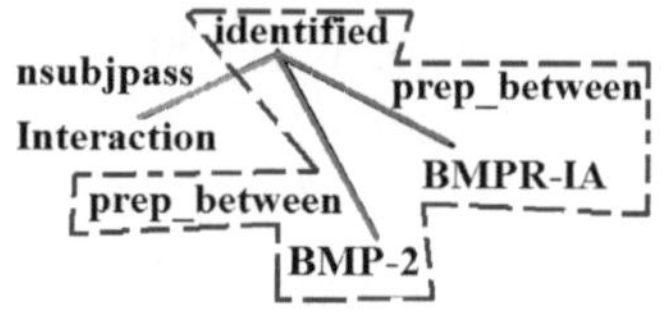

Figure 2: Part of the DT for the sentence "Interaction was identified between *BMP-2* and *BMPR-IA*". The dotted area indicates the minimal subtree.

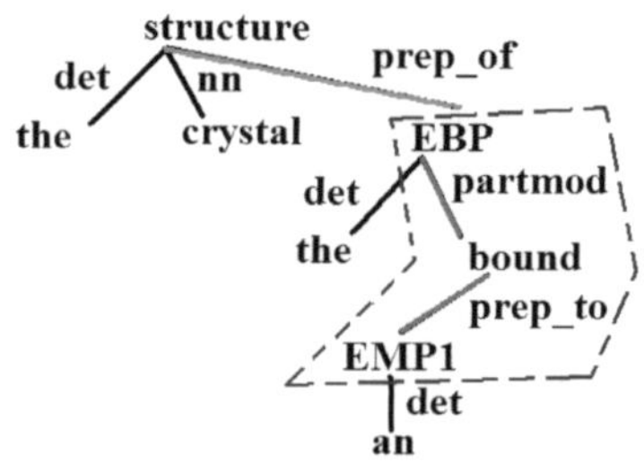

Figure 3: Part of the DT for the sentence "Phe93 forms extensive contacts with a peptide ligand in the crystal structure of the *EBP* bound to an *EMP1*". The dotted area indicates the minimal subtree.

- Ordered GRL (OGRL) or ordered GRW (OGRW) tree: in a GRW (or GRL) tree, the node containing the grammatical function of a word is inserted as the parent node of such word. So, if the word has a parent node containing its NE category, the newly inserted node with grammatical function becomes the child node of the node containing NE category, i.e. the order of the nodes is the following – *"NE category ⇒ grammatical relation ⇒ word (or lemma)"*. However, in OGRW (or OGRL), this ordering is modified as follows – *"grammatical relation ⇒ NE category ⇒ word (or lemma)"*.

- Ordered GRPL (OGRPL) tree: this is similar to the OGRL tree except for the order of the nodes, which is the following – *"grammatical relation ⇒ NE category ⇒ PoS ⇒ lemma"*.

- Simplified (S) tree: any tree structure would become an S tree if it contains simplified representations of the entity types, where all its parts except the head word of a multi-word entity are not considered in the minimal subtree.

The second objective is to extend DTKs to include important cue words or predicates that are missing in the minimal subtree. We do so by mildly expanding the minimal subtree, i.e. we define the mildly extended DT (MEDT) kernel. We propose three different expansion rules for three versions of MEDT as follows:

- Expansion rule for MEDT-1 kernel: *If the root of the minimal subtree is not a modifier (e.g. adjective) or a verb, then look for such node in its children or in its parent (in the original DT tree) to extend the subtree.*

  The following example shows a sentence where this rule would be applicable:

  > The binding epitopes of **BMP-2** for **BMPR-IA** was characterized using BMP-2 mutant proteins.

  Here, the cue word is "binding", the root of the minimal subtree is "epitopes" and the target entities are *BMP-2* and *BMPR-IA*. However, as shown in Figure 1, the minimal subtree does not contain the cue word.

- Expansion rule for MEDT-2 kernel: *If the root of the minimal subtree is a verb and its subject (or passive subject) in the original DT tree is not included in the subtree, then include it.*

  Consider the following sentence:

  > Interaction was identified between **BMP-2** and **BMPR-IA**.

  Here, the cue word is "Interaction", the root is "identified" and the entities are *BMP-2* and *BMPR-IA*. The passive subject "Interaction" does not belong to the minimal subtree (see Figure 2).

- Expansion rule for MEDT-3 kernel: *If the root of the minimal subtree is the head word of one of the interacting entities, then add the parent node (in the original DT tree) of the root node as the new root of the subtree.*

  This is an example sentence where this rule is applicable (see Figure 3):

  > Phe93 forms extensive contacts with a peptide ligand in the crystal structure of the **EBP** bound to an **EMP1**.

5 Experiments and results

We carried out several experiments with different dependency structures and tree kernels. Most importantly, we tested tree kernels on PST and our improved representations for DT.

5.1 Data and experimental setup

We used the AIMed corpus (Bunescu et al., 2005) converted using the software provided by Pyysalo et al. (2008). AIMed is the largest benchmark corpus (in terms of number of sentences) for the PPI task. It contains 1,955 sentences, in which are annotated 1,000 positive PPI and 4,834 negative pairs.

We use the Stanford parser[6] for parsing the data.[7] The SPECIALIST lexicon tool[8] is used to normalize words to avoid spelling variations and also to provide lemmas. For training and evaluating tree kernels, we use the SVM-LIGHT-TK toolkit[9] (Moschitti, 2006; Joachims, 1999). We tuned the parameters μ, λ and c following the approach described by Hsu et al. (2003), and used biased hyperplane.[10] All the other parameters are left as their default values.

Our experiments are evaluated with 10-fold cross validation using the same split of the AIMed corpus used by Bunescu et al. (2005).

5.2 Results and Discussion

The results of different tree kernels applied to different structures are shown in Tables 1 and 2. All the tree structures are tested with four different tree kernel types: SST, SST+bow, PTK and uPTK.

According to the empirical outcome, our new DT structures perform better than the existing tree structures. The highest result (F: 46.26) is obtained by applying uPTK to MEDT-3 (SOGRL). This is 6.68 higher than the best F-measure obtained by previous DT structures proposed in Nguyen et al. (2009), and 0.36 higher than the best F-measure obtained using PST (PET).

[6] http://nlp.stanford.edu/software/lex-parser.shtml

[7] For some of the positive PPI pairs, the connecting dependency tree could not be constructed due to parsing errors for the corresponding sentences. Such pairs are considered as false negative (FN) during precision and recall measurements.

[8] http://lexsrv3.nlm.nih.gov/SPECIALIST/index.html

[9] http://disi.unitn.it/moschitti/Tree-Kernel.htm

[10] Please refer to http://svmlight.joachims.org/ and http://disi.unitn.it/moschitti/Tree-Kernel.htm for details about parameters of the respective tools

	DT (GR)	DT (SGR)	DT (DW)	DT (SDW)	DT (GRW)	DT (SGRW)	DT (SGRL)	DT (SGRPL)	DT (OGRPL)
SST	P: 55.29 R: 23.5 F: 32.98	P: 54.22 R: 24.4 F: 33.66	P: 31.87 R: 27.5 F: 29.52	P: 30.74 R: 27.3 F: 28.92	P: 52.76 R: 33.4 F: 40.9	P: 52.47 R: 30.8 F: 38.82	P: 56.09 R: 33.6 F: 42.03	P: 56.03 R: 33.0 F: 41.54	P: 57.85 R: 31.7 F: 40.96
SST + bow	P: 57.87 R: 21.7 F: 31.56	P: 54.91 R: 23.5 F: 32.91	P: 30.71 R: 26.9 F: 28.68	P: 29.98 R: 25.9 F: 27.79	P: 52.98 R: 32.0 F: 39.9	P: 51.06 R: 31.3 F: 38.81	P: 51.99 R: 31.4 F: 39.15	P: 56.8 R: 28.8 F: 38.22	P: 61.73 R: 29.2 F: 39.65
PT	P: 60.0 R: 15.9 F: 25.14	P: 57.84 R: 16.6 F: 25.8	P: 40.44 R: 23.9 F: 30.04	P: 42.2 R: 26.5 F: 32.56	P: 53.35 R: 34.2 F: 41.68	P: 53.41 R: 36.0 F: 43.01	P: 51.29 R: 37.9 F: 43.59	P: 52.88 R: 33.0 F: 40.64	P: 53.55 R: 33.2 F: 40.99
uPT	P: 58.77 R: 23.8 F: 33.88	P: 59.5 R: 26.0 F: 36.19	P: 29.21 R: 30.2 F: 29.7	P: 29.52 R: 31.5 F: 30.48	P: 51.86 R: 32.0 F: 39.58	P: 52.17 R: 33.7 F: 40.95	P: 52.1 R: 36.0 F: 42.58	P: 54.64 R: 31.2 F: 39.72	P: 56.43 R: 30.7 F: 39.77

Table 1: Performance of DT (GR), DT (DW) and DT (GRW) (proposed by (Nguyen et al., 2009)) and their modified and improved versions on the *converted* AIMed corpus.

RE experiments carried out on newspaper text corpora (such as ACE 2004) have indicated that kernels based on PST obtain better results than kernels based on DT. Interestingly, our experiments on a biomedical text corpus indicate an opposite trend. Intuitively, this might be due to the different nature of the PPI task. PPI can be often identified by spotting cue words such as *interaction*, *binding*, etc, since the interacting entities (i.e. proteins) usually have direct syntactic dependency relation on such cue words. This might have allowed kernels based on DT to be more accurate.

Although tree kernels applied on DT and PST structures have produced high performance on corpora of news text (Zhou et al., 2007; Nguyen et al., 2009), in case of biomedical text the results that we obtained are relatively low. This may be due to the fact that biomedical texts are different from newspaper texts: more variation in vocabulary, more complex naming of (bio) entities, more diversity of the valency of verbs and so on.

One important finding of our experiments is the effectiveness of the mild extension of DT structures. MEDT-3 achieves the best result for all kernels (SST, SST+bow, PTK and uPTK). However, the other two versions of MEDT appear to be less effective.

In general, the empirical outcome suggests that uPTK can better exploit our proposed DT structures as well as PST. The superiority of uPTK on PTK demonstrates that single lexical features (i.e. features with flat structure) tend to overfit.

Finally, we have performed statistical tests to assess the significance of our results. For each kernel (i.e. SST, SST+bow, PTK, uPTK), the PPI predictions using the best structure (i.e. MEDT-3 applied to SOGRL) are compared against the predictions of the other structures. The tests were performed using the approximate randomization procedure (Noreen, 1989). We set the number of iterations to 1,000 and the confidence level to 0.01. According to the tests, for each kernel, our best structure produces significantly better results.

5.3 Comparison with previous work

To the best of our knowledge, the only work on tree kernel applied on dependency trees that we can directly compare to ours is reported by Sætre et al. (2007). Their DT kernel achieved an F-score of 37.1 on AIMed corpus which is lower than our best results. As discussed earlier, Miwa et al. (2009)) also used tree kernel on dependency analyses and achieved a much higher result. However, the tree structure they used contains multiple nodes for a single word and this does not comply with the constraints usually applied to dependency tree structures (refer to Section 2.3). It would be interesting to examine why such type of tree representation leads to

	DT (SOGRPL)	DT (OGRL)	DT (SOGRW)	DT (SOGRL)	MEDT-1 (SOGRL)	MEDT-2 (SOGRL)	MEDT-3 (SOGRPL)	PST (PET)
SST	P: 57.59	P: 54.38	P: 51.49	P: 54.08	P: 58.15	P: 54.46	P: 59.55	P: 52.72
	R: 33.0	R: 33.5	R: 31.2	R: 33.8	R: 34.6	R: 33.6	R: 37.1	R: 35.9
	F: 41.96	F: 41.46	F: 38.86	F: 41.6	F: 43.39	F: 41.56	F: 45.72	F: 42.71
SST	P: 60.31	P: 53.22	P: 50.08	P: 53.26	P: 58.84	P: 52.87	P: 59.35	P: 52.88
+	R: 30.7	R: 33.1	R: 30.9	R: 32.7	R: 32.6	R: 32.2	R: 34.9	R: 37.7
bow	F: 40.69	F: 40.82	F: 38.22	F: 40.52	F: 41.96	F: 40.02	F: 43.95	F: 44.02
PT	P: 55.45	P: 49.78	P: 51.05	P: 51.61	P: 52.94	P: 50.89	P: 54.1	P: 58.39
	R: 34.6	R: 34.6	R: 34.1	R: 36.9	R: 36.0	R: 37.0	R: 38.9	R: 36.9
	F: 42.61	F: 40.82	F: 40.89	F: 43.03	F: 42.86	F: 42.85	F: 45.26	F: 45.22
uPT	P: 56.2	P: 50.87	P: 50.0	P: 52.74	P: 55.0	P: 52.17	P: 56.85	P: 56.6
	R: 32.2	R: 35.0	R: 33.0	R: 35.6	R: 34.1	R: 34.8	R: 39.0	R: 38.6
	F: 40.94	F: 41.47	F: 39.76	F: 42.51	F: 42.1	F: 41.75	**F: 46.26**	F: 45.9

Table 2: Performance of the other improved versions of DT kernel structures (including MEDT kernels) as well as PST (PET) kernel (Moschitti, 2004; Nguyen et al., 2009) on the *converted* AIMed corpus.

a better result.

In this work, we compare the performance of tree kernels applied of DT with that of PST. Previously, Tikk et al. (2010) applied similar kernels on PST for exactly the same task and data set. They reported that SST and PTK (on PST) achieved F-scores of 26.2 and 34.6, respectively on the converted AIMed corpus (refer to Table 2 in their paper). Such results do not match our figures obtained with the same kernels on PST. We obtain much higher results for those kernels. It is difficult to understand the reason for such differences between our and their results. A possible explanation could be related to parameter settings. Another source of uncertainty is given by the tool for tree kernel computation, which in their case is not mentioned. Moreover, their description of PT and SST (in Figure 1 of their paper) appears to be imprecise: for example, in (partial or complete) phrase structure trees, words can only appear as leaves but in their figure they appear as non-terminal nodes.

The comparison with other kernel approaches (i.e. not necessarily tree kernels on DT or PST) shows that there are model achieving higher results (e.g. Giuliano et al. (2006), Kim et al. (2010), Airola et al. (2008), etc). State-of-the-art results on most of the PPI data sets are obtained by the hybrid kernel presented in Miwa et al. (2009). As noted earlier, our work focuses on the design of an effective DTK for PPI that can be combined with others and that can hopefully be used to design state-of-the-art hybrid kernels.

6 Conclusion

In this paper, we have proposed a study of PPI extraction from specific biomedical data based on tree kernels. We have modeled and experimented with new kernels and DT structures, which can be exploited for RE tasks in other domains too.

More specifically, we applied four different tree kernels on existing and newly proposed DT and PST structures. We have introduced some extensions of DT kernel structures which are linguistically motivated. We call these as mildly extended DT kernels. We have also shown that in PPI extraction lexical information can lead to overfitting as uPTK outperforms PTK. In general, the empirical results show that our DT structures perform better than the previously proposed PST and DT structures.

The ultimate objective of our work is to improve tree kernels applied to DT and then combine them with other types of kernels and data to produce more accurate models.

Acknowledgments

This work was carried out in the context of the project "eOnco - Pervasive knowledge and data management in cancer care". The authors would like to thank the anonymous reviewers for providing excellent feedback.

References

A Airola, S Pyysalo, J Bjorne, T Pahikkala, F Ginter, and T Salakoski. 2008. A graph kernel for protein-protein interaction extraction. In *Proceedings of BioNLP 2008*, pages 1–9, Columbus, USA.

R Bunescu and R Mooney. 2005. A shortest path dependency kernel for relation extraction. In *Proceedings of Human Language Technology Conference and Conference on Empirical Methods in Natural Language Processing*, pages 724–731, Vancouver, British Columbia, Canada, October. Association for Computational Linguistics.

R Bunescu and RJ Mooney. 2006. Subsequence kernels for relation extraction. In *Proceedings of the 19th Conference on Neural Information Processing Systems*, pages 171–178.

R Bunescu, R Ge, RJ Kate, EM Marcotte, RJ Mooney, AK Ramani, and YW Wong. 2005. Comparative experiments on learning information extractors for proteins and their interactions. *Artificial Intelligence in Medicine (Special Issue on Summarization and Information Extraction from Medical Documents)*, 33(2):139–155.

M Collins and N Duffy. 2001. Convolution kernels for natural language. In *Proceedings of Neural Information Processing Systems (NIPS'2001)*.

A Culotta and J Sorensen. 2004. Dependency tree kernels for relation extraction. In *Proceedings of the 42nd Annual Meeting of the Association for Computational Linguistics*, Barcelona, Spain.

G Erkan, A Ozgur, and DR Radev. 2007. Semi-Supervised Classification for Extracting Protein Interaction Sentences using Dependency Parsing. In *Proceedings of the 2007 Joint Conference on Empirical Methods in Natural Language Processing and Computational Natural Language Learning (EMNLP-CoNLL 2007)*, pages 228–237.

C Giuliano, A Lavelli, and L Romano. 2006. Exploiting shallow linguistic information for relation extraction from biomedical literature. In *Proceedings of the 11th Conference of the European Chapter of the Association for Computational Linguistics (EACL'2006)*, pages 401–408, Trento, Italy.

CW Hsu, CC Chang, and CJ Lin, 2003. *A practical guide to support vector classification*. Department of Computer Science and Information Engineering, National Taiwan University, Taipei, Taiwan.

T Joachims. 1999. Making large-scale support vector machine learning practical. In *Advances in kernel methods: support vector learning*, pages 169–184. MIT Press, Cambridge, MA, USA.

S Kim, J Yoon, J Yang, and S Park. 2010. Walk-weighted subsequence kernels for protein-protein interaction extraction. *BMC Bioinformatics*, 11(1).

H Lodhi, C Saunders, J Shawe-Taylor, N Cristianini, and C Watkins. 2002. Text classification using string kernels. *Journal of Machine Learning Research*, 2:419–444, March.

M Miwa, R Sætre, Y Miyao, T Ohta, and J Tsujii. 2009. Protein-protein interaction extraction by leveraging multiple kernels and parsers. *International Journal of Medical Informatics*, 78.

A Moschitti. 2004. A study on convolution kernels for shallow semantic parsing. In *Proceedings of the 42nd Annual Meeting of the Association for Computational Linguistics*, ACL '04, Barcelona, Spain.

A Moschitti. 2006. Efficient convolution kernels for dependency and constituent syntactic trees. In Johannes Fürnkranz, Tobias Scheffer, and Myra Spiliopoulou, editors, *Machine Learning: ECML 2006*, volume 4212 of *Lecture Notes in Computer Science*, pages 318–329. Springer Berlin / Heidelberg.

A Moschitti. 2006a. Making Tree Kernels Practical for Natural Language Learning. In *Proceedings of the 11th Conference of the European Chapter of the Association for Computational Linguistics*, Trento, Italy.

TT Nguyen, A Moschitti, and G Riccardi. 2009. Convolution kernels on constituent, dependency and sequential structures for relation extraction. In *Proceedings of the 2009 Conference on Empirical Methods in Natural Language Processing (EMNLP'2009)*, pages 1378–1387, Singapore, August.

EW Noreen. 1989. *Computer-Intensive Methods for Testing Hypotheses: An Introduction*. Wiley-Interscience.

S Pyysalo, A Airola, J Heimonen, J Björne, F Ginter, and T Salakoski. 2008. Comparative analysis of five protein-protein interaction corpora. *BMC Bioinformatics*, 9(Suppl 3):S6.

R Sætre, K Sagae, and J Tsujii. 2007. Syntactic features for protein-protein interaction extraction. In *Proceedings of the Second International Symposium on Languages in Biology and Medicine (LBM 2007)*, pages 6.1–6.14, Singapore.

A Severyn and A Moschitti. 2010. Fast cutting plane training for structural kernels. In *Proceedings of ECML-PKDD*.

D Tikk, P Thomas, P Palaga, J Hakenberg, and U Leser. 2010. A Comprehensive Benchmark of Kernel Methods to Extract Protein-Protein Interactions from Literature. *PLoS Computational Biology*, 6(7), July.

SVN Vishwanathan and AJ Smola. 2002. Fast kernels on strings and trees. In *Proceedings of Neural Information Processing Systems (NIPS'2002)*, pages 569–576, Vancouver, British Columbia, Canada.

D Zhang and WS Lee. 2003. Question classification using support vector machines. In *Proceedings of the*

26th annual international ACM SIGIR conference on Research and development in information retrieval, SIGIR '03, pages 26–32, Toronto, Canada.

M Zhang, J Su, D Wang, G Zhou, and CL Tan. 2005. Discovering relations between named entities from a large raw corpus using tree similarity-based clustering. In Robert Dale, Kam-Fai Wong, Jian Su, and Oi Yee Kwong, editors, *Natural Language Processing IJC-NLP 2005*, volume 3651 of *Lecture Notes in Computer Science*, pages 378–389. Springer Berlin / Heidelberg.

GD Zhou, M Zhang, DH Ji, and QM Zhu. 2007. Tree kernel-based relation extraction with context-sensitive structured parse tree information. In *Proceedings of the 2007 Joint Conference on Empirical Methods in Natural Language Processing and Computational Natural Language Learning (EMNLP-CoNLL)*, pages 728–736, June.

The CISP Annotation Schema Uncovers Hypotheses and Explanations in Full-Text Scientific Journal Articles

Elizabeth White, K. Bretonnel Cohen, and Lawrence Hunter
Department of Pharmacology, Computational Bioscience Program,
University of Colorado School of Medicine, Aurora, Colorado, USA
elizabeth.white@ucdenver.edu,
kevin.cohen@gmail.com,
larry.hunter@ucdenver.edu

Abstract

Increasingly, as full-text scientific papers are becoming available, scientific queries have shifted from looking for facts to looking for arguments. Researchers want to know when their colleagues are proposing theories, outlining evidentiary relations, or explaining discrepancies. We show here that sentence-level annotation with the CISP schema adapts well to a corpus of biomedical articles, and we present preliminary results arguing that the CISP schema is uniquely suited to recovering common types of scientific arguments about hypotheses, explanations, and evidence.

1 Introduction

In the scientific domain, the deluge of full-text publications is driving researchers to find better techniques for extracting or summarizing the main claims and findings in a paper. Many researchers have noted that the sentences of a paper play a small set of different rhetorical roles (Teufel and Moens, 1999; Blais et al., 2007; Agarwal and Yu, 2009). We are investigating the rhetorical roles of sentences in the CRAFT corpus, a set of 97 full-text papers that we have annotated using the CISP schema. Hand alignment of the resulting annotations suggests that patterns in these CISP-annotated sentences correspond to common argumentative gambits in scientific writing.

2 Methods

The CRAFT corpus is a set of 97 full-text papers describing the function of genes in the Mouse Genome Informatics database (Blake et al., 2011). These documents have already been annotated with syntactic information (parse trees and part-of-speech tags), linguistic phenomena (coreference), and semantic entities (genes, chemicals, cell lines, biological functions and molecular processes), making the corpus a rich resource for extracting or inferring information from full scientific papers.

The CISP schema (Soldatova and Liakata, 2007; Liakata et al., 2009) contains 11 categories, and several of the categories describe the intentions of the authors, making it well suited for markup of argumentation. We chose to narrow these down to 9 categories (excluding Model and Object) during annotation training; our guidelines are shown in Figure 1. We expect this schema to describe the pragmatics in the text well, while still offering the potential for high interannotator agreement due to a manageable number of categories. The process of marking the sentences in the CRAFT corpus according to the CISP guidelines took one annotator about four months.

3 Results and Discussion

Six of the 97 CRAFT papers do not follow the standard IMRaD paper structure (one was a review article, and five combined Results and Discussion); these documents were eliminated from this analysis. Annotation of the 91 remaining CRAFT papers resulted in 20676 sentences. The distribution of the annotated classes is shown in Table 1.

Our use of the CISP schema exposes an approach for recovering two types of explanatory arguments. The first sets the context with a sequence of Back-

Proceedings of the 2011 Workshop on Biomedical Natural Language Processing, ACL-HLT 2011, pages 134–135,
Portland, Oregon, USA, June 23-24, 2011.

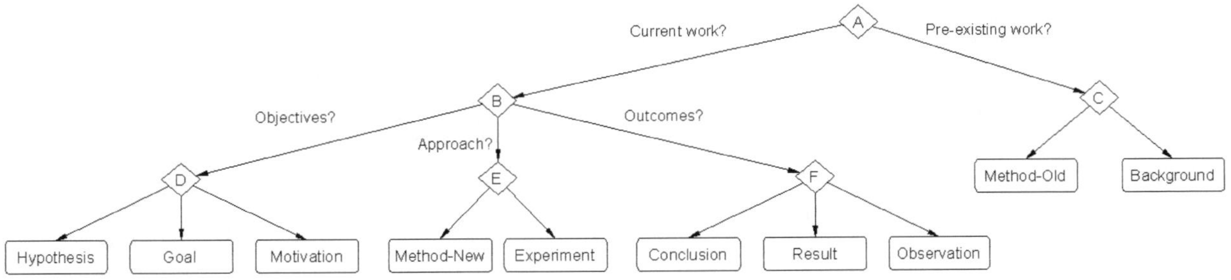

Figure 1: Flow chart for CISP annotation of the CRAFT corpus.

CISP Type	Count	Percentage
Hypothesis	1050	5.08
Goal	992	4.80
Motivation	928	4.49
Background	2838	13.73
Method	637	3.08
Experiment	5270	25.49
Result	5471	26.46
Observation	1168	5.65
Conclusion	2322	11.23
Total	20676	100.0

Table 1: Distribution of CISP sentence types annotated in 91 CRAFT articles.

ground sentences, followed by a Hypothesis, Motivation, or Goal; this echoes a motif found by Swales (1990) and Teufel and Moens (1999). We also find another pattern that consists of a combination of Results and Observations, either preceded or followed by a Conclusion; Teufel and Moens (1999) also find exemplars of this maneuver, and note that it parallels Swales' notion of occupying a niche in the research world. Hand alignment of CISP annotations in Introduction and Result sections suggests that a finite state machine may be capable of modeling the transitions between CISP sentence types in these arguments, and machine learning approaches to represent these and other patterns with hidden Markov models or conditional random fields are underway.

References

Shashank Agarwal and Hong Yu. 2009. Automatically classifying sentences in full-text biomedical articles into Introduction, Methods, Results, and Discussion. *Bioinformatics*, 25(23): 3174–3180.

Antoine Blais, Iana Atanassova, Jean-Pierre Desclés, Mimi Zhang, and Leila Zighem. 2007. Discourse automatic annotation of texts: an application to summarization. In *Proceedings of the Twentieth International Florida Artificial Intelligence Research Society Conference*, May 7-9, 2007, Key West, Florida, USA, 350–355. AAAI Press.

Judith A. Blake, Carol J. Bult, James A. Kadin, Joel E. Richardson, Janan T. Eppig, and the Mouse Genome Database Group 2011. The Mouse Genome Database (MGD): premier model organism resource for mammalian genomics and genetics. *Nucleic Acids Res.*, 39(Suppl. 1): D842–D848.

Maria Liakata, Claire Q, and Larisa N. Soldatova. Semantic Annotation of Papers: Interface & Enrichment Tool (SAPIENT). 2009. In *Proceedings of BioNLP 2009*, Boulder, Colorado,193–200.

Larisa Soldatova and Maria Liakata. 2007. An ontology methodology and CISP - the proposed Core Information about Scientific Papers. JISC intermediate project report.

John M. Swales. 1990. *Genre Analysis: English in academic and research settings*, 137–166. Cambridge University Press, Cambridge.

Simone Teufel and Marc Moens. 1999. Argumentative classification of extracted sentences as a first step towards flexible abstracting. In *Advances in Automatic Text Summarization*, I. Mani and D. Maybury, eds. MIT Press, Cambridge, MA.

SimSem: Fast Approximate String Matching in Relation to Semantic Category Disambiguation

Pontus Stenetorp[*†] **Sampo Pyysalo**[*] **and Jun'ichi Tsujii**[‡]
[*] Tsujii Laboratory, Department of Computer Science, The University of Tokyo, Tokyo, Japan
[†] Aizawa Laboratory, Department of Computer Science, The University of Tokyo, Tokyo, Japan
[‡] Microsoft Research Asia, Beijing, People's Republic of China
{pontus,smp}@is.s.u-tokyo.ac.jp
jtsujii@microsoft.com

Abstract

In this study we investigate the merits of fast approximate string matching to address challenges relating to spelling variants and to utilise large-scale lexical resources for semantic class disambiguation. We integrate string matching results into machine learning-based disambiguation through the use of a novel set of features that represent the distance of a given textual span to the closest match in each of a collection of lexical resources. We collect lexical resources for a multitude of semantic categories from a variety of biomedical domain sources. The combined resources, containing more than twenty million lexical items, are queried using a recently proposed fast and efficient approximate string matching algorithm that allows us to query large resources without severely impacting system performance. We evaluate our results on six corpora representing a variety of disambiguation tasks. While the integration of approximate string matching features is shown to substantially improve performance on one corpus, results are modest or negative for others. We suggest possible explanations and future research directions. Our lexical resources and implementation are made freely available for research purposes at: http://github.com/ninjin/simsem

1 Introduction

The use of dictionaries for boosting performance has become commonplace for Named Entity Recognition (NER) systems (Torii et al., 2009; Ratinov and Roth, 2009). In particular, dictionaries can give an initial improvement when little or no training data is available. However, no dictionary is perfect, and all resources lack certain spelling variants and lag behind current vocabulary usage and thus are unable to cover the intended domain in full. Further, due to varying dictionary curation and corpus annotation guidelines, the definition of what constitutes a semantic category is highly unlikely to precisely match for any two specific resources (Wang et al., 2009). Ideally, for applying a lexical resource to an entity recognition or disambiguation task to serve as a definition of a semantic category there would be a precise match between the definitions of the lexical resource and target domain, but this is seldom or never the case.

Most previous work studying the use of dictionary resources in entity mention-related tasks has focused on single-class NER, in particular this is true for BioNLP where it has mainly concerned the detection of proteins. These efforts include Tsuruoka and Tsujii (2003), utilising dictionaries for protein detection by considering each dictionary entry using a novel distance measure, and Sasaki et al. (2008), applying dictionaries to restrain the contexts in which proteins appear in text. In this work, we do not consider entity mention detection, but instead focus solely on the related task of disambiguating the semantic category for a given continuous sequence of characters (a textual span), doing so we side-step the issue of boundary detection in favour of focusing on novel aspects of semantic category disambiguation. Also, we are yet to see a high-performing multi-class biomedical NER system, this motivates our desire to include multiple semantic categories.

Proceedings of the 2011 Workshop on Biomedical Natural Language Processing, ACL-HLT 2011, pages 136–145,
Portland, Oregon, USA, June 23-24, 2011.

2 Methods

In this section we introduce our approach and the structure of our system.

2.1 SimSem

Many large-scale language resources are available for the biomedical domain, including collections of domain-specific lexical items (Ashburner et al., 2000; Bodenreider, 2004; Rebholz-Schuhmann et al., 2010). These resources present obvious opportunities for semantic class disambiguation. However, in order to apply them efficiently, one must be able to query the resources taking into consideration both lexical variations in dictionary entries compared to real-world usage and the speed of look-ups.

We can argue that each resource offers a different view of what constitutes a particular semantic category. While these views will not fully overlap between resources even for the same semantic category, we can expect a certain degree of agreement. When learning to disambiguate between semantic categories, a machine learning algorithm could be expected to learn to identify a specific semantic category from the similarity between textual spans annotated for the category and entries in a related lexical resource. For example, if we observe the text "Carbonic anhydrase IV" marked as PROTEIN and have an entry for "Carbonic anhydrase 4" in a lexical resource, a machine learning method can learn to associate the resource with the PROTEIN category (at specific similarity thresholds) despite syntactic differences.

In this study, we aim to construct such a system and to demonstrate that it outperforms strict string matching approaches. We refer to our system as SimSem, as in "Similarity" and "Semantic".

2.2 SimString

SimString[1] is a software library utilising the CP-Merge algorithm (Okazaki and Tsujii, 2010) to enable fast approximate string matching. The software makes it possible to find matches in a collection with over ten million entries using cosine similarity and a similarity threshold of 0.7 in approximately 1 millisecond with modest modern hardware. This makes it useful for querying a large collection of strings to find entries which may differ from the query string only superficially and may still be members of the same semantic category.

As an example, if we construct a SimString database using an American English wordlist[2] and query it using the cosine measure and a threshold of 0.7. For the query "reviewer" SimString would return the following eight entries: review, viewer, preview, reviewer, unreviewed, televiewer, and revieweress. We can observe that most of the retrieved entries share some semantic similarity with the query.

2.3 Machine Learning

For the machine learning component of our system we use the L2-regularised logistic regression implementation of the LIBLINEAR[3] software library (Fan et al., 2008). We do not normalise our feature vectors and optimise our models' penalty parameter using k-fold cross-validation on the training data. In order to give a fair representation of the performance of other systems, we use a rich set of features that are widely applied for NER (See Table 1).

Our novel SimString features are generated as follows. We query each SimString database using the cosine measure with a sliding similarity threshold, starting at 1.0 and ending at 0.7, lowering the threshold by 0.1 per query. If a query is matched, we generate a feature unique for that database and threshold, we also generate the same feature for each step from the current threshold to the cut-off of 0.7 (a match at e.g. 0.9 similarity also implies matches at 0.8 and 0.7).

The cut-off is motivated by the fact that very low thresholds introduces a large degree of noise. For example, for our American English wordlist the query "rejection" using threshold 0.1 and the cosine measure will return 13,455 results, among them "questionableness" which only have a single sequence "ion" in common.

It is worthwhile to note that during our preliminary experiments we failed to establish a consistent benefit from contextual features across our development sets. Thus, contextual features are not included in our feature set and instead our study focuses only

[1] http://www.chokkan.org/software/simstring/

[2] /usr/share/dict/web2 under FreeBSD 8.1-RELEASE, based on Webster's Second International dictionary from 1934

[3] We used version 1.7 of LIBLINEAR for our experiments

Feature	Type	Input	Value(s)
Text	Text	Flu	Flu
Lower-cased	Text	DNA	dna
Prefixes: sizes 3 to 5	Text	bull	bul, ...
Suffixes: sizes 3 to 5	Text	bull	ull, ...
Stem (Porter, 1993)	Text	performing	perform
Is a pair of digits	Bool	42	True
Is four digits	Bool	4711	True
Letters and digits	Bool	C4	True
Digits and hyphens	Bool	9-12	True
Digits and slashes	Bool	1/2	True
Digits and colons	Bool	3,1	True
Digits and dots	Bool	3.14	True
Upper-case and dots	Bool	M.C.	True
Initial upper-case	Bool	Pigeon	True
Only upper-case	Bool	PMID	True
Only lower-case	Bool	pure	True
Only digits	Bool	131072	True
Only non-alpha-num	Bool	#*$!	True
Contains upper-case	Bool	gAwn	True
Contains lower-case	Bool	After	True
Contains digits	Bool	B52	True
Contains non-alpha-num	Bool	B52;s	True
Date regular expression[4]	Bool	1989-01-30	True
Pattern	Text	1B-zz	0A-aa
Collapsed Pattern	Text	1B-zz	0A-a

Table 1: Basic features used for classification

the features that are generated solely from the textual span which has been annotated with a semantic category (span-internal features) and the comparison of approximate and strict string matching.

3 Resources

This section introduces and discusses the preprocessing and statistics of the lexical and corpus resources used in our experiments.

3.1 Lexical Resources

To generate a multitude of SimString databases covering a wide array of semantic categories we employ several freely available lexical resources (Table 2).

The choice of lexical resources was initially made with the aim to cover commonly annotated domain semantic categories: the CHEBI and CHEMICAL subsets of JOCHEM for chemicals, LINNAEUS for species, Entrez Gene and SHI for proteins. We then expanded the selection based on error analysis to increase our coverage of a wider array of semantic categories present in our development data.

We used the GO version from March 2011, extracting all non-obsolete terms from the ontology and separating them into the three GO subontologies: biological process (BP), cellular component (CC) and molecular function (MF). We then created an additional three resources by extracting all exact synonyms for each entry. Lastly, we expanded these six resources into twelve resources by applying the GO term variant generation technique described by Beisswanger et al. (2008).

UMLS, a collection of various resources, contain 135 semantic categories (e.g. Body Location or Region and Inorganic Chemical) which we use to create a database for each category.

For Entrez Gene we extracted all entries for the following types: gene locus, protein name, protein description, nomenclature symbol and nomenclature fullname, creating a SimString database for each. This leaves some parts of Entrez Gene unutilised, but we deemed these categories to be sufficient for our experiments.

The Turku Event Corpus is a resource created by applying an automated event extraction system on the full release of PubMed from 2009. As a precondition for the event extraction system to operate, protein name recognition is necessary; for this corpus, NER has been performed by the corpus curators using the BANNER (Leaman and Gonzalez, 2008) NER system trained on GENETAG (Tanabe et al., 2005). We created a database (PROT) containing all protein annotations, extracted all event triggers (TRIG) and created a database for each of the event types covered by the event extraction system.

For the AZDC corpus, we extracted each annotated textual span since the corpus covers only a single semantic category. Similarly, the LINNAEUS dictionary was converted into a single database since it covers the single category "species".

Table 3 contains the statistics per dictionary resource and the number of SimString databases created for each resource. Due to space requirements we leave out the full details for GO BP, GO CC, GO MF, UMLS, Entrez Gene and TURKU TRIG, and instead give the total entries for all the databases generated from these resources.

[4] A simple regular expression matching dates:
`^(19|20)\d\d[- /.](0[1-9]|1[012])[- /.](0[1-9]|[12][0-9]|3[01])$`
from http://www.regular-expressions.info/dates.html

Name	Abbreviation	Semantic Categories	Publication
Gene Ontology	GO	Multiple	Ashburner et al. (2000)
Protein Information Resource	PIR	Proteins	Wu et al. (2003)
Unified Medical Language System	UMLS	Multiple	Bodenreider (2004)
Entrez Gene	–	Proteins	Maglott et al. (2005)
Automatically generated dictionary	SHI	Proteins	Shi and Campagne (2005)
Jochem	JOCHEM	Multiple	Hettne et al. (2009)
Turku Event Corpus	TURKU	Proteins and biomolecular events	Björne et al. (2010)
Arizona Disease Corpus	AZDC	Diseases	Chowdhury and Lavelli (2010)
LINNAEUS Dictionary	LINNAEUS	Species	Gerner et al. (2010)
Webster's International Dictionary	WID	Multiple	–

Table 2: Lexical resources gathered for our experiments

Resource	Unique Entries	Databases
GO BP	67,411	4
GO CC	5,993	4
GO MF	55,595	4
PIR	691,577	1
UMLS	5,902,707	135
Entrez Gene	3,602,757	5
SHI	61,676	1
CHEBI	187,993	1
CHEMICAL	1,527,751	1
TURKU PROT	4,745,825	1
TURKU TRIG	130,139	10
AZDC	1,195	1
LINNAEUS	3,119,005	1
WID	235,802	1
Total:	20,335,426	170

Table 3: Statistics per dictionary resource

3.2 Corpora

To evaluate our approach we need a variety of corpora annotated with multiple semantic categories. For this purpose we selected the six corpora listed in Table 4.

The majority of our corpora are available in the common stand-off style format introduced for the BioNLP 2009 Shared Task (BioNLP'09 ST) (Kim et al., 2009). The remaining two, NLPBA and CALBC CII, were converted into the BioNLP'09 ST format so that we could process all resources in the same manner for our experimental set-up.

In addition to physical entity annotations, the GREC, EPI, ID and GENIA corpora incorporate event trigger annotations (e.g. Gene Regulatory Event (GRE) for GREC). These trigger expressions carry with them a specific semantic type (e.g. "interact" can carry the semantic type BINDING for GENIA), allowing us to enrich the data sets with additional semantic categories by including these types in our dataset as distinct semantic categories. This gave us the following increase in semantic categories: GREC one, EPI 15, ID ten, GENIA nine.

The original GREC corpus contains an exceptionally wide array of semantic categories. While this is desirable for evaluating the performance of our approach under different task settings, the sparsity of the data is a considerable problem; the majority of categories do not permit stable evaluation as they have only a handful of annotations each. To alleviate this problem we used the five ontologies defined in the GREC annotation guidelines[5], collapsing the annotations into five semantic super categories to create a resource we refer to as Super GREC. This preprocessing conforms with how the categories were used when annotating the GREC corpus (Thompson et al., 2009). This resource contains sufficient annotations for each semantic category to enable evaluation on a category-by-category basis. Also, for the purpose of our experiments we removed all "SPAN" type annotations since they themselves carry no semantic information (cf. GREC annotation guidelines).

CALBC CII contains 75,000 documents, which is more than enough for our experiments. In order to maintain balance in size between the resources in our experiments, we sampled a random 5,000 documents and used these as our CALBC CII dataset.

[5] http://www.nactem.ac.uk/download.php?target=GREC/Event_annotation_guidelines.pdf

Name	Abbreviation	Publication
BioNLP/NLPBA 2004 Shared Task Corpus	NLPBA	Kim et al. (2004)
Gene Regulation Event Corpus	GREC	Thompson et al. (2009)
Collaborative Annotation of a Large Biomedical Corpus	CALBC CII	Rebholz-Schuhmann et al. (2010)
Epigenetics and Post-Translational Modifications	EPI	Ohta et al. (2011)
Infectious Diseases Corpus	ID	Pyysalo et al. (2011)
Genia Event Corpus	GENIA	Kim et al. (2011)

Table 4: Corpora used for evaluation

3.3 Corpus Statistics

In this section we present statistics for each of our datasets. For resources with a limited number of semantic categories we use pie charts to illustrate their distribution (Figure 1). For the other corpora we use tables to illustrate this. Tables for the corpora for which pie charts are given has been left out due to space requirements.

The NLPBA corpus (Figure 1a) with 59,601 tokens annotated, covers five semantic categories, with a clear majority of protein annotations. While NLPBA contains several semantic categories, they are closely related, which is expected to pose challenges for disambiguation. This holds in particular for proteins, DNA and RNA, which commonly share names.

Our collapsed version of GREC, Super GREC (see Figure 1b), contains 6,777 annotated tokens and covers a total of six semantic categories: Regulatory Event (GRE), nucleic acids, proteins, processes, living system and experimental. GREC is an interesting resource in that its classes are relatively distinct and four of them are evenly distributed.

CALBC CII is balanced among its annotated categories, as illustrated in Figure 1c. The 6,433 tokens annotated are of the types: proteins and genes (PRGE), species (SPE), disorders (DISO) and chemicals and drugs (CHED). We note that we have introduced lexical resources covering each of these classes (Section 3.1).

For the BioNLP'11 ST resources EPI (Table 5), GENIA (Figure 1d and contains 27,246 annotated tokens) and ID (Table 6), we observe a very skewed distribution due to our decision to include event types as distinct classes; The dominating class for all the datasets are proteins. For several of these categories, learning accurate disambiguation is expected to be very challenging if not impossible due to sparsity: For example, Dehydroxylation in EPI has a single annotation.

Type	Ratio	Annotations
Acetylation	2.3%	294
Catalysis	1.4%	186
DNA demethylation	0.1%	18
DNA methylation	2.3%	301
Deacetylation	0.3%	43
Deglycosylation	0.2%	26
Dehydroxylation	0.0%	1
Demethylation	0.1%	12
Dephosphorylation	0.0%	3
Deubiquitination	0.1%	13
Entity	6.6%	853
Glycosylation	2.3%	295
Hydroxylation	0.9%	116
Methylation	2.5%	319
Phosphorylation	0.9%	112
Protein	77.7%	10,094
Ubiquitination	2.3%	297
Total:		12,983

Table 5: Semantic categories in EPI

ID is of particular interest since it contains a considerable amount of annotations for more than one physical entity category, including in addition to protein also organism and a minor amount of chemical annotations.

4 Experiments

In this section we introduce our experimental set-up and discuss the outcome of our experiments.

4.1 Experimental Set-up

To ensure that our results are not biased by overfitting on a specific set of data, all data sets were separated into training, development and test sets.

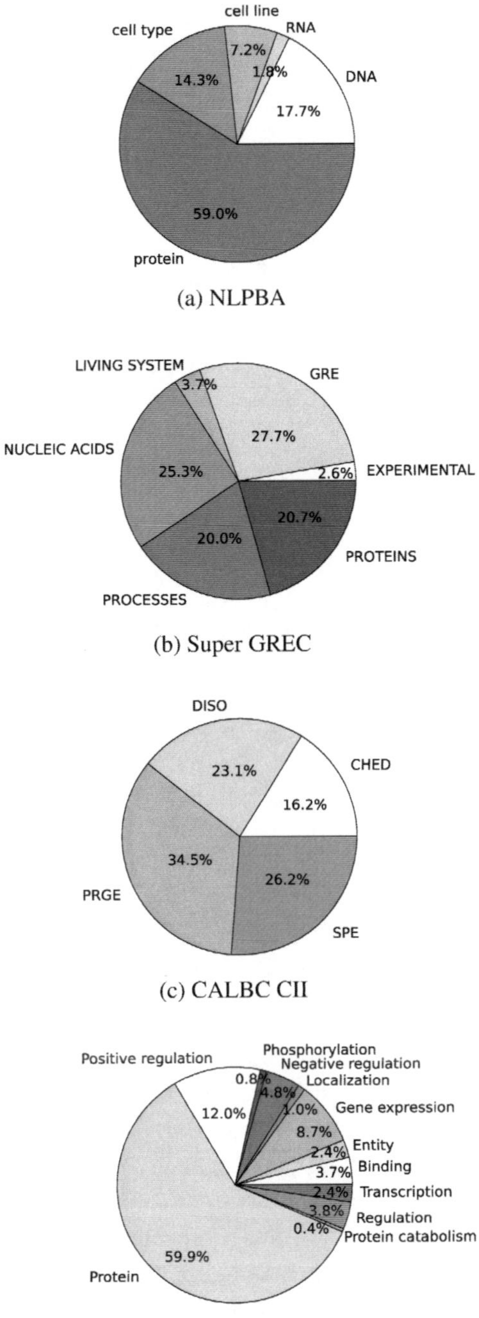

(a) NLPBA

(b) Super GREC

(c) CALBC CII

(d) GENIA

Figure 1: Semantic category distributions

NLPBA defines only a training and test set, GREC and CALBC CII are provided as resources and lack any given division, and for the BioNLP'11 ST data the test sets are not distributed. Thus, we combined all the available data for each dataset and separated the documents into fixed sets with the following ratios: 1/2 training, 1/4 development and 1/4 test.

Type	Ratio	Annotations
Binding	1.0%	102
Chemical	6.8%	725
Entity	0.4%	43
Gene expression	3.3%	347
Localization	0.3%	36
Negative regulation	1.6%	165
Organism	25.5%	2,699
Phosphorylation	0.5%	54
Positive regulation	2.5%	270
Process	8.0%	843
Protein	43.1%	4,567
Protein catabolism	0.0%	5
Regulation	1.8%	188
Regulon-operon	1.1%	121
Transcription	0.4%	47
Two-component-system	3.7%	387
Total:		10,599

Table 6: Semantic categories in ID

We use a total of six classifiers for our experiments. First, a naive baseline (Naive): a majority class voter with a memory based on the exact text of the textual span. The remaining five are machine learning classifiers trained using five different feature sets: gazetteer features constituting strict string matching towards our SimString databases (Gazetteer), SimString features generated from our SimString databases (SimString), the span internal features listed in Table 1 (Internal), the span internal and gazetteer features (Internal-Gazetteer) and the span internal and SimString features (Internal-SimString).

We evaluate performance using simple instance-level accuracy (correct classifications / all classifications). Results are represented as learning curves for each data set.

4.2 Results

From our experiments we find that – not surprisingly – the performance of the Naive, Gazetteer and SimString classifiers alone is comparatively weak. Their performance is illustrated in Figure 2. We can briefly summarize the results for these methods by noting that the SimString classifier outperforms the Gazetteer by a large margin for every dataset.[6] From

[6]Due to space restrictions we do not include further analysis or charts.

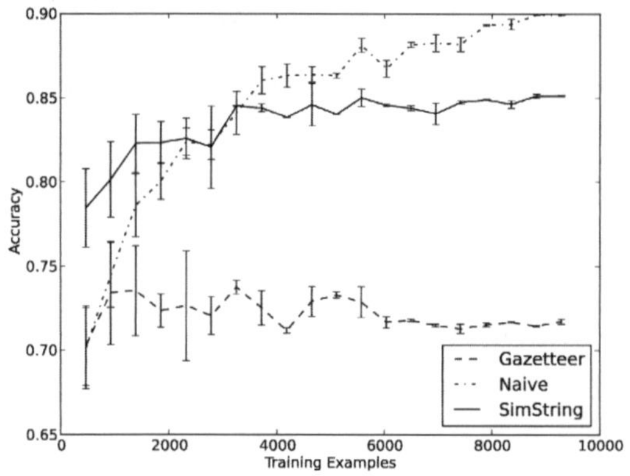

Figure 2: SimString, Gazetteer and Naive for ID

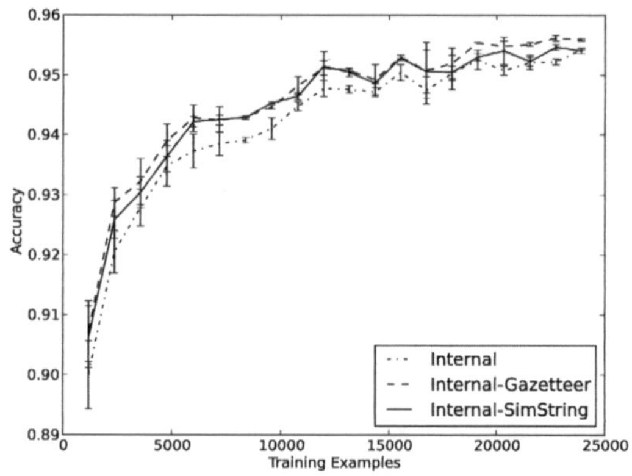

Figure 4: Learning curve for GENIA

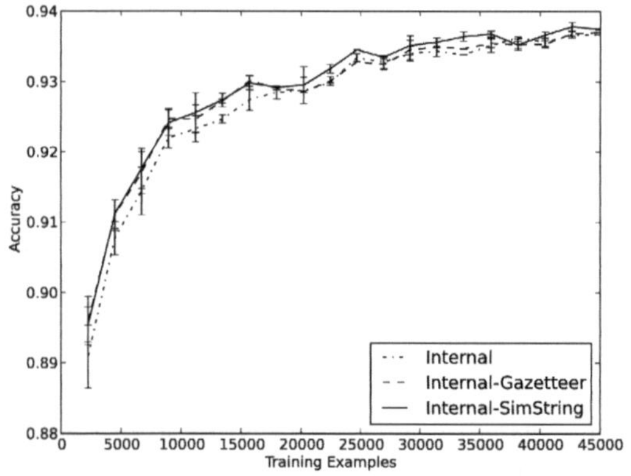

Figure 3: Learning curve for NLPBA

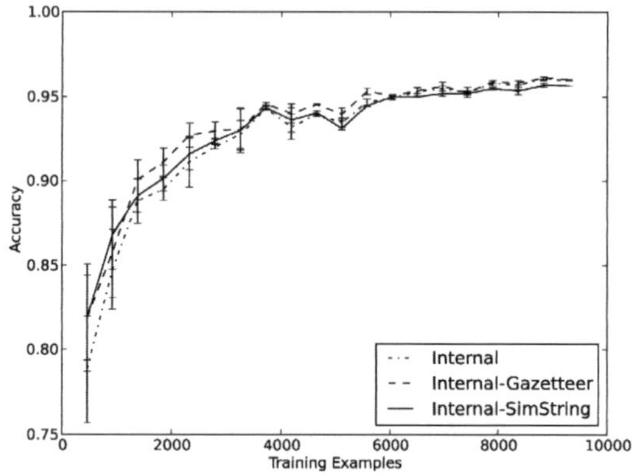

Figure 5: Learning curve for ID

here onwards we focus on the performance of the Internal classifier in combination with Gazetteer and SimString features.

For NLPBA (Figure 3), GENIA (Figure 4) and ID (Figure 5) our experiments show no clear systematic benefit from either SimString or Gazetteer features.

For Super GREC (Figure 6) and EPI (Figure 7) classifiers with Gazetteer and SimString features consistently outperform the Internal classifier, and the SimString classifier further shows some benefit over Gazetteer for EPI.

The only dataset for which we see a clear benefit from SimString features over Gazetteer and Internal is for CALBC CII (Figure 8).

5 Discussion and Conclusions

While we expected to see clear benefits from both using Gazetteers and SimString features, our experiments returned negative results for the majority of the corpora. For NLPBA, GENIA and ID we are aware that most of the instances are either proteins or belong to event trigger classes for which we may not have had adequate lexical resources for disambiguation. By contrast, for Super GREC there are several distinct classes for which we expected lexical resources to have fair coverage for SimString and Gazetteer features. While an advantage over Internal was observed for Super GREC, SimString features showed no benefit over Gazetteer features. The methods exhibited the expected result on only one of the six corpora, CALBC CII, where there is a clear advantage for Gazetteer over Internal and a further clear advantage for SimString over Gazetteer.

Disappointingly, we did not succeed in establishing a clear improvement for more than one of the six corpora. Although we have not been successful in

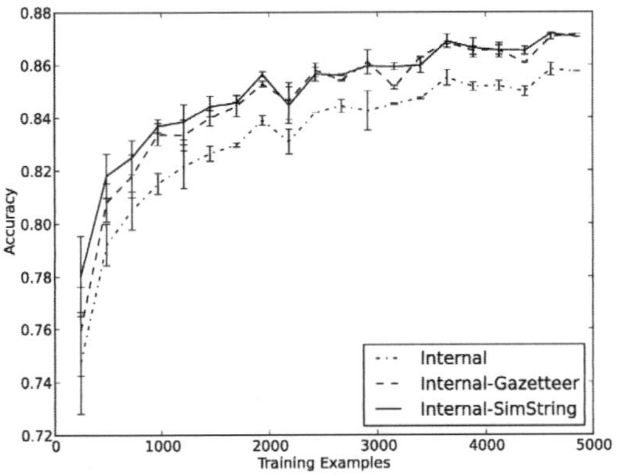

Figure 6: Learning curve for Super GREC

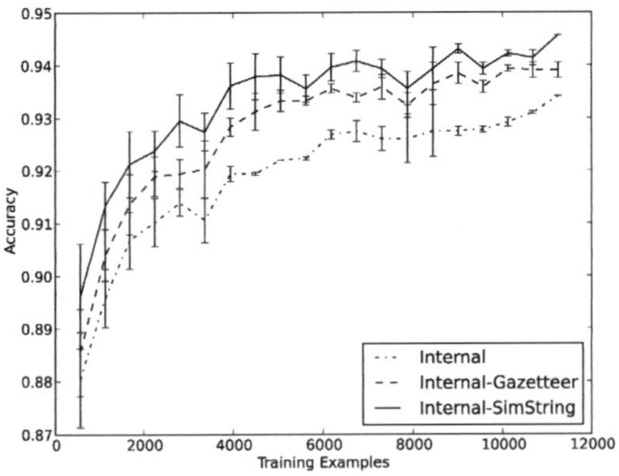

Figure 7: Learning curve for EPI

proving our initial hypothesis we argue that our results calls for further study due to several concerns raised by the results remaining unanswered. It may be that our notion of distance to lexical resource entries is too naive. A possible future direction would be to compare the query string to retrieved results using a method similar to that of Tsuruoka and Tsujii (2003). This would enable us to retain the advantage of fast approximate string matching, thus being able to utilise larger lexical resources than if we were to calculate sophisticated alignments for each lexical entry.

Study of the confusion matrices revealed that some event categories such as negative regulation, positive regulation and regulation for ID are commonly confused by the classifiers. Adding additional resources or contextual features may alleviate these problems.

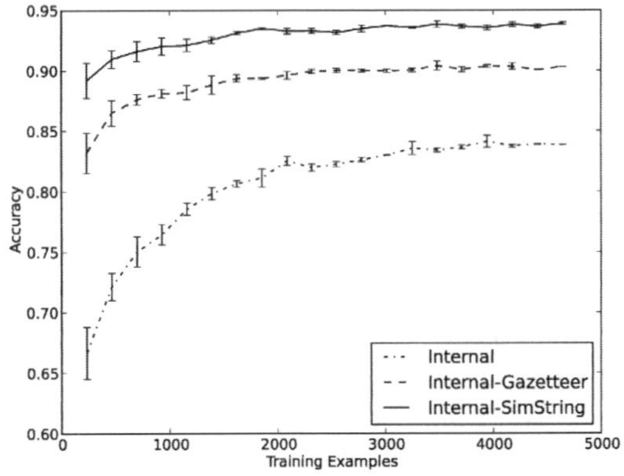

Figure 8: Learning curve for CALBC CII

To conclude, we have found a limited advantage but failed to establish a clear, systematic benefit from approximate string matching for semantic class disambiguation. However, we have demonstrated that approximate string matching can be used to generate novel features for classifiers and allow for the utilisation of large scale lexical resources in new and potentially interesting ways. It is our hope that by making our findings, resources and implementation available we can help the BioNLP community to reach a deeper understanding of how best to incorporate our proposed features for semantic category disambiguation and related tasks.

Our system and collection of resources are freely available for research purposes at http://github.com/ninjin/simsem

Acknowledgements

The authors would like to thank Dietrich Rebholz-Schuhmann and the CALBC organisers for allowing us the use of their data. and Jari Björne for answering questions regarding the Turku Event Corpus. We would also like to thank the anonymous reviewers and Luke McCrohon for their insightful and extensive feedback, which has considerably helped us to improve this work. Lastly the first author would like to thank Makoto Miwa and Jun Hatori for their timely and helpful advice on machine learning methods.

This work was supported by the Swedish Royal Academy of Sciences and by Grant-in-Aid for Specially Promoted Research (MEXT, Japan).

References

M. Ashburner, C.A. Ball, J.A. Blake, D. Botstein, H. Butler, J.M. Cherry, A.P. Davis, K. Dolinski, S.S. Dwight, J.T. Eppig, et al. 2000. Gene ontology: tool for the unification of biology. The Gene Ontology Consortium. *Nature genetics*, 25(1):25.

E. Beisswanger, M. Poprat, and U. Hahn. 2008. Lexical Properties of OBO Ontology Class Names and Synonyms. In *3rd International Symposium on Semantic Mining in Biomedicine*.

J. Björne, F. Ginter, S. Pyysalo, J. Tsujii, and T. Salakoski. 2010. Scaling up biomedical event extraction to the entire PubMed. In *Proceedings of the 2010 Workshop on Biomedical Natural Language Processing*, pages 28–36. Association for Computational Linguistics.

O Bodenreider. 2004. The unified medical language system (umls): integrating biomedical terminology. *Nucleic Acids Research*, 32:D267–D270.

M.F.M. Chowdhury and A. Lavelli. 2010. Disease Mention Recognition with Specific Features. *ACL 2010*, page 83.

Rong-En Fan, Kai-Wei Chang, Cho-Jui Hsieh, Xiang-Rui Wang, and Chih-Jen Lin. 2008. LIBLINEAR: A library for large linear classification. *Journal of Machine Learning Research*, 9:1871–1874.

M. Gerner, G. Nenadic, and C.M. Bergman. 2010. LINNAEUS: A species name identification system for biomedical literature. *BMC bioinformatics*, 11(1):85.

K.M. Hettne, R.H. Stierum, M.J. Schuemie, P.J.M. Hendriksen, B.J.A. Schijvenaars, E.M. Mulligen, J. Kleinjans, and J.A. Kors. 2009. A dictionary to identify small molecules and drugs in free text. *Bioinformatics*, 25(22):2983.

Jin-Dong Kim, Tomoko Ohta, Yoshimasa Tsuruoka, Yuka Tateisi, and Nigel Collier. 2004. Introduction to the bio-entity recognition task at JNLPBA. In *Proceedings of the International Joint Workshop on Natural Language Processing in Biomedicine and its Applications (JNLPBA)*, pages 70–75.

Jin-Dong Kim, Tomoko Ohta, Sampo Pyysalo, Yoshinobu Kano, and Jun'ichi Tsujii. 2009. Overview of BioNLP'09 Shared Task on Event Extraction. In *Proceedings of Natural Language Processing in Biomedicine (BioNLP) NAACL 2009 Workshop*, pages 1–9.

Jin-Dong Kim, Yue Wang, Toshihasi Takagi, and Akinori Yonezawa. 2011. Overview of genia event task in bionlp shared task 2011. In *Proceedings of the BioNLP 2011 Workshop Companion Volume for Shared Task*, Portland, Oregon, June. Association for Computational Linguistics.

R. Leaman and G. Gonzalez. 2008. BANNER: an executable survey of advances in biomedical named entity recognition. In *Pacific Symposium on Biocomputing*, volume 13, pages 652–663. Citeseer.

D. Maglott, J. Ostell, K.D. Pruitt, and T. Tatusova. 2005. Entrez Gene: gene-centered information at NCBI. *Nucleic Acids Research*, 33(suppl 1):D54.

Tomoko Ohta, Sampo Pyysalo, and Jun'ichi Tsujii. 2011. Overview of the Epigenetics and Post-translational Modifications (EPI) task of BioNLP Shared Task 2011. In *Proceedings of the BioNLP 2011 Workshop Companion Volume for Shared Task*, Portland, Oregon, June. Association for Computational Linguistics.

Naoaki Okazaki and Jun'ichi Tsujii. 2010. Simple and efficient algorithm for approximate dictionary matching. In *Proceedings of the 23rd International Conference on Computational Linguistics (Coling 2010)*, pages 851–859, Beijing, China, August.

M.F. Porter. 1993. An algorithm for suffix stripping. *Program: electronic library and information systems*, 14(3):130–137.

Sampo Pyysalo, Tomoko Ohta, Rafal Rak, Dan Sullivan, Chunhong Mao, Chunxia Wang, Bruno Sobral, Jun'ichi Tsujii, and Sophia Ananiadou. 2011. Overview of the Infectious Diseases (ID) task of BioNLP Shared Task 2011. In *Proceedings of the BioNLP 2011 Workshop Companion Volume for Shared Task*, Portland, Oregon, June. Association for Computational Linguistics.

L. Ratinov and D. Roth. 2009. Design challenges and misconceptions in named entity recognition. In *Proceedings of the Thirteenth Conference on Computational Natural Language Learning*, pages 147–155. Association for Computational Linguistics.

D. Rebholz-Schuhmann, A.J.J. Yepes, E.M. Van Mulligen, N. Kang, J. Kors, D. Milward, P. Corbett, E. Buyko, E. Beisswanger, and U. Hahn. 2010. CALBC silver standard corpus. *Journal of bioinformatics and computational biology*, 8(1):163–179.

Y. Sasaki, Y. Tsuruoka, J. McNaught, and S. Ananiadou. 2008. How to make the most of NE dictionaries in statistical NER. *BMC bioinformatics*, 9(Suppl 11):S5.

L. Shi and F. Campagne. 2005. Building a protein name dictionary from full text: a machine learning term extraction approach. *BMC bioinformatics*, 6(1):88.

L. Tanabe, N. Xie, L. Thom, W. Matten, and W.J. Wilbur. 2005. GENETAG: a tagged corpus for gene/protein named entity recognition. *BMC bioinformatics*, 6(Suppl 1):S3.

P. Thompson, S.A. Iqbal, J. McNaught, and S. Ananiadou. 2009. Construction of an annotated corpus to support biomedical information extraction. *BMC bioinformatics*, 10(1):349.

M. Torii, Z. Hu, C.H. Wu, and H. Liu. 2009. BioTagger-GM: a gene/protein name recognition system. *Journal of the American Medical Informatics Association*, 16(2):247.

Y. Tsuruoka and J. Tsujii. 2003. Boosting precision and recall of dictionary-based protein name recognition. In *Proceedings of the ACL 2003 workshop on Natural language processing in biomedicine-Volume 13*, pages 41–48. Association for Computational Linguistics.

Yue Wang, Jin-Dong Kim, Rune Saetre, Sampo Pyysalo, and Jun'ichi Tsujii. 2009. Investigating heterogeneous protein annotations toward cross-corpora utilization. *BMC Bioinformatics*, 10(1):403.

C.H. Wu, L.S.L. Yeh, H. Huang, L. Arminski, J. Castro-Alvear, Y. Chen, Z. Hu, P. Kourtesis, R.S. Ledley, B.E. Suzek, et al. 2003. The protein information resource. *Nucleic Acids Research*, 31(1):345.

Building Timelines from Narrative Clinical Records: Initial Results Based-on Deep Natural Language Understanding

Hyuckchul Jung, James Allen, Nate Blaylock, Will de Beaumont, Lucian Galescu, Mary Swift

Florida Institute for Human and Machine Cognition
40 South Alcaniz Street, Pensacola, Florida, USA
{hjung,blaylock,jallen,wbeaumont,lgalescu,mswift}@ihmc.us

Abstract

We present an end-to-end system that processes narrative clinical records, constructs timelines for the medical histories of patients, and visualizes the results. This work is motivated by real clinical records and our general approach is based on deep semantic natural language understanding.

1 Introduction

It is critical for physicians and other healthcare providers to have complete and accurate knowledge of the medical history of patients that includes disease/symptom progression over time and related tests/treatments in chronological order. While various types of clinical records (e.g., discharge summaries, consultation notes, etc.) contain comprehensive medical history information, it can be often challenging and time-consuming to comprehend the medical history of patients when the information is stored in multiple documents in different formats and the relations among various pieces of information is not explicit.

For decades, researchers have investigated temporal information extraction and reasoning in the medical domain (Zhou and Hripcsak, 2007). However, information extraction in the medical domain typically relies on shallow NLP techniques (e.g., pattern matching, chunking, templates, etc.), and most temporal reasoning techniques are based on structured data with temporal tags (Augusto, 2005; Stacey and McGregor, 2007).

In this paper, we present our work on developing an end-to-end system that (i) extracts interesting medical concepts (e.g., medical conditions/ tests/treatments), related events and temporal expressions from raw clinical text records, (ii) constructs timelines of the extracted information; and (iii) visualizes the timelines, all using deep semantic natural language understanding (NLU).

Our deep NLU system extracts rich semantic information from narrative text records and builds logical forms that contain ontology types as well as linguistic features. Ontology- and pattern-based extraction rules are used on the logical forms to retrieve time points/intervals, medical concepts/ events and their temporal/causal relations that are pieced together by our system's temporal reasoning component to create comprehensive timelines.

Our system is an extension to a well-proven general-purpose NLP system (Allen et al., 2000) rather than a system specialized to the clinical domain, and the temporal reasoning in our system is tightly integrated into the NLP system's deep semantic analysis. We believe this approach will allow us to process a broader variety of documents and complex forms of temporal expressions.

In the coming sections, we first present a motivating example, a real clinical record of a cancer patient. Next, we give an overview of our NLU system including how medical ontology is integrated into our system. The overview section is followed by detailed description of our information extraction and temporal reasoning approach. Then, we discuss our results and conclude.

2 Motivating Example

Our work is carried out as a collaboration with the Moffitt Cancer Center (part of the NCI Comprehensive Cancer Centers), who have provided us with access to clinical records for over 1500 patients. Figure 1 shows a (de-identified) "History of Present Illness" (HPI) section of a Thoracic Consultation Note from this data set.

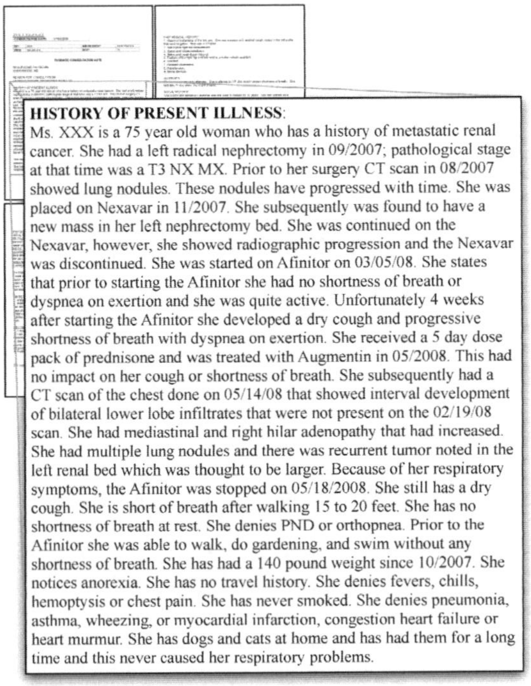

HISTORY OF PRESENT ILLNESS:
Ms. XXX is a 75 year old woman who has a history of metastatic renal cancer. She had a left radical nephrectomy in 09/2007; pathological stage at that time was a T3 NX MX. Prior to her surgery CT scan in 08/2007 showed lung nodules. These nodules have progressed with time. She was placed on Nexavar in 11/2007. She subsequently was found to have a new mass in her left nephrectomy bed. She was continued on the Nexavar, however, she showed radiographic progression and the Nexavar was discontinued. She was started on Afinitor on 03/05/08. She states that prior to starting the Afinitor she had no shortness of breath or dyspnea on exertion and she was quite active. Unfortunately 4 weeks after starting the Afinitor she developed a dry cough and progressive shortness of breath with dyspnea on exertion. She received a 5 day dose pack of prednisone and was treated with Augmentin in 05/2008. This had no impact on her cough or shortness of breath. She subsequently had a CT scan of the chest done on 05/14/08 that showed interval development of bilateral lower lobe infiltrates that were not present on the 02/19/08 scan. She had mediastinal and right hilar adenopathy that had increased. She had multiple lung nodules and there was recurrent tumor noted in the left renal bed which was thought to be larger. Because of her respiratory symptoms, the Afinitor was stopped on 05/18/2008. She still has a dry cough. She is short of breath after walking 15 to 20 feet. She has no shortness of breath at rest. She denies PND or orthopnea. Prior to the Afinitor she was able to walk, do gardening, and swim without any shortness of breath. She has had a 140 pound weight since 10/2007. She notices anorexia. She has no travel history. She denies fevers, chills, hemoptysis or chest pain. She has never smoked. She denies pneumonia, asthma, wheezing, or myocardial infarction, congestion heart failure or heart murmur. She has dogs and cats at home and has had them for a long time and this never caused her respiratory problems.

Figure 1: A sample medical record -- Thoracic Consultation Note[1]

The text of this section provides a very detailed description of what problems/tests/treatments an anonymous cancer patient went through over a period. Such narrative text is common in clinical notes and, because such notes are carefully created by physicians, they tend to have only relevant information about patient medical history.

Nonetheless, there are lots of challenges in constructing complete and accurate medical history because of complex temporal expressions/ relations, medical language specific grammar/ jargons, implicit information and domain-specific medical knowledge (Zhou and Hripcsak, 2007).

In this paper, as an initial step towards constructing complete timelines from narrative text, we focus on sentences with explicit temporal expressions listed below (tagged as *Line 1 ~ 11*) plus a sentence in the present tense (*Line 12*):

- *Line 1*: She had a left radical nephrectomy **in 09/ 2007**; pathological stage **at that time** was a T3 NX MX.
- *Line 2*: **Prior to** her surgery CT scan **in 08/2007** showed lung nodules.
- *Line 3*: She was placed on Nexavar **in 11/2007**.
- *Line 4*: She was started on Afinitor **on 03/05/08**.
- *Line 5*: She states that **prior to** starting the Afinitor she had no shortness of breath or dyspnea on exertion and she was quite active.
- *Line 6*: Unfortunately **4 weeks after** starting the Afinitor she developed a dry cough and progressive shortness of breath with dyspnea on exertion.
- *Line 7*: She received a 5 day dose pack of prednisone and was treated with Augmentin **in 05/2008**.
- *Line 8*: She subsequently had a CT scan of the chest done **on 05/14/08** that showed interval development of bilateral lower lobe infiltrates that were not present on the **02/19/08** scan.
- *Line 9*: Because of her respiratory symptoms, the Afinitor was stopped **on 05/18/2008**.
- *Line 10*: **Prior to** the Afinitor she was able to walk, do gardening, and swim without any shortness of breath.
- *Line 11*: She has had a 140 pound weight **since 10/2007**.
- *Line 12*: She denies fevers, chills, hemoptysis or chest pain.

In these 12 sentences, there are instances of 10 treatments (e.g., procedures such as "nephrectomy" and drugs such as "Nexavar"), 3 tests (e.g., CT-scan), 13 problems/symptoms (e.g., lung nodules) and 2 other types of clinical findings (e.g., the cancer stage level "T3 NX MX"). There are also 23 events of various types represented with verbs such as "had", "was", "showed", and "was started".

While there are simple expressions such as "on 03/05/08" in *Line 3*, there are also temporal expressions in more complex forms with time relations (e.g., "prior to"), time references (e.g., "at that time") or event references (e.g., "4 weeks after starting Afinitor"). Throughout this paper, we will use *Line 1 ~ 12* as a concrete example based on which we develop general techniques to construct timelines.

[1] For privacy, identities of patients/physicians were concealed and the dates/time-spans in the original sources were altered while maintaining their chronological order. Some measurements and geographic names were also modified.

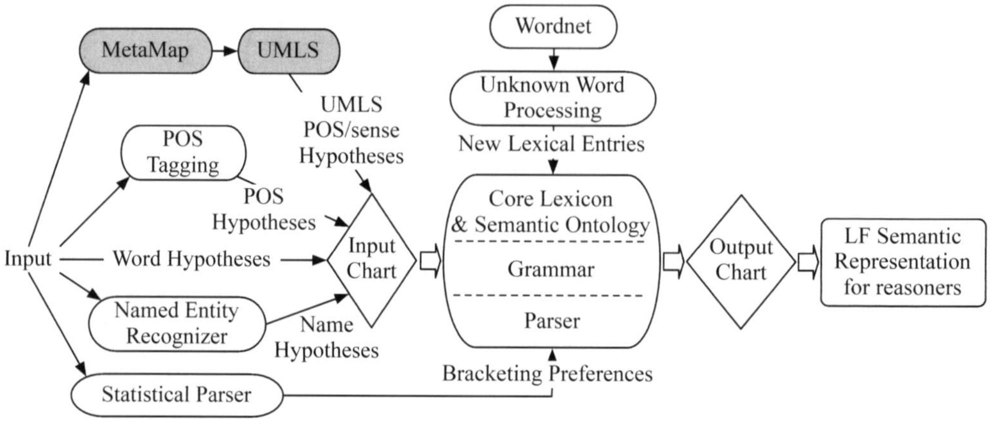

Figure 2: Front-end language processing components with MetaMap and UMLS

3 Natural Language Understanding (NLU) System

Our system is an extension to an existing NLU system that is the result of a decade-long research effort in developing generic natural language technology. The system uses a "deep" understanding approach, attempting to find a linked, overall meaning for all the words in a paragraph. An architectural view of the system is shown in Figure 2.

3.1 Core NLU Components

At the core of the system is a packed-forest chart parser which builds constituents bottom-up using a best-first search strategy. The core grammar is a hand-built, lexicalized context-free grammar, augmented with feature structures and feature unification. The parser draws on a general purpose semantic lexicon and ontology which define a range of word senses and lexical semantic relations. The core semantic lexicon was constructed by hand and contains more than 7000 lemmas. It can be also dynamically augmented for unknown words by consulting WordNet (Miller, 1995).

To support more robust processing as well as domain configurability, the core system is informed by a variety of statistical and symbolic preprocessors. These include several off-the-shelf statistical NLP tools such as the Stanford POS tagger (Toutanova and Manning, 2000), the Stanford named-entity recognizer (NER) (Finkel et al., 2005) and the Stanford Parser (Klein and Manning, 2003). The output of these and other specialized preprocessors (such as a street address recognizer) are sent to the parser as advice. The parser then can include or not include this advice (e.g., that a certain phrase is a named entity) as it searches for the optimal parse of the sentence.

The result of parsing is a frame-like semantic representation that we call the Logical Form (LF). The LF representation includes semantic types, semantic roles for predicate arguments, and dependency relations. Figure 3 shows an LF example for the sentence *"She had a left radical nephrectomy in 09/2007"*. In the representation, elements that start with colons (e.g., :THEME) are semantic roles of ontological concepts, and role values can be a variable to refer to another LF term.

3.2 UMLS Integration

By far the most critical aspect of porting our generic NLU components to the task of understanding clinical text is the need for domain-specific lexical and ontologic information. One widely used comprehensive resource that can provide both is the National Library of Medicine's Unified Medical Language System (UMLS) (Bodenreider, 2004). UMLS was integrated into or system via MetaMap (Aronson and Lang, 2010), a tool also developed by NLM, that can identify and rank UMLS concepts in text.

Specifically, we added MetaMap as a special kind of named entity recognizer feeding advice into the Parser's input chart (see Figure 2). We run MetaMap twice on the input text to obtain UMLS information both for the maximal constituents, and for individual words in those constituents (e.g., "lung cancer", as well as "lung" and "cancer").

The lexicon constructs representations for the new words and phrases on the fly. Our general approach for dealing with how the corresponding concepts fit in our system ontology uses an ontol-

(F *V1* (:* ONT::HAVE W::HAVE) :AFFECTED *V2* :THEME *V3* :MOD *V4* :TENSE W::PAST)
(PRO *V2* (:* ONT::PERSON W::SHE) :PROFORM ONT::SHE :CO-REFERENCE *V0*)
(A *V3* (:* ONT::TREATMENT W::LEFT-RADICAL-NEPHRECTOMY) :DOMAIN-INFO (UMLS .....)
(F *V4* (:* ONT::TIME-SPAN-REL W::IN) :OF *V1* :VAL *V5*)
(THE *V5* ONT::TIME-LOC :YEAR 2007 :MONTH 9)

Figure 3: LF semantic representation for "*She had a left radical nephrectomy in 09/2007*"

(*?x1 ?y2* (*? type1* ONT::HAVE) :AFFECTED *?y2* :THEME *?y3* :MOD *?y4*) (*?x2 ?y2* (*? type2* ONT::PERSON))) (*?x3 ?y3* (*? type3* ONT::TREATMENT ONT::MEDICAL-DIAGNOSTIC) :DOMAIN-INFO ***?!info***)	*List of LF patterns*
-extract-person-has-treatment-or-medical-diagnostic>	*Unique rule ID*
(EVENT :type ?type1 :class occurrence :subject ?y2 :object ?y3)	*Output Specification*

Figure 4: An example extraction rule

ogy specialization mechanism which we call ontology grafting, whereby new branches are created from third party ontological sources, and attached to appropriate leaf nodes in our ontology.

The UMLS Semantic Network and certain vocabularies included in the UMLS Metathesaurus define concept hierarchies along multiple axes. First, we established links between the 15 UMLS semantic groups and corresponding concepts in our ontology. Second, we selected a list of nodes from the SNOMED-CT and NCI hierarchies (27 and 11 nodes, respectively) and formed ontological branches rooted in these nodes that we grafted onto our ontology.

Based on these processes, UMLS information gets integrated into our LF representation. In Figure 3, the 3rd term has a role called :domain-info and, in fact, its value is *(UMLS :CUI C2222800 :CONCEPT "left nephrectomy" :PREFERRED "nephrectomy of left kidney (treatment)" :SEMANTIC-TYPES (TOPP) :SEMANTIC-GROUPS (PROC) :SOURCES (MEDCIN MTH))* that provides detailed UMLS concept information. Here, the semantic type "TOPP" is a UMLS abbreviation for "Therapeutic or Preventive Procedure". More details about complex issues surrounding UMLS integration into our system can be found in (Swift et al., 2010).

4 Information Extraction (IE) from Clinical Text Records

In this section, we describe how to extract basic elements that will be used as a foundation to construct timelines. We first describe our general approach to extracting information from LF graphs. Then we give details specific to the various types of information we extract in our system: various clinical concepts, temporal concepts (points as well as intervals), events and temporal relations.

4.1 LF Pattern-based Extraction

Given LF outputs from the NLU system described in Section 3, we use LF pattern-based rules for information extraction. The basic structure of an extraction rule is a list of LF patterns followed by a unique rule ID and the output specification.

Each LF-pattern specifies a pattern against an LF. Variables can appear anywhere except as role names in different formats:

- ?x - (unconstrained) match anything
- ?!x - match any non-null value
- (? x V1 V2 ...) - (constrained) match one of the specified values V1, V2, ...

As an example, the extraction rule in Figure 4 will match LFs that mean a person had a treatment or a medical-diagnostic with explicit UMLS information (i.e., part of LFs in Figure 3 matches). The output specification records critical information from the extraction to be used by other reasoners.

The extraction rules have all been developed by hand. Nevertheless, they are quite general, since a) LF patterns abstract away from lexical and syntactic variability in the broad class of expressions of interest (however, lexical and syntactic features may be used if needed); and b) LF patterns make heavy use of ontological categories, which provides abstraction at the semantic level.

4.2 Clinical Concept Extraction

Among various types of concepts included in clinical records, we focus on concepts related to problems/tests/treatments to build a medical his-

```
((?x1 ?y1 (? type1 ONT::SUBSTANCE) :domain-info
?info :quantifier ?quan)
-extract-substance>
(extraction :type substance :concept ?type1 :umlsinfo
?info :ont-term ?y1 :quantifier ?quan))
```

Figure 5: A rule to extract substances

```
(?x1 ?y1 (:* ont::event-time-rel w::until) :val ?val)
(?x2 ?val (? type2 ont::time-loc) :mod ?mod)
(?x3 ?mod (? type3 ont::event-time-rel) :displacement
?displacement)
(?x4 ?displacement (? type4 ont::quantity) :unit ?unit
:amount ?amount)
(?x5 ?amount ont::number :value ?num)
```

Figure 6: LF patterns to extract a time-span

tory and extract them using extraction rules as described above. Figure 5 shows a rule to extract substances by matching any LF with a substance concept (as mentioned already, subclasses such as pharmacologic substances, would also match).

The rule in Figure 5 checks the :quantifier role and its value (e.g., none) is used to infer the presence or the absence of concepts. Using similar rules, we extract additional concepts such as medical-disorders-and-conditions, physical-symptom, treatment, medical-diagnostic, medical-action and clinical-finding. Here, medical-action and clinical-finding are to extract concepts in a broader sense.[2] To cover additional concepts, we can straightforwardly update extraction rules.

4.3 Temporal Expression Extraction

Temporal expressions are also extracted in the same way but using different LF patterns. We have 14 rules to extract dates and time-spans of varying levels of complexity; for the example in Figure 1 six of these rules were applied. Figure 6 shows LF patterns for a rule to extract temporal expressions of the form "until X days/months/years ago"; for example, here is what the rule extracts for "until 3 days ago":

```
(extraction :type time-span :context-rel (:*
ont::event-time-rel w::until) :reference (time-position
:context-rel (:* ont::event-time-rel w::ago) :amount 3
:unit (:* ont::time-unit ont::day)))
```

From this type of output, other reasoners can easily access necessary information about given temporal expressions without investigating the whole LF representation on their own.

4.4 Event Extraction

To construct timelines, the concepts of interest (Section 4.2) and the temporal expressions (Section 4.3) should be pieced together. For that purpose, it is critical to extract events because they not only describe situations that happen or occur but also represent states or circumstances where something holds. Furthermore, event features provide useful cues to reason about situations surrounding extracted clinical concepts.

Here, we do not formally define events, but refer to (Sauri et al., 2006) for detailed discussion about events. While events can be expressed by multiple means (e.g., verbs, nominalizations, and adjectives), our extraction rules for events focus on verbs and their features such as class, tense, aspect, and polarity. Figure 7 shows a rule to extract an event with the verb "start" like the one in *Line 4*, "She was *started* on Afinitor on 03/05/08". The output specification from this rule for *Line 4* will have the :class, :tense, and :passive roles as *(aspectual initiation)*, *past*, and *true* respectively.

These event features play a critical role in constructing timelines (Section 5). For instance, the event class *(aspectual initiation)* from applying the rule in Figure 7 to *Line 4* implies that the concept "Afinitor" (a pharmacologic-substance) is not just something tried on the given date, 03/05/08, but something that continued from that date.

4.5 Relation Information Extraction

The relations among extracted concepts (namely, conjoined relations between events and set relations between clinical concepts) also play a key role in our approach. When events or clinical concepts are closely linked with such relations, heuristically, they tend to share similar properties that are exploited in constructing timelines as described in Section 5.

5 Building Timelines from Extracted Results

Extracted clinical concepts, temporal expressions, events, and relations (Section 4) are used as a

---

[2] While concept classification into certain categories is a very important task in the medical domain, sophisticated concept categorization like the one specified in the 2010 i2b2/VA Challenge (https://www.i2b2.org/NLP/Relations/) is not the primary goal of this paper. We rather focus on how to associate extracted concepts with other events and temporal expressions to build timelines.

```
(?x1 ?ev (? type1 ont::start) :affected ?affected :tense ?tense :passive ?passive :progressive ?progresive
  :perfective ?perfective :negation ?negation)
-extract-start-event>
(EVENT :type ?type1 :class (aspectual initiation) :subject ?affected :object null :tense ?tense :passive
  ?passive :progressive ?progresive :perfective ?perfective :negation ?negation :ont-term ?ev)
```

Figure 7: An event extraction rule example

foundation to construct timelines that represent patients' medical history. In this section, we present timeline construction processes (as shown in Figure 8), using example sentences from Section 2.

Inputs: Clinical concepts, Temporal Expressions, Events, Relations, LFs
Outputs: Clinical concepts with associated dates or timespans.

Steps:
1. Build links between events and clinical concepts
2. Find associated temporal expressions for events
3. Propagate temporal expressions through relations between events when applicable
4. Compute concrete time values for temporal expressions, taking into account the context of clinical records
5. Compute time values for clinical concepts based on their associated events

Figure 8: Pseudocode for Timeline Construction

Step 1: We first make connections between events and clinical concepts. In the current system, events and clinical concepts are extracted in separate rules and their relations are not always explicit in the output specification of the rules applied. For instance, Figure 9 shows LFs for the sentence in *Line 7* in a graph format, using simplified LF terms for illustration. The clinical concept "prednisone" and the event "received" get extracted by different rules and the relation between them is not explicit in their output specifications.

To address such a case, for a pair of an event and a clinical concept, we traverse LF graphs and decide that a relation between them exists if there is a path that goes through certain pre-defined concepts that do not separate them semantically and syntactically (e.g., concepts of measure-units, evidence/history, development, and some propositions).

Step 2: Second, we find temporal expressions associated with events. This step is relatively straightforward. While temporal expressions and events get extracted separately, by investigating their LFs, we can decide if a given temporal expression is a modifier of an event. In Figure 9, the time-span-relation (i.e., "in") in the dotted-line box is a direct modifier of the event "was treated".

Step 3: Next, we propagate the association between events and temporal expressions. That is, when the relation between an event and a temporal expression is found, we check if the temporal expression can be associated with additional events related to the event (esp. when the related events do not have any associated temporal expression).

In Figure 9, the event "received" does not have a temporal expression as a modifier. However, it is conjoined with the event "was treated" in the same past tense under the same speech act. Thus, we let the event "received" share the same temporal expression with its conjoined event. Here, the conjoined relation was extracted with relation rules described in Section 4.5, which allows us to focus on only related events.

Step 4: When temporal expressions do not have concrete time values within the expressions, we need to designate times for them by looking into information in their LFs:

- *Event references*: The system needs to find the referred event and gets its time value. For instance, in "4 weeks after starting Afinitor" (*Line 6*), "starting Afinitor" refers to a previous event in *Line 4*. The system investigates all events with a verb with the same- or sub-type of ont::start and Afinitor as its object (active verbs) or its subject (passive verbs). After resolving event references, additional time reference or relation computation may be required (e.g., computation for "4 weeks after").
- *Time references*: Concrete times for expressions like the above example "N weeks after <reference-time>" can be easily computed by checking the time displacement information in LFs with the reference time. However, expressions such as "N days ago" are based on the context of clinical records (e.g., record creation

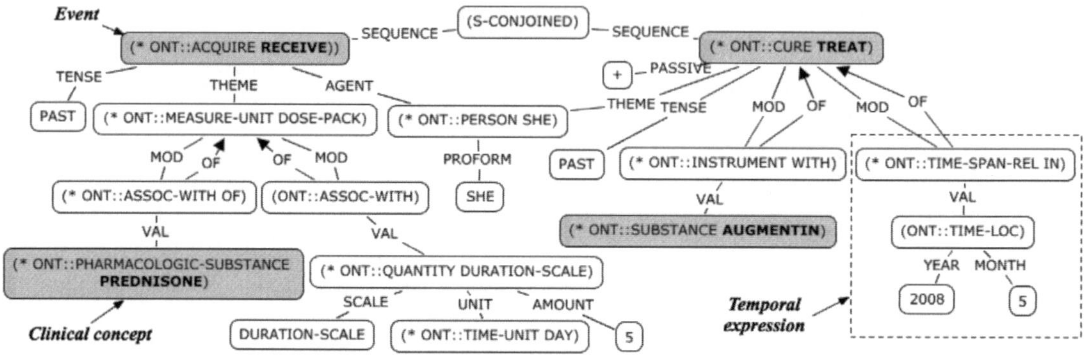

Figure 9: Graph format LFs of the sentence in *Line 7* -- "She received a 5 day dose pack of prednisone and was treated with Augmentin in 05/2008."

time). Document creation time is usually represented as metadata attached to the document itself, or it could be retrieved from a database where clinical records are stored. In addition, previously mentioned dates or time-spans can be referred to using pronouns (e.g., "at that/this time"). For such expressions, we heuristically decide that it refers to the most recent temporal expression.

- *Time relation*: Some temporal expressions have directional time relations (e.g., "until", "prior to", and "after") specifying intervals with open ends. When the ending time of a time span is not specified (e.g., "since 10/2007" in *Line 10*). We heuristically set it from the context of the clinical record such as the document creation time.

Step 5: Finally, we designate or compute times on or during which the presence or the absence of each clinical concept is asserted. Since temporal expressions are associated with events, to find time values for clinical concepts, we first check the relations between events and clinical concepts. When an event with a concrete time is found for a clinical concept, the event's class is examined. For classes such as state and occurrence, the concrete time value of the event is used. In contrast, for an aspectual event, we check its feature (e.g., initiation or termination) and look for other aspectual events related to the clinical concept and compute a time span. For instance, regarding "Afinitor", *Line 4* and *Line 9* have events with classes (aspectual initiation) and (aspectual termination) respectively, which leads to a time span between the two dates in *Line 4* and *Line 9*. Currently, we do not resolve conflicting hypotheses.

Assertion of Presence or Absence of Clinical Concepts: To check if a certain concept is present or not, we take into account quantifier information (e.g., none), the negation role values of events, and the verb types of events (e.g., "deny" indicates the absence assertion). In addition to such information readily available in the output specifications of the clinical concept- and event-extraction rules, we also check the path (as in Step 1) that relates the clinical concepts and the events, and the quantifiers of the concepts in the path are used to compute negation values. For instance, given "The scan shows no evidence of lung nodules", the quantifier of the concept "evidence" indicates the absence of the clinical finding "lung nodules".

6 Timeline Results and Discussion

For the example in Section 2 (*Line 1 ~ 12*), we extract all the instances of the clinical concepts and the temporal expressions. Out of 23 events, 17 were extracted. While we missed events such as *state/was* (*Line 5*), *done* (*Line 8*), and *walk/do/swim* (*Line 10*), our event extraction rules can be extended to cover them if need be.

Figure 10 visualizes the extraction results of the example. We use a web widget tool called Simile Timeline (www.simile-widgets.org/timeline/). Some property values (that were also extracted by rules) are shown alongside some concepts (e.g., weight measurement). Note that not all extracted clinical findings are displayed in Figure 10 because we visualize clinical concepts only when they are associated with temporal expressions in our LFs. For instance, the CT-scan on 05/14/08 in *Line 8* is not shown because the date was not associated with it due to fragmented LFs from the Parser.

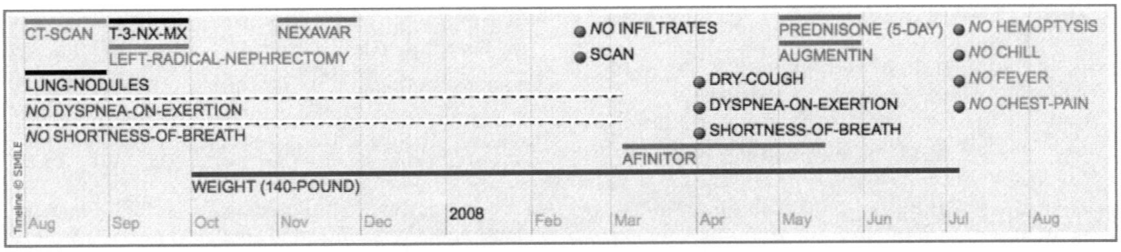

Figure 10: Visualization of timeline results

However, we were still able to extract "no infiltrates" and "scan" from a meaningful fragment.

In addition to the fragmented LF issue, we plan to work on temporal reasoning for concepts in the sentences without explicit temporal expressions, and the current limited event reference resolution will be improved. We are also working on evaluation with 48 clinical records from 10 patients. Annotated results will be created as a gold-standard and precision/recall will be measured.

7 Related Work

Temporal information is of crucial importance in clinical applications, which is why it has attracted a lot interest over the last two decades or more (Augusto, 2005). Since so much clinical information is still residing in unstructured form, in particular as text in the patient's health record, the last decade has seen a number of serious efforts in medical NLP in general (Meystre et al., 2008) and in extracting temporal information from clinical text in particular.

Some of this surge in interest has been spurred by dedicated competitions on extraction of concepts and events from clinical text (such as the i2b2 NLP challenges). At the same time, the evolution of temporal markup languages such as TimeML (Sauri et al., 2006), and temporal extraction/inference competitions (such as the two TempEval challenges, Verhagen et al., 2009) in the general area of NLP have led to the development of tools such as TARSQI (Verhagen et al., 2005) that could be adapted to the clinical domain.

Although the prevailing paradigm in this area is to use superficial methods for extracting and classifying temporal expressions, it has long been recognized that higher level semantic processing, including discourse-level analysis, would have to be performed to get past the limits of the current approaches (cf. Zhou and Hripcsak, 2007).

Recent attempts to use deeper linguistic features include the work of Bethard et al. (2007), who used syntactic structure in addition to lexical and some minor semantic features to classify temporal relations of the type we discussed in Section 4.3. Savova and her team have also expressed interest in testing off-the-shelf deep parsers and semantic role labelers for aiding in temporal relation identification and classification (Savova et al., 2009); although we are not aware of any temporal extraction results yet, we appreciate their effort in expanding the TimeML annotation schema for the clinical domain, as well as their efforts in developing corpora of clinical text annotated with temporal information.

The work of Mulkar-Mehta et al. (2009) also deserves a mention, even though they apply their techniques to biomedical text rather than clinical text. They obtain a shallow logical form that represents predicate-argument relations implicit in the syntax by post-processing the results of a statistical parser. Temporal relations are obtained from the shallow LF based on a set of hand-built rules by an abductive inference engine.

To our knowledge, however, our system is the first general-purpose NLU system that produces a full, deep syntactic and semantic analysis of the text as a prerequisite to the extraction and analysis of relevant clinical and temporal information.

8 Conclusion

In this paper, we presented a prototype deep natural language understanding system to construct timelines for the medical histories of patients. Our approach is generic and extensible to cover a variety of narrative clinical text records. The results from our system are promising and they can be used to support medical decision making.

9 Acknowledgement

This work was supported by the National Cancer Institute and the H. Lee Moffitt Cancer Center and Research Institute (Award # RC2CA1488332).

References

James Allen, Donna Byron, Myroslava Dzikovska, George Ferguson, Lucian Galescu, and Amanda Stent. 2000. An architecture for a generic dialogue shell. Journal of Natural Language Engineering 6(3):1–16.

Mary Swift, Nate Blaylock, James Allen, Will de Beaumont, Lucian Galescu, and Hyuckchul Jung. 2010. Augmenting a Deep Natural Language Processing System with UMLS. Proceedings of the Fourth International Symposium on Semantic Mining in Biomedicine (poster abstract)

Alan R. Aronson and François-Michel Lang. 2010. An overview of MetaMap: historical perspective and recent advances. Journal of the American Medical Informatics Association. 17:229-236.

Juan C. Augusto. 2005. Temporal reasoning for decision support in medicine. Artificial Intelligence in Medicine, 33(1): 1-24.

Steven Bethard, James H. Martin, and Sara Klingenstein. 2007. Timelines from Text: Identification of Syntactic Temporal Relations. In Proceedings of the International Conference on Semantic Computing (ICSC '07), 11-18.

Olivier Bodenreider. 2004. The Unified Medical Language System (UMLS): integrating biomedical terminology. Nucleic Acids Research, Vol. 32.

Jenny Rose Finkel, Trond Grenager, and Christopher Manning. 2005. Incorporating Non-local Information into Information Extraction Systems by Gibbs Sampling. Proceedings of the Annual Meeting of the Association for Computational Linguistics.

Dan Klein and Christopher D. Manning. 2003. Fast Exact Inference with a Factored Model for Natural Language Parsing. In Advances in Neural Information Processing Systems 15 (NIPS 2002), Cambridge, MA: MIT Press.

S. M. Meystre, G. K. Savova, K. C. Kipper-Schuler, J. F. Hurdle. 2008. Extracting information from textual documents in the electronic health record: a review of recent research. IMIA Yearbook of Medical Informatics.

George A. Miller. 1995. WordNet: A lexical database for English. Communications of the ACM, 38(5).

R. Mulkar-Mehta, J.R. Hobbs, C.-C. Liu, and X.J. Zhou. 2009. Discovering causal and temporal relations in biomedical texts. In AAAI Spring Symposium, 74-80.

Roser Sauri, Jessica Littman, Bob Knippen, Robert Gaizauskas, Andrea Setzer, and James Pustejovsky. 2006. TimeML annotation guidelines. (available at http://www.timeml.org/site/publications/timeMLdocs/annguide_1.2.1.pdf)

G. Savova, S. Bethard, W. Styler, J. Martin, M. Palmer, J. Masanz, and W. Ward. 2009. Towards temporal relation discovery from the clinical narrative. Proceedings of the Annual AMIA Symposium, 568-572.

Michael Stacey and Carolyn McGregor. 2007. Temporal abstraction in intelligent clinical data analysis: A survey. Artificial Intelligence in Medicine, 39.

Kristina Toutanova and Christopher D. Manning. 2000. Enriching the Knowledge Sources Used in a Maximum Entropy Part-of-Speech Tagger. In Proceedings of the Joint SIGDAT Conference on Empirical Methods in Natural Language Processing and Very Large Corpora (EMNLP/VLC-2000).

M. Verhagen, I. Mani, R. Sauri, R. Knippen, S.B. Jang, J. Littman, A. Rumshisky, J. Phillips, and J. Pustejovsky. 2005. Automating temporal annotation with TARSQI. In Proceedings of the ACL 2005 on Interactive poster and demonstration sessions (ACLdemo '05), 81-84.

M. Verhagen, R. Gaizauskas, F. Schilder, M. Hepple, J. Moszkowicz, and J. Pustejovsky. 2009. The TempEval challenge: identifying temporal relations in text. Language Resources and Evaluation 43(2):161-179.

Li Zhou, Carol Friedman, Simon Parsons and George Hripcsak. 2005. System Architecture for Temporal Information Extraction, Representation and Reasoning in Clinical Narrative Reports. Proceedings of the Annual AMIA Symposium.

Li Zhou and George Hripcsak. 2007. Temporal reasoning with medical data - A review with emphasis on medical natural language processing. Journal of Biomedical Informatics, 40.

Text Mining Techniques for Leveraging Positively Labeled Data

Lana Yeganova*, Donald C. Comeau, Won Kim, W. John Wilbur
National Center for Biotechnology Information, NLM, NIH, Bethesda, MD 20894 USA
{yeganova, comeau, wonkim, wilbur}@mail.nih.gov
* Corresponding author. Tel.:+1 301 402 0776

Abstract

Suppose we have a large collection of documents most of which are unlabeled. Suppose further that we have a small subset of these documents which represent a particular class of documents we are interested in, i.e. these are labeled as positive examples. We may have reason to believe that there are more of these positive class documents in our large unlabeled collection. What data mining techniques could help us find these unlabeled positive examples? Here we examine machine learning strategies designed to solve this problem. We find that a proper choice of machine learning method as well as training strategies can give substantial improvement in retrieving, from the large collection, data enriched with positive examples. We illustrate the principles with a real example consisting of multiword UMLS phrases among a much larger collection of phrases from Medline.

1 Introduction

Given a large collection of documents, a few of which are labeled as interesting, our task is to identify unlabeled documents that are also interesting. Since the labeled data represents the data we are interested in, we will refer to it as the positive class and to the remainder of the data as the negative class. We use the term negative class, however, documents in the negative class are not necessarily negative, they are simply unlabeled and the negative class may contain documents relevant to the topic of interest. Our goal is to retrieve these unknown relevant documents.

A naïve approach to this problem would simply take the positive examples as the positive class and the rest of the collection as the negative class and apply machine learning to learn the difference and rank the negative class based on the resulting scores. It is reasonable to expect that the top of this ranking would be enriched for the positive class. But an appropriate choice of methods can improve over the naïve approach.

One issue of importance would be choosing the most appropriate machine learning method. Our problem can be viewed from two different perspectives: the problem of learning from imbalanced data as well as the problem of recommender systems. In terms of learning from imbalanced data, our positive class is significantly smaller than the negative, which is the remainder of the collection. Therefore we are learning from imbalanced data. Our problem is also a recommender problem in that based on a few examples found of interest to a customer we seek similar positive examples amongst a large collection of unknown status. Our bias is to use some form of wide margin classifier for our problem as such classifiers have given good performance for both the imbalanced data problem and the recommender problem (Zhang and Iyengar 2002; Abkani, Kwek et al. 2004; Lewis, Yang et al. 2004).

Imbalanced data sets arise very frequently in text classification problems. The issue with imbalanced learning is that the large prevalence of negative documents dominates the decision process and harms classification performance. Several approaches have been proposed to deal with the problem including sampling methods and cost-sensitive learning methods and are described in (Chawla, Bowyer et al. 2002; Maloof 2003; Weiss, McCarthy et al. 2007). These studies have shown that there is no clear advantage of one approach versus another. Elkan (2001) points out that cost-sensitive methods and sampling methods are related in the sense that altering the class distribution of training data is equivalent to altering misclassification cost. Based on these studies we examine cost-sensitive learning in which the cost on the positive set is increased, as a useful approach to consider when using an SVM.

In order to show how cost-sensitive learning for an SVM is formulated, we write the standard equations for an SVM following (Zhang 2004).

Given training data $\{(x_i, y_i)\}$ where y_i is 1 or −1 depending on whether the data point x_i is classified as positive or negative, an SVM seeks that vector w_i which minimizes

$$\sum_i h\left(y_i(\vec{x}_i \cdot \vec{w} + \theta)\right) + \frac{\lambda}{2}\|\vec{w}\|^2 \quad (1)$$

where the loss function is defined by

$$h(z) = \begin{cases} 1-z, & z \le 1 \\ 0, & z > 1. \end{cases} \quad (2)$$

The cost-sensitive version modifies (1) to become

$$r_+ \cdot \sum_{i \in C_+} h\left(y_i(\vec{x}_i \cdot \vec{w} + \theta)\right) + r_- \cdot \sum_{i \in C_-} h\left(y_i(\vec{x}_i \cdot \vec{w} + \theta)\right) + \frac{\lambda}{2}\|\vec{w}\|^2 \quad (3)$$

and now we can choose r_+ and r_- to magnify the losses appropriately. Generally we take r_- to be 1, and r_+ to be some factor larger than 1. We refer to this formulation as CS-SVM. Generally, the same algorithms used to minimize (1) can be used to minimize (3).

Recommender systems use historical data on user preferences, purchases and other available data to predict items of interest to a user. Zhang and Iyengar (2002) propose a wide margin classifier with a quadratic loss function as very effective for this purpose (see appendix). It is used in (1) and requires no adjustment in cost between positive and negative examples. It is proposed as a better method than varying costs because it does not require searching for the optimal cost relationship between positive and negative examples. We will use for our wide margin classifier the modified Huber loss function (Zhang 2004). The modified Huber loss function is quadratic where this is important and has the form

$$h(z) = \begin{cases} -4 \cdot z, & z \le -1 \\ (1-z)^2, & -1 < z \le 1 \\ 0, & z > 1. \end{cases} \quad (4)$$

We also use it in (1). We refer to this approach as the Huber method (Zhang 2004) as opposed to SVM. We compare it with SVM and CS-SVM. We used our own implementations for SVM, CS-SVM, and Huber that use gradient descent to optimize the objective function.

The methods we develop are related to semi-supervised learning approaches (Blum and Mitchell 1998; Nigam, McCallum et al. 1999) and active learning (Roy and McCallum 2001; Tong and Koller 2001). Our method differs from active learning in that active learning seeks those unlabeled examples for which labels prove most informative in improving the classifier. Typically these examples are the most uncertain. Some semi-supervised learning approaches start with labeled examples and iteratively seek unlabeled examples closest to already labeled data and impute the known label to the nearby unlabeled examples. Our goal is simply to retrieve plausible members for the positive class with as high a precision as possible. Our method has value even in cases where human review of retrieved examples is necessary. The imbalanced nature of the data and the presence of positives in the negative class make this a challenging problem.

In Section 2 we discuss additional strategies proposed in this work, describe the data used and design of experiments, and provide the evaluation measure used. In Section 3 we present our results, in Sections 4 and 5 we discuss our approach and draw conclusions.

2 Methods

2.1 Cross Training

Let D represent our set of documents, and C_+ those documents that are known positives in D. Generally C_+ would be a small fraction of D and for the purposes of learning we assume that $C_- = D \setminus C_+$.

We are interested in the case when some of the negatively labeled documents actually belong to the positive class. We will apply machine learning to learn the difference between the documents in the class C_+ and documents in the class C_- and use the weights obtained by training to score the documents in the negative class C_-. The highest scoring documents in set C_- are candidate mislabeled documents. However, there may be a problem with this approach, because the classifier is based on partially mislabeled data. Candidate mislabeled documents are part of the C_- class. In the process of training, the algorithm purposely learns to score them low. This effect can be magnified by any overtraining that takes place. It will also be promoted by a large number of features, which makes it more likely that any positive point in the negative class is in some aspect different from any member of C_+.

Another way to set up the learning is by excluding documents from directly participating in the training used to score them. We first divide the negative set into disjoint pieces

$$C_- = Z_1 \cup Z_2$$

Then train documents in C_+ versus documents in Z_1 to rank documents in Z_2 and train documents in C_+ versus documents in Z_2 to rank documents in Z_1. We refer to this method as cross training (CT). We will apply this approach and show that it confers benefit in ranking the false negatives in C_-.

2.2 Data Sources and Preparation

The databases we studied are *MeSH25*, *Reuters*, *20NewsGroups*, and *MedPhrase*.

MeSH25. We selected 25 MeSH® terms with occurrences covering a wide frequency range: from 1,000 to 100,000 articles. A detailed explanation of MeSH can be found at http://www.nlm.nih.gov/mesh/.

For a given MeSH term *m*, we treat the records assigned that MeSH term *m* as positive. The remaining MEDLINE® records do not have *m* assigned as a MeSH and are treated as negative. Any given MeSH term generally appears in a small minority of the approximately 20 million MEDLINE documents making the data highly imbalanced for all MeSH terms.

Reuters. The data set consists of 21,578 Reuters newswire articles in 135 overlapping topic categories. We experimented on the 23 most populated classes.

For each of these 23 classes, the articles in the class of interest are positive, and the rest of 21,578 articles are negative. The most populous positive class contains 3,987 records, and the least populous class contains 112 records.

20NewsGroups. The dataset is a collection of messages from twenty different newsgroups with about one thousand messages in each newsgroup. We used each newsgroup as the positive class and pooled the remaining nineteen newsgroups as the negative class.

Text in the *MeSH25* and *Reuters* databases has been preprocessed as follows: all alphabetic characters were lowercased, non-alphanumeric characters replaced by blanks, and no stemming was done. Features in the *MeSH25* dataset are all single nonstop terms and all pairs of adjacent nonstop terms that are not separated by punctuation. Features in the *Reuters* database are single nonstop terms only. Features in the *20Newsgroups* are extracted using the Rainbow toolbox (McCallum 1996).

MedPhrase. We process MEDLINE to extract all multiword UMLS® (http://www.nlm.nih.gov/research/umls/) phrases that are present in MEDLINE. From the resulting set of strings, we drop the strings that contain punctuation marks or stop words. The remaining strings are normalized (lowercased, redundant white space is removed) and duplicates are removed. We denote the resulting set of 315,679 phrases by $U_{phrases}$.

For each phrase in $U_{phrases}$, we randomly sample, as available, up to 5 MEDLINE sentences containing it. We denote the resulting set of 728,197 MEDLINE sentences by $S_{phrases}$. From $S_{phrases}$ we extract all contiguous multiword expressions that are not present in $U_{phrases}$. We call them n-grams, where n>1. N-grams containing punctuation marks and stop words are removed and remaining n-grams are normalized and duplicates are dropped. The result is 8,765,444 n-grams that we refer to as M_{ngram}. We believe that M_{ngram} contains many high quality biological phrases. We use $U_{phrases}$, a known set of high quality biomedical phrases, as the positive class, and M_{ngram} as the negative class.

In order to apply machine learning we need to define features for each n-gram. Given an n-gram *grm* that is composed of *n* words, $grm = w_1 w_2 \cdots w_n$, we extract a set of 11 numbers

$\{f_i\}_{i=1}^{11}$ associated with the n-gram *grm*. These are as follows:

f_1: number of occurrences of *grm* throughout Medline;

f_2: -(number of occurrences of $w_2 \ldots w_n$ not following w_1 in documents that contain *grm*)/ f_1;

f_3: -(number of occurrences of $w_1 \ldots w_{n-1}$ not preceding w_n in documents that contain *grm*)/ f_1;

f_4: number of occurrences of (n+1)-grams of the form $xw_1 \ldots w_n$ throughout Medline;

f_5: number of occurrences of (n+1)-grams of the form $w_1 \ldots w_n x$ throughout Medline;

f_6: $\log\left(\frac{p(w_1 \mid w_2)(1-p(w_1 \mid \neg w_2))}{(1-p(w_1 \mid w_2))p(w_1 \mid \neg w_2)}\right)$

f_7: mutual information between w_1 and w_2;

f_8: $\log\left(\frac{p(w_{n-1} \mid w_n)(1-p(w_{n-1} \mid \neg w_n))}{(1-p(w_{n-1} \mid w_n))p(w_{n-1} \mid \neg w_n)}\right)$

f_9: mutual information between w_{n-1} and w_n;

f_{10}: -(number of different multiword expressions beginning with w_1 in Medline);

f_{11}: -(number of different multiword expressions ending with w_n in Medline).

We discretize the numeric values of the $\{f_i\}_{i=1}^{11}$ into categorical values.

In addition to these features, for every n-gram *grm*, we include the part of speech tags predicted by the MedPost tagger (Smith, Rindflesch et al. 2004). To obtain the tags for a given n-gram *grm* we randomly select a sentence from $S_{phrases}$ containing *grm*, tag the sentence, and consider the tags $t_0 t_1 t_2 \cdots t_{n-1} t_n t_{n+1}$ where t_0 is the tag of the word preceding word w_1 in n-gram *grm*, t_1 is the tag of word w_1 in n-gram *grm*, and so on. We construct the features

$$\begin{cases} \text{if } n > 2: \ \{(t_0,1),(t_1,2)(t_n,3)(t_{n+1},4),t_2,\ldots,t_{n-1}\} \\ \text{otherwise: } \{(t_0,1),(t_1,2)(t_n,3)(t_{n+1},4)\}. \end{cases}$$

These features emphasize the left and right ends of the n-gram and include parts-of-speech in the middle without marking their position. The resulting features are included with $\{f_i\}_{i=1}^{11}$ to represent the n-gram.

2.3 Experimental Design

A standard way to measure the success of a classifier is to evaluate its performance on a collection of documents that have been previously classified as positive or negative. This is usually accomplished by randomly dividing up the data into training and test portions which are separate. The classifier is then trained on the training portion, and is tested on test portion. This can be done in a cross-validation scheme or by randomly re-sampling train and test portions repeatedly.

We are interested in studying the case where only some of the positive documents are labeled. We simulate that situation by taking a portion of the positive data and including it in the negative training set. We refer to that subset of positive documents as *tracer data (Tr)*. The tracer data is then effectively mislabeled as negative. By introducing such an artificial supplement to the negative training set we are not only certain that the negative set contains mislabeled positive examples, but we know exactly which ones they are. Our goal is to automatically identify these mislabeled documents in the negative set and knowing their true labels will allow us to measure how successful we are. Our measurements will be carried out on the negative class and for this purpose it is convenient to write the negative class as composed of true negatives and tracer data (false negatives)

$$C_-' = C_- \cup Tr.$$

When we have trained a classifier, we evaluate performance by ranking C_-' and measuring how well tracer data is moved to the top ranks. The challenge is that *Tr* appears in the negative class and will interact with the training in some way.

2.4 Evaluation

We evaluate performance using Mean Average Precision (MAP) (Baeza-Yates and Ribeiro-Neto 1999). The mean average precision is the mean value of the average precisions computed for all topics in each of the datasets in our study. *Average precision* is the average of the precisions at each rank that contains a true positive document.

3 Results

3.1 *MeSH25*, *Reuters*, and *20NewsGroups*

We begin by presenting results for the *MeSH25* dataset. Table 1 shows the comparison between Huber and SVM methods. It also compares the performance of the classifiers with different levels of tracer data in the negative set. We set aside 50% of C_+ to be used as tracer data and used the remaining 50% of C_+ as the positive set for training. We describe three experiments where we have different levels of tracer data in the negative set at training time. These sets are C_-, $C_- \cup Tr_{20}$, and $C_- \cup Tr_{50}$ representing no tracer data, 20% of C_+ as tracer data and 50% of C_+ as tracer data, respectively. The test set $C_- \cup Tr_{20}$ is the same for all of these experiments. Results indicate that on average Huber outperforms SVM on these highly imbalanced datasets. We also observe that performance of both methods deteriorates with increasing levels of tracer data.

Table 2 shows the performance of Huber and SVM methods on negative training sets with tracer data $C_- \cup Tr_{20}$ and $C_- \cup Tr_{50}$ as in Table 1, but with cross training. As mentioned in the Methods section, we first divide each negative training set into two disjoint pieces Z_1 and Z_2. We then train documents in the positive training set versus documents in Z_1 to score documents in Z_2 and train documents in the positive training set versus documents in Z_2 to score documents in Z_1. We then merge Z_1 and Z_2 as scored sets and report measurements on the combined ranked set of documents. Comparing with Table 1, we see a significant improvement in the MAP when using cross training.

Table 1: MAP scores trained with three levels of tracer data introduced to the negative training set.

No Cross Training	No Tracer Data		Tr_{20} in training		Tr_{50} in training	
MeSH Terms	Huber	SVM	Huber	SVM	Huber	SVM
celiac disease	0.694	0.677	0.466	0.484	0.472	0.373
lactose intolerance	0.632	0.635	0.263	0.234	0.266	0.223
myasthenia gravis	0.779	0.752	0.632	0.602	0.562	0.502
carotid stenosis	0.466	0.419	0.270	0.245	0.262	0.186
diabetes mellitus	0.181	0.181	0.160	0.129	0.155	0.102
rats, wistar	0.241	0.201	0.217	0.168	0.217	0.081
myocardial infarction	0.617	0.575	0.580	0.537	0.567	0.487
blood platelets	0.509	0.498	0.453	0.427	0.425	0.342
serotonin	0.514	0.523	0.462	0.432	0.441	0.332
state medicine	0.158	0.164	0.146	0.134	0.150	0.092
urinary bladder	0.366	0.379	0.312	0.285	0.285	0.219
drosophila melanogaster	0.553	0.503	0.383	0.377	0.375	0.288
tryptophan	0.487	0.480	0.410	0.376	0.402	0.328
laparotomy	0.186	0.173	0.138	0.101	0.136	0.066
crowns	0.520	0.497	0.380	0.365	0.376	0.305
streptococcus mutans	0.795	0.738	0.306	0.362	0.218	0.306
infectious mononucleosis	0.622	0.614	0.489	0.476	0.487	0.376
blood banks	0.283	0.266	0.170	0.153	0.168	0.115
humeral fractures	0.526	0.495	0.315	0.307	0.289	0.193
tuberculosis, lymph node	0.385	0.397	0.270	0.239	0.214	0.159
mentors	0.416	0.420	0.268	0.215	0.257	0.137
tooth discoloration	0.499	0.499	0.248	0.215	0.199	0.151
pentazocine	0.710	0.716	0.351	0.264	0.380	0.272
hepatitis e	0.858	0.862	0.288	0.393	0.194	0.271
genes, p16	0.278	0.313	0.041	0.067	0.072	0.058
Avg	0.491	0.479	0.321	0.303	0.303	0.238

Table 2: MAP scores for Huber and SVM trained with two levels of tracer data introduced to the negative training set using cross training technique.

2-fold Cross Training	Tr_{20} in training		Tr_{50} in training	
MeSH Terms	Huber	SVM	Huber	SVM
celiac disease	0.550	0.552	0.534	0.521
lactose intolerance	0.415	0.426	0.382	0.393
myasthenia gravis	0.652	0.643	0.623	0.631
carotid stenosis	0.262	0.269	0.241	0.241
diabetes mellitus	0.148	0.147	0.144	0.122
rats, wistar	0.212	0.186	0.209	0.175
myocardial infarction	0.565	0.556	0.553	0.544
blood platelets	0.432	0.435	0.408	0.426
serotonin	0.435	0.447	0.417	0.437
state medicine	0.135	0.136	0.133	0.132
urinary bladder	0.295	0.305	0.278	0.280
drosophila melanogaster	0.426	0.411	0.383	0.404
tryptophan	0.405	0.399	0.390	0.391
laparotomy	0.141	0.128	0.136	0.126
crowns	0.375	0.376	0.355	0.353
streptococcus mutans	0.477	0.517	0.448	0.445
infectious mononucleosis	0.519	0.514	0.496	0.491
blood banks	0.174	0.169	0.168	0.157
humeral fractures	0.335	0.335	0.278	0.293
tuberculosis, lymph node	0.270	0.259	0.262	0.244
mentors	0.284	0.278	0.275	0.265
tooth discoloration	0.207	0.225	0.209	0.194
pentazocine	0.474	0.515	0.495	0.475
hepatitis e	0.474	0.499	0.482	0.478
genes, p16	0.102	0.101	0.083	0.093
Avg	0.350	0.353	0.335	0.332

We performed similar experiments with the *Reuters* and *20NewsGroups* datasets, where 20% and 50% of the good set is used as tracer data. We report MAP scores for these datasets in Tables 3 and 4.

3.2 Identifying high quality biomedical phrases in the MEDLINE Database

We illustrate our findings with a real example of detecting high quality biomedical phrases among M_{ngram}, a large collection of multiword expressions from Medline. We believe that M_{ngram} contains many high quality biomedical phrases. These examples are the counterpart of the mislabeled positive examples (tracer data) in the previous tests.

Table 3: MAP scores for Huber and SVM trained with 20% and 50% tracer data introduced to the negative training set for *Reuters* dataset.

Reuters	Tr_{20} in training		Tr_{50} in training	
	Huber	SVM	Huber	SVM
No CT	0.478	0.451	0.429	0.403
2-Fold CT	0.662	0.654	0.565	0.555

Table 4: MAP scores for Huber and SVM trained with 20% and 50% tracer data introduced to the negative training set for *20NewsGroups* dataset.

20News Groups	Tr_{20} in training		Tr_{50} in training	
	Huber	SVM	Huber	SVM
No CT	0.492	0.436	0.405	0.350
2-Fold CT	0.588	0.595	0.502	0.512

To identify these examples, we learn the difference between the phrases in $U_{phrases}$ and M_{ngram}. Based on the training we rank the n-grams in M_{ngram}. We expect the n-grams that cannot be separated from UMLS phrases are high quality biomedical phrases. In our experiments, we perform 3-fold cross validation for training and testing. This insures we obtain any possible benefit from cross training. The results shown in figure 1 are MAP values for these 3 folds.

Figure 1. Huber, CS-SVM, and naïve Bayes classifiers applied to the *MedPhrase* dataset.

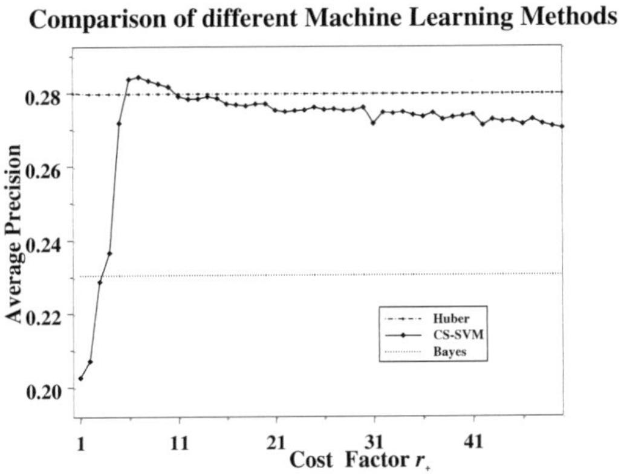

We trained naïve Bayes, Huber, and CS-SVM with a range of different cost factors. The results are presented in Figure 1. We observe that the Huber classifier performs better than naïve Bayes. CS-SVM with the cost factor of 1 (standard SVM) is quite ineffective. As we increase the cost factor, the performance of CS-SVM improves until it is comparable to Huber. We believe that the quality of ranking is better when the separation of $U_{phrases}$ from M_{ngram} is better.

Because we have no tracer data we have no direct way to evaluate the ranking of M_{ngram}. However, we selected a random set of 100 n-grams from M_{ngram}, which score as high as top-scoring 10% of phrases in $U_{phrases}$. Two reviewers manually reviewed that list and identified that 99 of these 100 n-grams were high quality biomedical phrases. Examples are: *aminoshikimate pathway, berberis aristata, dna hybridization, subcellular distribution, acetylacetoin synthase*, etc. One false-positive example in that list was *congestive heart*.

4 Discussion

We observed that the Huber classifier performs better than SVM on imbalanced data with no cross training (see appendix). The improvement of Huber over SVM becomes more marked as the percentage of tracer data in the negative training set is increased. However, the results also show that cross training, using either SVM or Huber (which are essentially equivalent), is better than using Huber without cross training. This is demonstrated in our experiments using the tracer data. The results are consistent over the range of different data sets. We expect cross training to have benefit in actual applications.

Where does cost-sensitive learning fit into this picture? We tested cost-sensitive learning on all of our corpora using the tracer data. We observed small and inconsistent improvements (data not shown). The optimal cost factor varied markedly between cases in the same corpus. We could not conclude this was a useful approach and instead saw better results simply using Huber. This conclusion is consistent with (Zhang and Iyengar 2002) which recommend using a quadratic loss function. It is also consistent with results reported in (Lewis, Yang et al. 2004) where CS-SVM is compared with SVM on multiple imbalanced text classification problems and no benefit is seen using CS-SVM. Others have reported a benefit with CS-SVM (Abkani, Kwek et al. 2004; Eitrich and Lang 2005). However, their datasets involve relatively few features and we believe this is an important aspect where cost-sensitive learning has proven effective. We hypothesize that this is the case because with few features the positive data is more likely to be duplicated in the negative set. In our case, the *MedPhrase* dataset involves relatively few features (410) and indeed we see a dramatic improvement of CS-SVM over SVM.

One approach to dealing with imbalanced data is the artificial generation of positive examples as seen with the SMOTE algorithm (Chawla, Bowyer et al. 2002). We did not try this method and do not know if this approach would be beneficial for textual data or data with many features. This is an area for possible future research.

Effective methods for leveraging positively labeled data have several potential applications:

- Given a set of documents discussing a particular gene, one may be interested in finding other documents that talk about the same gene but use an alternate form of the gene name.
- Given a set of documents that are indexed with a particular MeSH term, one may want to find new documents that are candidates for being indexed with the same MeSH term.
- Given a set of papers that describe a particular disease, one may be interested in other diseases that exhibit a similar set of symptoms.
- One may identify incorrectly tagged web pages.

These methods can address both removing incorrect labels and adding correct ones.

5 Conclusions

Given a large set of documents and a small set of positively labeled examples, we study how best to use this information in finding additional positive examples. We examine the SVM and Huber classifiers and conclude that the Huber classifier provides an advantage over the SVM classifier on such imbalanced data. We introduce a technique which we term cross training. When this technique is applied we find that the SVM and Huber classifiers are essentially equivalent and superior to applying either method without cross training. We confirm this on three different corpora. We also analyze an example where cost-sensitive learning is effective. We hypothesize that with datasets having few features, cost-sensitive learning can be beneficial and comparable to using the Huber classifier.

Appendix: Why Huber Loss Function works better for problems with Unbalanced Class Distributions.

The drawback of the standard SVM for the problem with an unbalanced class distribution results from the shape of $h(z)$ in (2). Consider the initial condition at $w=0$ and also imagine that there is a lot more C_- training data than C_+ training data. In this case, by choosing $\theta=-1$, we can achieve the minimum value of the loss function in (1) for the initial condition $w=0$. Under these conditions, all C_- points yield $z=1$ and $h(z)=0$ and all C_+ points yield $z=-1$ and $h(z)=2$. The change of the loss function $\Delta h(z)$ in (2) with a change Δw is given by

$$\Delta h(z)=\frac{dh(z)}{dz}\nabla_w z\cdot\Delta w=-y_i x_i\cdot\Delta w \quad (5).$$

In order to reduce the loss at a C_+ data point (x_i, y_i), we must choose Δw such that $x_i\cdot\Delta w>0$. But we assume that there are significantly more C_- class data points than C_+ and many such points x' are mislabeled and close to x_i such that $x'\cdot\Delta w>0$. Then $h(z)$ is likely be increased by $x'\cdot\Delta w(>0)$ for these mislabeled points. Clearly, if there are significantly more C_- class data than those of C_+ class and the C_- set contains a lot of mislabeled points, it may be difficult to find Δw that can result in a net effect of decreasing the right hand side of (2). The above analysis shows why the standard support vector machine formulation in (2) is vulnerable to an unbalanced and noisy training data set. The problem is clearly caused by the fact that the SVM loss function $h(z)$ in (2) has a constant slope for $z\le 1$. In order to alleviate this problem, Zhang and Iyengar (2002) proposed the loss function $h^2(z)$ which is a smooth non-increasing function with slope 0 at $z=1$. This allows the loss to decrease while the positive points move a small distance away from the bulk of the negative points and take mislabeled points with them. The same argument applies to the Huber loss function defined in (4).

Acknowledgments

This research was supported by the Intramural Research Program of the NIH, National Library of Medicine.

References

Abkani, R., S. Kwek, et al. (2004). Applying Support Vector Machines to Imballanced Datasets. ECML.

Baeza-Yates, R. and B. Ribeiro-Neto (1999). Modern Information Retrieval. New York, ACM Press.

Blum, A. and T. Mitchell (1998). "Combining Labeled and Unlabeled Data with Co-Training." COLT: Proceedings of the Workshop on Computational Learning Theory: 92-100.

Chawla, N. V., K. W. Bowyer, et al. (2002). "SMOTE: Synthetic Minority Over-sampling Technique." Journal of Artificial Intelligence Research **16**: 321-357.

Eitrich, T. and B. Lang (2005). "Efficient optimization of support vector machine learning parameters for unbalanced datasets." Journal of Computational and Applied Mathematics **196**(2): 425-436.

Elkan, C. (2001). The Foundations of Cost Sensitive Learning. Proceedings of the Seventeenth International Joint Conference on Artificial Intelligence.

Lewis, D. D., Y. Yang, et al. (2004). "RCV1: A New Benchmark Collection for Text Categorization Research." Journal of Machine Learning Research **5**: 361-397.

Maloof, M. A. (2003). Learning when data sets are imbalanced and when costs are unequal and unknown. ICML 2003, Workshop on Imballanced Data Sets.

McCallum, A. K. (1996). "Bow: A toolkit for statistical language modeling, text retrieval, classification and clustering. http://www.cs.cmu.edu/~mccallum/bow/."

Nigam, K., A. K. McCallum, et al. (1999). "Text Classification from Labeled and Unlabeled Documents using EM." Machine Learning: 1-34.

Roy, N. and A. McCallum (2001). Toward Optimal Active Learning through Sampling Estimation of Error Reduction. Eighteenth International Conference on Machine Learning.

Smith, L., T. Rindflesch, et al. (2004). "MedPost: A part of speech tagger for biomedical text." Bioinformatics **20**: 2320-2321.

Tong, S. and D. Koller (2001). "Support vector machine active learning with applications to text classification." Journal of Machine Learning Research **2**: 45-66.

Weiss, G., K. McCarthy, et al. (2007). Cost-Sensitive Learning vs. Sampling: Which is Best for Handling Unbalanced Classes with Unequal Error Costs? Proceedings of the 2007 International Conference on Data Mining.

Zhang, T. (2004). Solving large scale linear prediction problems using stochastic gradient descent algorithms. Twenty-first International Conference on Machine learning, Omnipress.

Zhang, T. and V. S. Iyengar (2002). "Recommender Systems Using Linear Classifiers." Journal of Machine Learning Research **2**: 313-334.

Parsing Natural Language Queries for Life Science Knowledge

Tadayoshi Hara
National Institute of Informatics
2-1-2 Hitotsubashi, Chiyoda-ku,
Tokyo 101-8430, JAPAN
harasan@nii.ac.jp

Yuka Tateisi
Faculty of Informatics, Kogakuin University
1-24-2 Nishi-shinjuku, Shinjuku-ku,
Tokyo 163-8677, JAPAN
yucca@cc.kogakuin.ac.jp

Jin-Dong Kim
Database Center for Life Science
2-11-16 Yayoi, Bunkyo-ku,
Tokyo 113-0032, JAPAN
jdkim@dbcls.rois.ac.jp

Yusuke Miyao
National Institute of Informatics
2-1-2 Hitotsubashi, Chiyoda-ku,
Tokyo 101-8430, JAPAN
yusuke@nii.ac.jp

Abstract

This paper presents our preliminary work on adaptation of parsing technology toward natural language query processing for biomedical domain. We built a small treebank of natural language queries, and tested a state-of-the-art parser, the results of which revealed that a parser trained on Wall-Street-Journal articles and Medline abstracts did not work well on query sentences. We then experimented an adaptive learning technique, to seek the chance to improve the parsing performance on query sentences. Despite the small scale of the experiments, the results are encouraging, enlightening the direction for effective improvement.

1 Introduction

Recent rapid progress of life science resulted in a greatly increased amount of life science knowledge, e.g. genomics, proteomics, pathology, therapeutics, diagnostics, etc. The knowledge is however scattered in pieces in diverse forms over a large number of databases (DBs), e.g. PubMed, Drugs.com, Therapy database, etc. As more and more knowledge is discovered and accumulated in DBs, the need for their integration is growing, and corresponding efforts are emerging (BioMoby[1], BioRDF[2], etc.).

Meanwhile, the need for a query language with high expressive power is also growing, to cope with the complexity of accumulated knowledge. For example, SPARQL[3] is becoming an important query language, as RDF[4] is recognized as a standard interoperable encoding of information in databases. SPARQL queries are however not easy for human users to compose, due to its complex vocabulary, syntax and semantics. We propose natural language (NL) query as a potential solution to the problem. Natural language, e.g. English, is the most straightforward language for human beings. Extra training is not required for it, yet the expressive power is very high. If NL queries can be automatically translated into SPARQL queries, human users can access their desired knowledge without learning the complex query language of SPARQL.

This paper presents our preliminary work for NL query processing, with focus on syntactic parsing. We first build a small treebank of natural language queries, which are from Genomics track (Hersh et al., 2004; Hersh et al., 2005; Hersh et al., 2006; Hersh et al., 2007) topics (Section 2 and 3). The small treebank is then used to test the performance of a state-of-the-art parser, Enju (Ninomiya et al., 2007; Hara et al., 2007) (Section 4). The results show that a parser trained on Wall-Street-Journal (WSJ) articles and Medline abstracts will not work well on query sentences. Next, we experiment an adaptive learning technique, to seek the chance to improve the parsing performance on query sentences. Despite the small scale of the experiments, the results enlighten directions for effective

[1] http://www.biomoby.org/
[2] http://esw.w3.org/HCLSIG_BioRDF_Subgroup
[3] http://www.w3.org/TR/rdf-sparql-query/
[4] http://www.w3.org/RDF/

Proceedings of the 2011 Workshop on Biomedical Natural Language Processing, ACL-HLT 2011, pages 164–173,

	GTREC			
	04	05	06	07
Declarative	1	0	0	0
Imperative	22	60	0	0
Infinitive	1	0	0	0
Interrogative				
- WP/WRB/WDT	3 / 1 / 11	0 / 0 / 0	6 / 22 / 0	0 / 0 / 50
- Non-*wh*	5	0	0	0
NP	14	0	0	0
Total	58	60	28	50

Table 1: Distribution of sentence constructions

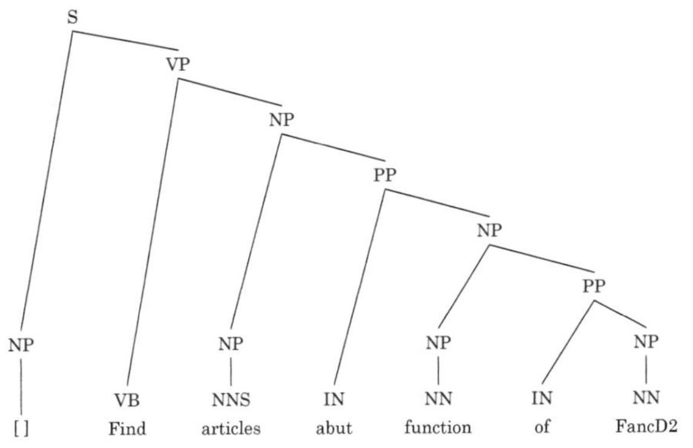

Figure 1: The tree structure for an imperative sentence

improvement (Section 5).

2 Syntactic Features of Query Sentences

While it is reported that the state-of-art NLP technology shows reasonable performance for IR or IE applications (Ohta et al., 2006), NLP technology has long been developed mostly for declarative sentences. On the other hand, NL queries include wide variety of sentence constructions such as interrogative sentences, imperative sentences, and noun phrases. Table 1 shows the distribution of the constructions of the 196 query sentences from the topics of the ad hoc task of Genomics track 2004 (GTREC04) and 2005 (GTREC05) in their narrative forms, and the queries for the passage retrieval task of Genomics track 2006 (GTREC06) and 2007 (GTREC07).

GTREC04 set has a variety of sentence constructions, including noun phrases and infinitives, which are not usually considered as full sentences. In the 2004 track, the queries were derived from interviews eliciting information needs of real biologists, without any control on the sentence constructions.

GTREC05 consists only of imperative sentences. In the 2005 track, a set of templates were derived from an analysis of the 2004 track and other known biologist information needs. The derived templates were used as the commands to find articles describing biological interests such as methods or roles of genes. Although the templates were in the form "Find articles describing ...", actual obtained imperatives begin with "Describe the procedure or method for" (12 sentences), "Provide information about" (36 sentences) or "Provide information on" (12 sentences).

GTREC06 consists only of *wh*-questions where a *wh*-word constitutes a noun phrase by itself (i.e. its part-of-speech is the WP in Penn Treebank (Marcus et al., 1994) POS tag set) or is an adverb (WRB). In the 2006 track, the templates for the 2005 track were reformulated into the constructions of questions and were then utilized for deriving the questions. For example, the templates to find articles describing the role of a gene involved in a given disease is reformulated into the question "What is the role of gene in disease?"

GTREC07 consists only of *wh*-questions where a *wh*-word serves as a pre-nominal modifier (WDT). In the 2007 track, unlike in those of last two years, questions were not categorized by the templates, but were based on biologists' information needs where the answers were lists of named entities of a given type. The obtained questions begin with "what + *entity type*" (45 sentences), "which + *entity type*" (4 sentences), or "In what + *entity type*" (1 sentence).

In contrast, the GENIA Treebank Corpus (Tateisi et al., 2005)[5] is estimated to have no imperative sentences and only seven interrogative sentences (see Section 5.2.2). Thus, the sentence constructions in GTREC04–07 are very different from those in the GENIA treebank.

3 Treebanking GTREC query sentences

We built a treebank (with POS) on 196 query sentences following the guidelines of the GENIA Treebank (Tateisi and Tsujii, 2006). The queries were first parsed using the Stanford Parser (Klein and Manning, 2003), and manual correction was made

[5] http://www-tsujii.is.s.u-tokyo.ac.jp/GENIA/home/wiki.cgi?page=GENIA+Treebank

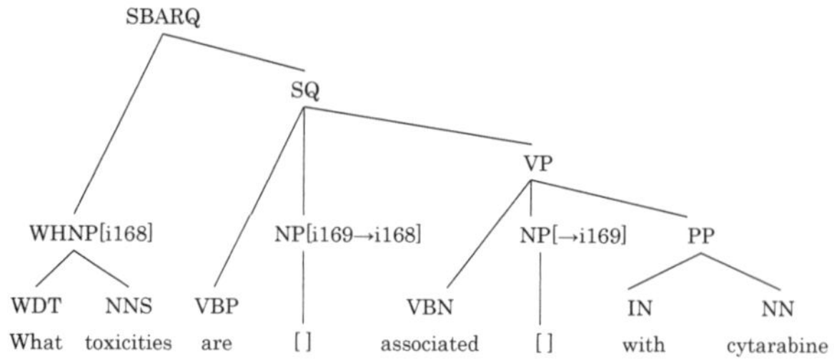

Figure 2: The tree structure for an interrogative sentence

by the second author. We tried to follow the guideline of the GENIA Treebank as closely as possible, but for the constructions that are rare in GENIA, we used the ATIS corpus in Penn Treebank (Bies et al., 1995), which is also a collection of query sentences, for reference.

Figure 1 shows the tree for an imperative sentence. A leaf node with [ ] corresponds to a null constituent. Figure 2 shows the tree for an interrogative sentence. Coindexing is represented by assigning an ID to a node and a reference to the ID to the node which is coindexed. In Figure 2, WHNP[i168] means that the WHNP node is indexed as i168, NP[i169→i168] means that the NP node is indexed as i169 and coindexed to the i168 node, and NP[→i169] means that the node is coindexed to the i169 node. In this sentence, which is a passive *wh*-question, it is assumed that the logical object (*what toxicities*) of the verb (*associate*) is moved to the subject position (the place of i169) and then moved to the sentence-initial position (the place of i168).

As most of the query sentences are either imperative or interrogative, there are more null constituents compared to the GENIA Corpus. In the GTREC query treebank, 184 / 196 (93.9%) sentences contained one or more null constituents, whereas in GENIA, 12,222 / 18,541 (65.9%) sentences did. We expected there are more sentences with multiple null constituents in GTREC compared to GENIA, due to the frequency of passive interrogative sentences, but on the contrary the number of sentences containing more than one null constituents are 65 (33.1%) in GTREC, and 6,367 (34.5%) in GENIA. This may be due to the frequency of relative clauses in GENIA.

4 Parsing system and extraction of imperative and question sentences

We introduce the parser and the POS tagger whose performances are examined, and the extraction of imperative or question sentences from GTREC treebank on which the performances are measured.

4.1 HPSG parser

The Enju parser (Ninomiya et al., 2007)[6] is a deep parser based on the HPSG formalism. It produces an analysis of a sentence that includes the syntactic structure (i.e., parse tree) and the semantic structure represented as a set of predicate-argument dependencies. The grammar is based on the standard HPSG analysis of English (Pollard and Sag, 1994). The parser finds a best parse tree scored by a maxent disambiguation model using a Cocke-Kasami-Younger (CKY) style algorithm.

We used a toolkit distributed with the Enju parser for training the parser with a Penn Treebank style (PTB-style) treebank. The toolkit initially converts the PTB-style treebank into an HPSG treebank and then trains the parser on it. We used a toolkit distributed with the Enju parser for extracting a HPSG lexicon from a PTB-style treebank. The toolkit initially converts the PTB-style treebank into an HPSG treebank and then extracts the lexicon from it.

The HPSG treebank converted from the test section was used as the gold-standard in the evaluation. As the evaluation metrics of the Enju parser, we used labeled and unlabeled precision/recall/F-score of the predicate-argument dependencies produced by the parser. A predicate-argument dependency is represented as a tuple of $\langle w_p, w_a, r\rangle$, where w_p is the predicate word, w_a is the argument word, and r is the label of the predicate-argument relation, such as `verb-ARG1` (semantic subject of a verb) and `prep-ARG1` (modifiee of a prepositional phrase).

4.2 POS tagger

The Enju parser assumes that the input is already POS-tagged. We use a tagger in (Tsuruoka et al., 2005). It has been shown to give a state-of-the-art accuracy on the standard Penn WSJ data set and also on a different text genre (biomedical literature) when trained on the combined data set of the WSJ data and

[6]http://www-tsujii.is.s.u-tokyo.ac.jp/enju

the target genre (Tsuruoka et al., 2005). Since our target is biomedical domain, we utilize the tagger adapted to the domain as a baseline, which we call "the GENIA tagger".

4.3 Extracting imperative and question sentences from GTREC treebank

In GTREC sentences, two major constructions of sentences can be observed: imperative and question sentences. These two types of sentences have different sentence constructions and we will observe the impact of each or both of these constructions on the performances of parsing or POS-tagging. In order to do so, we collected imperative and question sentences from our GTREC treebank as follows:

- **GTREC imperatives** - Most of the imperative sentences in GTREC treebank begin with empty subjects "(NP-SBJ */-NONE-)". We extracted such 82 imperative sentences.

- **GTREC questions** - Interrogative sentences are annotated with the phrase label "SBARQ" or "SQ", where "SBARQ" and "SQ" respectively denote a *wh*-question and an yes/no question. We extracted 98 interrogative sentences whose top phrase labels were either of them.

5 Experiments

We examine the POS-tagger and the parser for the sentences in the GTREC corpus. They are adapted to each of GTREC overall, imperatives, and questions. We then observe how the parsing or POS-tagging accuracies are improved and analyze what is critical for parsing query sentences.

5.1 Experimental settings

5.1.1 Dividing corpora

We prepared experimental datasets for the following four domains:

- **GENIA Corpus (GENIA) (18,541 sentences)**
  Divided into three parts for training (14,849 sentences), development test (1,850 sentences), and final test (1,842 sentences).

- **GTREC overall (196 sentences)**
  Divided into two parts: one for ten-folds cross validation test (17-18 × 10 sentences) and the other for error analysis (17 sentences)

Target	GENIA tagger	Adapted tagger
GENIA	99.04%	-
GTREC (overall)	89.98%	96.54%
GTREC (imperatives)	90.32%	97.30%
GRREC (questions)	89.25%	94.77%

Table 2: Accuracy of the POS tagger for each domain

- **GTREC imperatives (82 sentences)**
  Divided into two parts: one for ten-folds cross validation test (7-8 × 10 sentences) and the other for error analysis (7 sentences)

- **GTREC questions (98 sentences)**
  Divided into two parts: one for ten-folds cross validation test (9 × 10 sentences) and the other for error analysis (8 sentences)

5.1.2 Adaptation of POS tagger and parser

In order to adapt the POS tagger and the parser to a target domain, we took the following methods.

- **POS tagger** - For the GTREC overall / imperatives / questions, we replicated the training data for 100,000 times and utilized the concatenated replicas and GENIA training data in (Tsuruoka et al., 2005) for training. For POS tagger, the number of replicas of training data was determined among $10^n (n = 0, \ldots, 5)$ by testing these numbers on development test sets in three of ten datasets of cross validation.

- **Enju parser** - We used a toolkit in the Enju parser (Hara et al., 2007). As a baseline model, we utilized the model adapted to the GENIA Corpus. We then attempted to further adapt the model to each domain. In this paper, the baseline model is called "the GENIA parser".

5.2 POS tagger and parser performances

Table 2 and 3 respectively show the POS tagging and the parsing accuracies for the target domains, and Figure 3 and 4 respectively show the POS tagging and the parsing accuracies for the target domains given by changing the size of the target training data.

The POS tagger could output for each word either of one-best POS or POS candidates with probabilities, and the Enju parser could take either of the two output types. The bracketed numbers in Table 3 and

Parser	GENIA			Adapted		
POS	Gold	GENIA tagger	Adapted tagger	Gold	GENIA tagger	Adapted tagger
For GENIA	88.54	88.07 (88.00)	-	-	-	-
For GTREC overall	84.37	76.81 (72.43)	83.46 (81.96)	89.00	76.98 (74.44)	86.98 (85.42)
For GTREC imperatives	85.19	78.54 (77.75)	85.71 (85.48)	89.42	74.40 (74.84)	88.97 (88.67)
For GTREC questions	85.45	76.25 (67.27)	83.55 (80.46)	87.33	81.41 (71.90)	84.87 (82.70)

[using POS candidates with probabilities (using only one best POS)]

Table 3: Accuracy of the Enju parser for GTREC

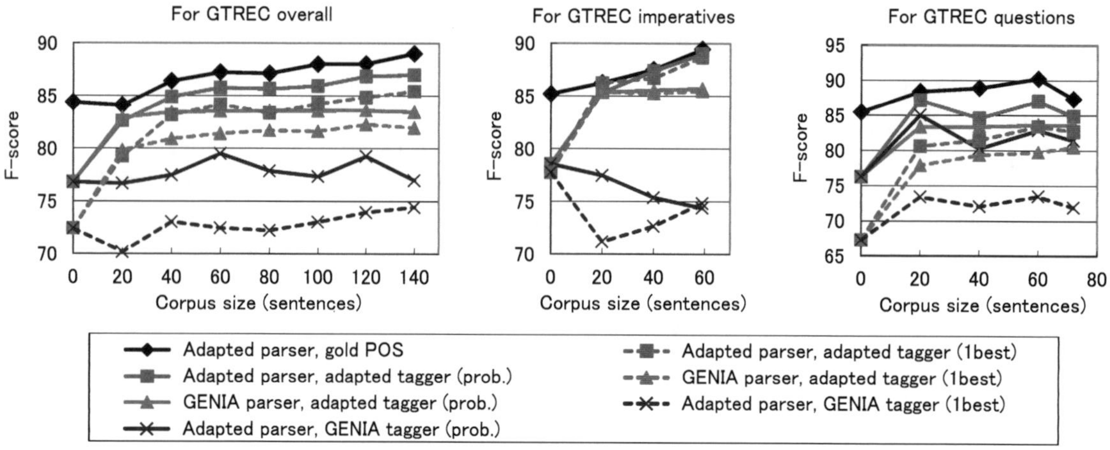

Figure 4: Parsing accuracy vs. corpus size

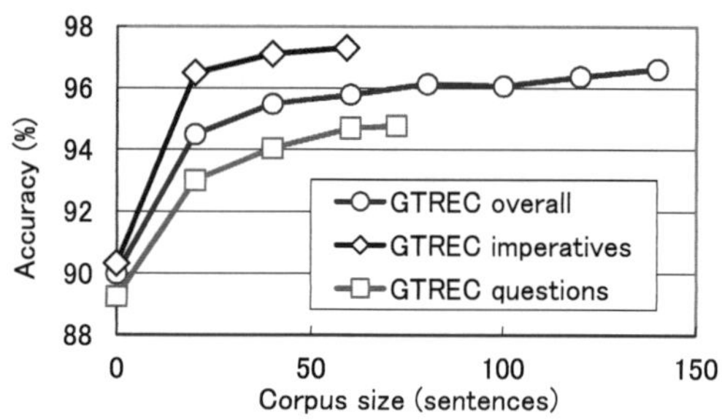

Figure 3: POS tagging accuracy vs. corpus size

Correct → Error	GENIA tagger	Adapted tagger
For GTREC overall (17 sentences)		
NN → NNP	4	0.6
VB → NN	4	0
WDT → WP	4	0
NN → JJ	1	1.9
For GTREC imperative (seven sentences)		
FW → NNP / NN / JJ	7	4
VB → NN	4	0
NN → NNP	2	0
For GTREC question (eight sentences)		
WDT → WP	3	0
VB → VBP	2	1
NNS → VBZ	2	0

(The table shows only error types observed more than once for either of the taggers)

Table 4: Tagging errors for each of the GTREC corpora

the dashed lines in Figure 4 show the parsing accuracies when we utilized one-best POS given by the POS tagger, and the other numbers and lines show the accuracies given by POS candidates with probabilities. In the rest of this section, when we just say "POS tagger", the tagger's output is POS candidates with probabilities.

Table 4 and 5 respectively compare the types of POS tagging and parsing errors for each domain between before and after adapting the POS tagger, and Table 6 compares the types of parsing errors for each domain between before and after adapting the parser. The numbers of errors for the rightmost column in each of the tables were given by the average of the ten-folds cross validation results.

In the following sections, we examine the impact of the performances of the POS taggers or the parsers on parsing the GTREC documents.

Error types	GENIA parser	
	GENIA tagger	Adapted tagger
For GTREC overall (17 sentences)		
Failure in detecting verb	12	0.2
Root selection	6	0
Range of NP	5	5
PP-attachment	4	3
Determiner / pronoun	4	1
Range of verb subject	4	4
Range of verb object	3	3
Adjective / modifier noun	2	3
For GTREC imperatives (seven sentences)		
Failure in detecting verb	8	0
Root selection	4	0
Range of NP	3	4
PP-attachment	3	1.8
Range of PP	2	2
For GTREC questions (eight sentences)		
Range of coordination	5	3
Determiner / pronoun	3	0
PP-attachment	3	1
Range of PP	2	2
Subject for verb	2	1

(The table shows only the types of parsing errors observed more than once for either of the parsers)

Table 5: Impact of adapting POS tagger on parsing errors

5.2.1 Impact of POS tagger on parsing

In Table 2, for each of the GTREC corpora, the GENIA tagger dropped its tagging accuracy by around nine points, and then recovered five to seven points by the adaptation. According to this behavior of the tagger, Table 3 shows that the GENIA and the adapted parsers with the GENIA tagger dropped their parsing accuracies by 6–15 points in F-score from the accuracies with the gold POS, and then recovered the accuracies within two points below the accuracies with the gold POS. The performance of the POS tagger would thus critically affect the parsing accuracies.

In Figure 3, we can observe that the POS tagging accuracy for each corpus rapidly increased only for first 20–30 sentences, and after that the improvement speed drastically declined. Accordingly, in Figure 4, the line for the adapted parser with the adapted tagger (the line with triangle plots) rose rapidly for the first 20–30 sentences, and after that slowed down.

We explored the tagging and parsing errors, and analyze the cause of the initial accuracy jump and the successive improvement depression.

Error types	Gold POS	
	GENIA parser	Adapted parser
For GTREC overall (17 sentences)		
Range of NP	5	1.3
Range of verb subject	3	2.6
PP-attachment	3	2.7
Whether verb takes object & complement	3	2.9
Range of verb object	2	1
For GTREC imperatives (seven sentences)		
Range of NP	4	1.1
PP-attachment	2	1.6
Range of PP	2	0.3
Preposition / modifier	2	2
For GTREC questions (eight sentences)		
Coordination / conjunction	2	2.2
Auxiliary / normal verb	2	2.6
Failure in detecting verb	2	2.6

(The table shows only the types of parsing errors observed more than once for either of the parsers)

Table 6: Impact of adapting parser on parsing errors

Cause of initial accuracy jump

In Table 4, "VB $\rightarrow$ NN" tagging errors were observed only in imperative sentences and drastically decreased by the adaptation. In a imperative sentence, a verb (VB) usually appears as the first word. On the other hand, the GENIA tagger was trained mainly on the declarative sentences and therefore would often take the first word in a sentence as the subject of the sentence, that is, noun (NN). When the parser received a wrong NN-tag for a verb, the parser would attempt to believe the information ("failure in detecting verb" in Table 6) and could then hardly choose the NN-tagged word as a main verb ("root selection" in Table 6). By adapting the tagger, the correct tag was given to the verb and the parser could choose the verb as a main verb.

"WDT $\rightarrow$ WP" tagging errors were observed only in the question sentences and also drastically decreased. For example, in the sentence "What toxicities are associated with cytarabine?", "What" works as a determiner (WDT) which takes "toxicities", while the GENIA tagger often took this "What" as a pronoun (WP) making a phrase by itself. This would be because the training data for the GENIA tagger would contain 682 WP "what" and only 27 WDT "what". WP "what" could not make a noun phrase by taking a next noun, and then the parsing of the parsing would corrupt ("determiner / pronoun" in Table 5). By adapting the tagger, "WDT" tag was given to "What", and the parser correctly made a phrase "What toxicities".

Since the variation of main verbs in GTREC imperatives is very small (see Section 2) and that of interrogatives is also very small, in order to correct the above two types of errors, we would require only small training data. In addition, these types of errors widely occurred among imperatives or questions, the accuracy improvement by correcting the errors was very large. The initial rapid improvement would thus occur.

Cause of improvement depression

"NN → NNP" tagging errors would come from the description style of words. In the GTREC queries, technical terms, such as the names of diseases or proteins, sometimes begin with capital characters. The GENIA tagger would take the capitalized words not as a normal noun (NN) but as a proper noun (NNP). By adaptation, the tagger would have learned the capital usage for terms and the errors then decreased.

However, in order to achieve such improvement, we would have to wait until a target capitalized term is added to the training corpus. "FW → NNP / NN / JJ", "NN → JJ", and several other errors would be similar to this type of errors in the point that, they would be caused by the difference in annotation policy or description style between the training data for the GENIA tagger and the GTREC queries.

"VB → VBP" errors were found in questions. For example, "affect" in the question "How do mutations in Sonic Hedgehog genes affect developmental disorders?" was base form (VB), while the GENIA tagger took it as a present tense (VBP) since the GENIA tagger would be unfamiliar with such verb behavior in questions. By adaptation, the tagger would learn that verbs in the domain tend to take base forms and the errors then decreased.

However, the tagger model based on local context features could not substantially solve the problem. VBP of course could appear in question sentences. We observed that a verb to be VBP was tagged with VB by the adapted tagger. In order to distinguish VB from VBP, we should capture longer distance dependencies between auxiliary and main verbs.

In tagging, the fact that the above two types of errors occupied most of the errors other than the errors involved in the initial jump, would be related to why the accuracy improvement got so slowly, which would lead to the improvement depression of the parsing performances. With the POS candidates with probabilities, the possibilities of correct POSs would increase, and therefore the parser would give higher parsing performances than using only one-best POSs (see Table 3 and Figure 4).

Anyway, the problems were not substantially solved. For these tagging problems, just adding the training data would not work. We might need reconstruct the tagging system or re-consider the feature designs of the model.

5.2.2 Impact of parser itself on parsing

For the GTREC corpora, the GENIA parser with gold POSs lowered the parsing accuracy by more than three points than for the GENIA Corpus, while the adaptation of the parser recovered a few points for each domain (second and fifth column in Table 3). Figure 4 would also show that we could improve the parser's performance with more training data for each domain. For GTREC questions, the parsing accuracy dropped given the maximum size of the training data. Our training data is small and therefore small irregular might easily make accuracies drop or rise. [7] We might have to prepare more corpora for confirming our observation.

Table 6 would imply that the major errors for all of these three corpora seem not straightforwardly associated with the properties specific to imperative or question sentences. Actually, when we explored the parse results, errors on the sentence constructions specific to the two types of sentences would hardly be observed. ("Failure in detecting verb" errors in GTREC questions came from other causes.) This would mean that the GENIA parser itself has potential to parse the imperative or question sentences.

The training data of the GENIA parser consists of the WSJ Penn Treebank and the GENIA Corpus. As long as we searched with our extraction method in Section 4.3, the WSJ and GENIA Corpus seem respectively contain 115 and 0 imperative, and 432

[7] This time we could not analyze which training data affected the decrease, because through the cross validation experiments each sentence was forced to be once final test data. However, we would like to find the reason for this accuracy decrease in some way.

and seven question sentences. Unlike the POS tagger, the parser could convey more global sentence constructions from these sentences.

Although the GENIA parser might understand the basic constructions of imperative or question sentences, by adaptation of the parser to the GTREC corpora, we could further learn more local construction features specific to GTREC, such as word sequence constructing a noun phrase, attachment preference of prepositions or other modifiers. The error reduction in Table 6 would thus be observed.

However, we also observed that several types of errors were still mostly unsolved after the adaptation. Choosing whether to add complements for verbs or not, and distinguishing coordinations from conjunctions seems to be difficult for the parser. If two question sentences were concatenated by conjunctions into one sentence, the parser would tend to fail to analyze the sentence construction for the latter sentence. The remaining errors in Table 6 would imply that we should also re-consider the model designs or the framework itself for the parser in addition to just increasing the training data.

6 Related work

Since domain adaptation has been an extensive research area in parsing research (Nivre et al., 2007), a lot of ideas have been proposed, including un-/semi-supervised approaches (Roark and Bacchiani, 2003; Blitzer et al., 2006; Steedman et al., 2003; McClosky et al., 2006; Clegg and Shepherd, 2005; McClosky et al., 2010) and supervised approaches (Titov and Henderson, 2006; Hara et al., 2007). Their main focus was on adapting parsing models trained with a specific genre of text (in most cases PTB-WSJ) to other genres of text, such as biomedical research papers. A major problem tackled in such a task setting is the handling of unknown words and domain-specific ways of expressions. However, as we explored, parsing NL queries involves a significantly different problem; even when all words in a sentence are known, the sentence has a very different construction from declarative sentences.

Although sentence constructions have gained little attention, a notable exception is (Judge et al., 2006). They pointed out low accuracy of state-of-the-art parsers on questions, and proposed supervised parser adaptation by manually creating a treebank of questions. The question sentences are annotated with phrase structure trees in the PTB scheme, although function tags and empty categories are omitted. An LFG parser trained on the treebank then achieved a significant improvement in parsing accuracy. (Rimell and Clark, 2008) also worked on question parsing. They collected question sentences from TREC 9-12, and annotated the sentences with POSs and CCG (Steedman, 2000) lexical categories. They reported a significant improvement in CCG parsing without phrase structure annotations.

On the other hand, (Judge et al., 2006) also implied that just increasing the training data would not be enough. We went further from their work, built a small but complete treebank for NL queries, and explored what really occurred in HPSG parsing.

7 Conclusion

In this paper, we explored the problem in parsing queries. We first attempted to build a treebank on queries for biological knowledge and successfully obtained 196 annotated GTREC queries. We next examined the performances of the POS tagger and the HPSG parser on the treebank. In the experiments, we focused on the two dominant sentence constructions in our corpus: imperatives and questions, extracted them from our corpus, and then also examined the parser and tagger for them.

The experimental results showed that the POS tagger's mis-tagging to main verbs in imperatives and *wh*-interrogatives in questions critically decreased the parsing performances, and that our small corpus could drastically decrease such mis-tagging and consequently improve the parsing performances. The experimental results also showed that the parser itself could improve its own performance by increasing the training data. On the other hand, the experimental results suggested that the POS tagger or the parser performance would stagnate just by increasing the training data.

In our future research, on the basis of our findings, we would like both to build more training data for queries and to reconstruct the model or reconsider the feature design for the POS tagger and the parser. We would then incorporate the optimized parser and tagger into NL query processing applications.

References

Ann Bies, Mark Ferguson, Karen Katz, and Robert MacIntyre. 1995. Bracketing guidelines for Treebank II style — Penn Treebank project. Technical report, Department of Linguistics, University of Pennsylvania.

John Blitzer, Ryan Mcdonald, and Fernando Pereira. 2006. Domain adaptation with structural correspondence learning. In *Proceedings of the 2006 Conference on Empirical Methods in Natural Language Processing*, pages 120–128, Sydney, Australia.

A. B. Clegg and A. Shepherd. 2005. Evaluating and integrating treebank parsers on a biomedical corpus. In *Proceedings of the ACL 2005 Workshop on Software*, Ann Arbor, Michigan.

Tadayoshi Hara, Yusuke Miyao, and Jun'ichi Tsujii. 2007. Evaluating impact of re-training a lexical disambiguation model on domain adaptation of an hpsg parser. In *Proceedings of 10th International Conference on Parsing Technologies (IWPT 2007)*, pages 11–22.

William R. Hersh, Ravi Teja Bhupatiraju, L. Ross, Aaron M. Cohen, Dale Kraemer, and Phoebe Johnson. 2004. TREC 2004 Genomics Track Overview. In *Proceedings of the Thirteenth Text REtrieval Conference, TREC 2004*.

William R. Hersh, Aaron M. Cohen, Jianji Yang, Ravi Teja Bhupatiraju, Phoebe M. Roberts, and Marti A. Hearst. 2005. TREC 2005 Genomics Track Overview. In *Proceedings of the Fourteenth Text REtrieval Conference, TREC 2005*.

William R. Hersh, Aaron M. Cohen, Phoebe M. Roberts, and Hari Krishna Rekapalli. 2006. TREC 2006 Genomics Track Overview. In *Proceedings of the Fifteenth Text REtrieval Conference, TREC 2006*.

William R. Hersh, Aaron M. Cohen, Lynn Ruslen, and Phoebe M. Roberts. 2007. TREC 2007 Genomics Track Overview. In *Proceedings of The Sixteenth Text REtrieval Conference, TREC 2007*.

John Judge, Aoife Cahill, and Josef van Genabith. 2006. Questionbank: Creating a Corpus of Parsing-Annotated Questions. In *Proceedings of the 21st International Conference on Computational Linguistics and 44th Annual Meeting of the ACL*, pages 497–504.

Dan Klein and Christopher D. Manning. 2003. Accurate Unlexicalized Parsing. In *Proceedings of the 41st Annual Meeting of the Association for Computational Linguistics*, pages 423–430.

Mitchell Marcus, Grace Kim, Mary Ann Marcinkiewicz, Robert Macintyre, Ann Bies, Mark Ferguson, Karen Katz, and Britta Schasberger. 1994. The Penn Treebank: Annotating predicate argument structure. In *Proceedings of ARPA Human Language Technology Workshop*.

David McClosky, Eugene Charniak, and Mark Johnson. 2006. Reranking and self-training for parser adaptation. In *Proceedings of the 21st International Conference on Computational Linguistics and the 44th annual meeting of the Association for Computational Linguistics*, pages 337–344, Sydney, Australia.

David McClosky, Eugene Charniak, and Mark Johnson. 2010. Automatic Domain Adaptation for Parsing. In *Proceedings of the 2010 Annual Conference of the North American Chapter of the ACL*, pages 28–36, Los Angeles, California.

Takashi Ninomiya, Takuya Matsuzaki, Yusuke Miyao, and Jun'ichi Tsujii. 2007. A log-linear model with an n-gram reference distribution for accurate hpsg parsing. In *Proceedings of 10th International Conference on Parsing Technologies (IWPT 2007)*, pages 60–68.

Joakim Nivre, Johan Hall, Sandra Kübler, Ryan McDonald, Jens Nilsson, Sebastian Riedel, and Deniz Yuret. 2007. The CoNLL 2007 shared task on dependency parsing. In *Proceedings of the CoNLL Shared Task Session of EMNLP-CoNLL 2007*, pages 915–932, Prague, Czech Republic, June. Association for Computational Linguistics.

Carl Pollard and Ivan A. Sag. 1994. *Head-Driven Phrase Structure Grammar*. University of Chicago Press.

Laura Rimell and Stephen Clark. 2008. Adapting a Lexicalized-Grammar Parser to Contrasting Domains. In *Proceedings of the 2008 Conference on Empirical Methods in Natural Language Processing*, pages 475–584.

Brian Roark and Michiel Bacchiani. 2003. Supervised and unsupervised PCFG adaptation to novel domains. In *Proceedings of the 2003 Conference of the North American Chapter of the Association for Computational Linguistics on Human Language Technology*, pages 126–133, Edmonton, Canada.

Mark Steedman, Miles Osborne, Anoop Sarkar, Stephen Clark, Rebecca Hwa, Julia Hockenmaier, Paul Ruhlen, Steven Baker, and Jeremiah Crim. 2003. Bootstrapping statistical parsers from small datasets. In *Proceedings of the tenth conference on European chapter of the Association for Computational Linguistics*, pages 331–338, Budapest, Hungary.

Mark Steedman. 2000. *The Syntactic Process*. THE MIT Press.

Yuka Tateisi and Jun'ichi Tsujii. 2006. GENIA Annotation Guidelines for Treebanking. Technical Report TR-NLP-UT-2006-5, Tsujii Laboratory, University of Tokyo.

Yuka Tateisi, Akane Yakushiji, Tomoko Ohta, and Jun'ichi Tsujii. 2005. Syntax Annotation for the GENIA corpus. In *Proceedings of the Second International Joint Conference on Natural Language Process-*

ing (IJCNLP 2005), Companion volume, pages 222–227.

Ivan Titov and James Henderson. 2006. Porting statistical parsers with data-defined kernels. In *Proceedings of the Tenth Conference on Computational Natural Language Learning*, pages 6–13, New York City.

Yoshimasa Tsuruoka, Yuka Tateishi, Jin-Dong Kim, Tomoko Ohta, John McNaught, Sophia Ananiadou, and Jun'ichi Tsujii. 2005. Developing a robust part-of-speech tagger for biomedical text. In *Advances in Informatics - 10th Panhellenic Conference on Informatics*, volume LNCS 3746, pages 382–392, Volos, Greece, November. ISSN 0302-9743.

Unlocking Medical Ontologies for Non-Ontology Experts

Shao Fen Liang
School of Computer Science,
The University of Manchester, Oxford Road,
Manchester, M13 9PL, UK
Fennie.Liang@cs.man.ac.uk

Donia Scott
School of Informatics,
The University of Sussex, Falmer,
Brighton, BN1 9QH, UK
D.R.Scott@sussex.ac.uk

Robert Stevens
School of Computer Science,
The University of Manchester, Oxford Road,
Manchester, M13 9PL, UK
Robert.Stevens@cs.man.ac.uk

Alan Rector
School of Computer Science,
The University of Manchester, Oxford Road,
Manchester, M13 9PL, UK
Rector@cs.man.ac.uk

Abstract

Ontology authoring is a specialised task requiring amongst other things a deep knowledge of the ontology language being used. Understanding and reusing ontologies can thus be difficult for domain experts, who tend not to be ontology experts. To address this problem, we have developed a Natural Language Generation system for transforming the axioms that form the definitions of ontology classes into Natural Language paragraphs. Our method relies on deploying ontology axioms into a top-level Rhetorical Structure Theory schema. Axioms are ordered and structured with specific rhetorical relations under rhetorical structure trees. We describe here an implementation that focuses on a sub-module of SNOMED CT. With some refinements on articles and layout, the resulting paragraphs are fluent and coherent, offering a way for subject specialists to understand an ontology's content without need to understand its logical representation.

1 Introduction

SNOMED CT (Spackman and Campbell, 1998) is widely mandated and promoted as a controlled vocabulary for electronic health records in several countries including the USA, UK, Canada and Australia. It is managed by the International Health Terminology Standards Development Organisation (IHTSDO) [1]. SNOMED describes diagnoses, procedures, and the necessary anatomy, biological process (morphology[2]) and the relevant organisms that cause disease for over 400,000 distinct concepts. It is formulated using a Description Logic (DL) (Baader et al., 2005). Description logics, usually in the form of the Web Ontology Language (OWL)[3] have become a common means of representing ontologies. Description logics in general and SNOMED in particular have been recognised as difficult to understand and reuse (Namgoong and Kim, 2007;Power et al., 2009). Even with the more or less human readable, Manchester OWL Syntax (Horridge et al., 2006) and using tools such as Protégé (Knublauch et al., 2004) the task of understanding ontologies remains non-trivial for most domain experts.

Consider, for example, a clinician seeking information about the concept of *thoracic cavity structure* [4] (i.e., anything in the chest cavity). SNOMED provides the following six axioms:

1. <Structure of thoracic viscus> SubClassOf <Thoracic cavity structure>
2. <Intrathoracic cardiovascular structure> SubClassOf <Thoracic cavity structure>
3. <Mediastinal structure> SubClassOf <Thoracic cavity structure>
4. <Thoracic cavity structure> SubClassOf <Structure of respiratory system and/or intrathoracic structure>
5. <Thoracic cavity structure> SubClassOf <Thoracic structure>
6. <Thoracic cavity structure> SubClassOf <Body cavity structure>

[1] http://www.w3.org/TR/owl-features/
[2] Literally, the altered structure as seen by the pathologist, but usually the evidence for the process that gave rise to it.
[3] http://www.w3.org/TR/owl-features/
[4] The SNOMED identifier for this class is ID: SCT_43799004

Proceedings of the 2011 Workshop on Biomedical Natural Language Processing, ACL-HLT 2011, pages 174–181,
Portland, Oregon, USA, June 23-24, 2011.

Although these axioms are shown with the more readable Manchester OWL syntax, the represented meaning of *Thoracic cavity structure* will not be easy for the typical clinician to decode.

Ontology concepts can be much more complex than those shown above. Not only can there be more axioms, but there can be nested axioms to an arbitrary depth. So the comprehension problem facing the typical clinician is even greater than that just described. It should be reduced, however, if the ontological content were presented in a more coherent, fluent and natural way – for example as:

> A *thoracic cavity structure* is a kind of *structure of* the *respiratory system and/or intrathoracic structure*, *thoracic structure* and *body cavity structure*. It includes a *structure of* the *thoracic viscus*, an *intrathoracic cardiovascular structure* and a *mediastinal structure*.

or, with added layout, as:

A *thoracic cavity structure* is a kind of
- *structure of* the *respiratory system and/or intrathoracic structure,*
- *thoracic structure,*

and
- *body cavity structure.*

It includes
- a *structure of* the *thoracic viscus,*
- an *intrathoracic cardiovascular structure*

and
- a *mediastinal structure.*

In these (human-generated) texts, the author has chosen to retain the general form of the anatomical terms as they appear in SNOMED, signalling them through the use of italics and introducing in places a definite article (e.g., "*structure of* the *thoracic viscus*". While these terms (particularly in the peculiar form they take in SNOMED names[5]) still present a barrier to non-subject-specialists, nevertheless the ontological content rendered as natural language is now much more accessible to non-ontology specialists.

Using natural language descriptions is obviously one way of improving the transparency of ontologies. However, authoring such descriptions is tedious and time-consuming to achieve by hand. This is clearly an area where automatic generation could be beneficial. With this in mind, we have built a verbaliser that renders SNOMED concepts as fluent natural language paragraphs.

2 Mapping SNOMED to a Representation of Coherent Discourse

Our goal is to use standard techniques for natural language generation (NLG) to generate fluent paragraph-sized texts for SNOMED concepts automatically.

Verbalisation is a two-staged process of deciding *what to say* and then *how to say it*. In our work the first of these is a non-issue: the content of our verbalisation will be SNOMED concepts. Our focus is therefore on deciding how to express the content.

As with any NLG system, our task begins by organising the input content in such a way as to provide a structure that will lead to coherent text, as opposed to a string of apparently disconnected sentences. Given the nature of our problem, we need to focus on the semantics of the discourse that can accommodate the nature of ontology axioms. For this purpose, we have chosen to use Rhetorical Structure Theory (RST) (Mann and Thompson, 1987;Mann and Thompson, 1988), as a mechanism for organising the ontological content of the SNOMED input.

RST is a theory of discourse that addresses issues of semantics, communication and the nature of the coherence of texts, and plays an important role in computational methods for generating natural language texts (Hovy, 1990;Scott and Souza, 1990;Mellish et al., 1998;Power et al., 2003). According to the theory, a text is coherent when it can be described as a hierarchical structure composed of text spans linked by rhetorical relations that represent the relevance relation that holds between them (among the set of 23 relations are EVIDENCE, MOTIVATION, CONTRAST, ELABORATION, RESULT, CAUSE, CONDITION, ANTITHESIS, ALTERNATIVE, LIST, CONCESSION and JUSTIFICATION). Relations can be left implicit in the text, but are more often signalled through *discourse markers* – words or phrases such as "because" for EVIDENCE, "in order to" for ENABLEMENT, "although" for ANTITHESIS,

[5] To reduce this problem somewhat, we use here the 'preferred term' for given SNOMED names, but even these can be quite peculiar, e.g., "*renal hypertension complicating pregnancy, childbirth and the puerperium - delivered with postnatal complication*".

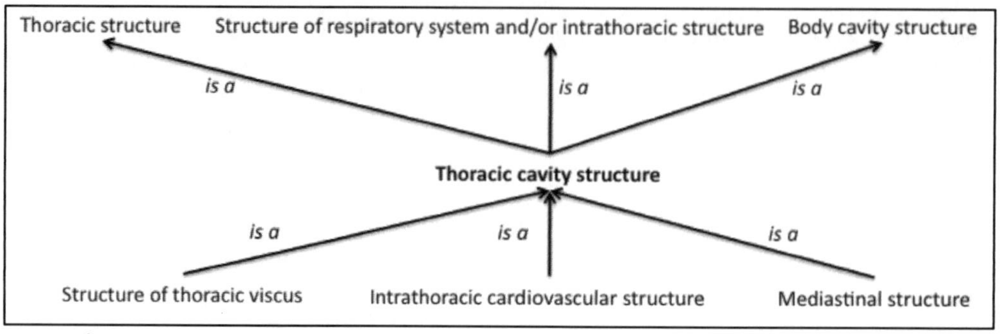

Figure 1: axioms and their relations to the class *Thoracic cavity structure*

"but" for CONCESSION, "and" for LIST, "or" for ALTERNATIVE, etc. (Sporleder and Lascarides, 2008;Callaway, 2003). They can also be signalled by punctuation (e.g., a colon for ELABORATION, comma between the elements of LIST, etc.).

In RST, text spans are divided into a schema, containing either a nucleus (N) and satellite (S), or two or more nuclei. Nuclei contain the information that is critical to the communicative message; satellites contain less critical information, which support the statements of their nuclei. The relations among nuclei and satellites are often expressed as:

RELATION(N,N)

RELATION(N,S)

These expressions conveniently take the same form as those expressing the types of ontology axiom, e.g.:

SubClassOf(A, B)
EquivalentClasses(C, D)

where, SubClassOf and EquivalentClasses express relations between A and B, and C and D. This suggests that with careful selection of RST relations, and applying appropriate discourse markers, ontologies can be represented as RST structures, and generated as natural language paragraphs that are not far from human written text.

To investigate the feasibility of this proposal, we have experimented with feeding axioms into RST trees, and have achieved a positive outcome. For example, the six axioms of the *thoracic cavity structure* concept that we have seen earlier can be organised into two groups of relations as shown in Figure 1. In the upper group are the super-classes of the *thoracic cavity structure* class, and in the lower are the sub-classes. This way of grouping the axioms can better present their relations to the class.

This structure can now be transformed into the RST tree shown in Figure 2, where the most important element of the message is the class *Thoracic cavity structure*, and this forms the main nucleus of the RST tree. The remaining content is related to this through an ELABORATION relation, the satellite of which is composed of two items of a multinucleus LIST, each of which is itself a LIST. This structure can be expressed textually as (among others) the two natural language descriptions we have shown earlier. These texts satisfy the requirement of coherence (as defined by RST), since each part bears a rhetorical relation to the other, and the entire text is itself spanned by a single rhetorical relation.

Our exploration of RST has shown that some relations map well to the characteristic features of ontology axioms. For example:

- the LIST relation captures well those cases where a group of axioms in the ontology bear the same level of relation to a given class;
- the ELABORATION relation applies generally to connect different notions of axioms to a class (i.e., super-, sub- and defining- classes), in order to provide additional descriptive information to the class;
- the CONDITION relation generally applies in cases where an axiom has property restrictions.

We also found that some rhetorical relations appear to bear a one-to-one mapping with logical forms of axioms, such as ALTERNATIVE to the logical `or`, and LIST to the logical `and`.

Our experience and the evidence over many practical cases have indicated that the full set of rhetorical relations is unlikely to be applied for ontology verbalisation. In particular, the set of so-called *presentational* relations are unlikely to apply, as ontology authors do not normally

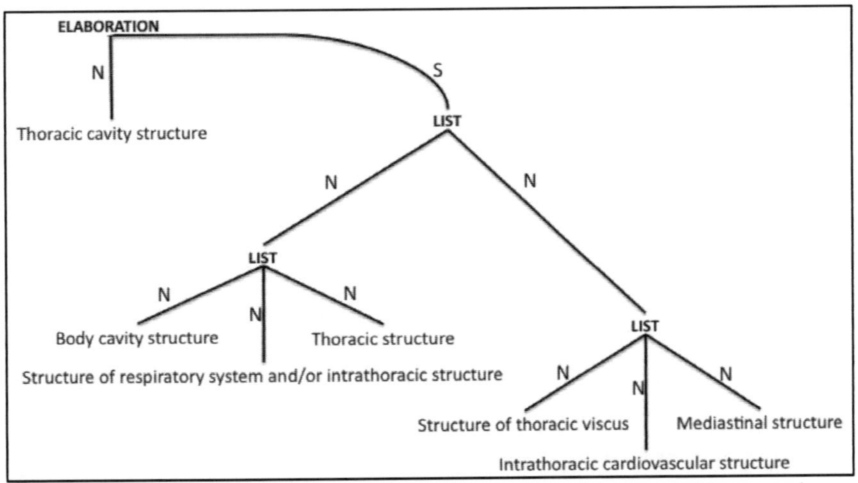

Figure 2: RST tree of the class *Thoracic cavity structure* with six axioms

create comparisons or attempt to state preferences amongst classes. (For example, SNOMED has no comparison operator between different treatments of diseases).

In addition, even within the set of *informational* relations (Moser and Moore, 1996), there are several that will not be found in ontologies. For example, since each axiom in an ontology is assumed to be true, using one axiom as an EVIDENCE of another axiom would be redundant. Similarly, using one axiom to JUSTIFY another axiom is not a conventional way of building ontologies.

3 Applying RST

Our investigations have shown that it is possible to build a top-level RST schema to cover all axioms with different meanings related to a class (see Figure 3). In SNOMED, axioms relating to a concept (i.e., class) can be either *direct* or *indirect*. Direct axioms describe the topic class directly, in which the topic class is the first class appearing in those axioms. Indirect axioms provide extra information, typically about how a class is used with other classes. For example, the axiom

<Structure of thoracic viscus>
SubClassOf(<Structure of viscus> and
<Thoracic cavity structure>)

can be placed as direct information about *structure of thoracic viscus*; it can also be placed as indirect information about *Structure of viscus* or *Thoracic cavity structure*.

Within the categories of direct and indirect information, axioms are also classified as either *simple* or *complex*. This distinction allows us to control the length of the verbalisation, since most complex axioms tend to be translated into longer sentences, involving as they do more properties and value restrictions. Simple axioms, on the other hand, describe only class relations, the length of which can be better controlled.

For a given SNOMED class, our verbalisation process starts with its super-, sub- and equivalent-classes, within an ELABORATION relation. The use of the ELABORATION relation allows the first part of the text to connect all classes relating to the topic class; the second part then starts to introduce more complex information directly related to the topic class. The ELABORATION relation is used until all the direct information has been included. Next the CONCESSION relation is applied to connect direct and indirect information.

Additionally, each indirect axiom should have its own subject, and therefore, they cannot be combined smoothly into a single sentence. We therefore use LIST as the relation for these axioms, since they are equally weighted, and changing the order among them does not affect the meaning of the whole paragraph.

Every complex axiom is translated using a CONDITION relation. This is because complex axioms contain conditional information to their subject class. For example:

<Disorder of soft tissue of thoracic cavity>
EquivalentTo(<Disorder of soft tissue of
body cavity>
and (<RoleGroup> some

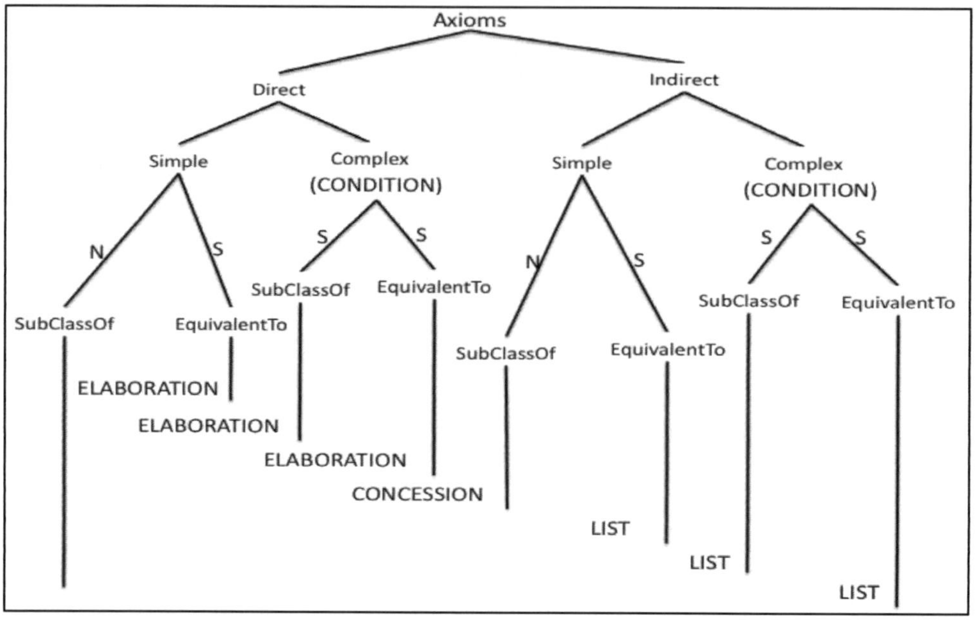

Figure 3: Top-level RST schema for SNOMED

(<Finding site> some
<Thoracic cavity structure>))
and (<RoleGroup> some
(<Finding site> some
<Soft tissues>)))

The condition in this axiom starts from the first "and" in the fourth line and extends to the end of the axiom. This condition needs to be attached to the class *Disorder of soft tissue of body cavity* to be equivalent to the *Disorder of soft tissue of thoracic cavity* class. We apply this rule to all complex axioms in an ontology.

4 Verbalising Individual Axioms

We use a template-based technique for verbalising the axioms as sentences. We have carefully selected translations of the SNOMED expressions. Our choice has been driven by an attempt to translate each axiom so as to preserve the meaning in the ontology and to avoid introducing misleading information. For example, the convention within ontologies is to conceptualise super-classes as an "is a" relation. However, translating this term as the English string "is a" can lead to misunderstanding, since the English expression can also be used to mean "equal to". Clearly, though, a class is not equal to its super-class. In this context, a more accurate translation is "is a kind of". We show some of these translations in Table 1.

Relation to the topic class X	**Translation wording**
With its simple super-class	X is a kind of …
With its complex super-class	X is a kind of … that …
With its simple sub-class	X includes …
With its simple equivalent class	X is defined as …
With its complex equivalent class	X is defined as … that

Table 1: Translations for axiom types

Consider for example, the SNOMED content:

<Benign hypertensive renal disease>
SubClassOf <Hypertensive renal disease>

<Benign arteriolar nephrosclerosis>
SubClassOf <Benign hypertensive renal disease>

<Benign hypertensive heart AND renal disease>
SubClassOf <Benign hypertensive renal disease>

<Benign hypertensive renal disease>
SubClassOf <Hypertensive renal disease>

and (<Finding site> some
<Kidney structure>)

<Benign arteriolar nephrosclerosis>
SubClassOf <Benign hypertensive renal disease>
and <Arteriolar nephrosclerosis>

<Benign hypertensive heart AND renal disease>
EquivalentTo<Benign hypertensive renal disease>
and <Benign hypertensive heart disease>
and <Hypertensive heart AND renal disease>

Our generator describes *Benign hypertensive renal disease* with its super-class as

> "*Benign hypertensive renal disease* is a kind of *hypertensive renal disease*."

and with its sub-classes as

> "*Benign hypertensive renal disease* includes *benign arteriolar nephrosclerosis* and *benign hypertensive heart and renal disease*."

There are two sub-classes in the above sentence, and we have signalled their connection (in a LIST relation) with "and" as the discourse marker. In those cases where there are more than two sub-classes, we use instead a comma "," except for the last mentioned, where we introduce "and". The same approach is applied to super-classes.

In those cases where a class has both super- and sub-classes to describe, we introduce the second sentence with "It" thus achieving better linguistic cohesion by avoiding having to repeat the same subject from the first sentence.

To bridge simple-direct and complex-direct axioms, we use "Additionally" to signal the introduction of more information relevant to the topic. For example to continue from the above two sentences, we have

> "Additionally, *benign hypertensive renal disease* is a kind of *hypertensive renal disease* that has a *finding site* in a *kidney structure*."

All direct information should have been consumed at this point, and we now need some bridging expression to signal the introduction of the indirect axioms. For this we use "Another relevant aspect of" or "Other relevant aspects of", depending on the number of axioms in the set. Continuing with our example, we now have

> "Other relevant aspects of *benign hypertensive renal disease* include the following: *benign arteriolar nephrosclerosis* is a kind of *benign hypertensive renal disease* and *arteriolar nephrosclerosis*; *benign hypertensive heart and renal disease* is defined as *benign hypertensive renal disease, benign hypertensive heart disease* and *hypertensive heart and renal disease*."

The improved transparency of the underlying ontological content can be clearly demonstrated by comparison with the SNOMED input from which it is derived.

The output that we have shown so far has all been generated as running text with minimal formatting except for the use of italic face for SNOMED labels. This works well for simple examples, but as can be seen from the previous example, readability becomes increasingly challenged as the expressions become longer. For this reason, we have also included in our system the facility to use layout to convey the logical structure of the ontological content. For example, the content shown above can also be generated as

> "*Benign hypertensive renal disease* is a kind of *hypertensive renal disease*. It includes
> - *benign arteriolar nephrosclerosis*
>
> and
> - *benign hypertensive heart and renal disease*.
>
> Additionally, *benign hypertensive renal disease* is a kind of *hypertensive renal disease* that has a *finding site* in a *kidney structure*. Other relevant aspects of *benign hypertensive renal disease* include the following:
> - *benign arteriolar nephrosclerosis* is defined as *benign hypertensive renal disease and arteriolar nephrosclerosis*;
> - *benign hypertensive heart and renal disease* is defined as *benign hypertensive renal disease, benign hypertensive heart disease* and *hypertensive heart and renal disease*."

5 Issues Related to Fluency

The quality of a text, whether human- or machine-generated, is to a large extent determined by its fitness for purpose. For example, the characteristics of a scientific article, a newspaper article or a twitter will be rather different, even though they may convey the same "message". The same is true for natural language descriptions of ontological content, which can range from the fully-fluent to the closely literal (e.g., something likely to be thought of as a kind of "SNOMED-ese"), depending on whether it is intended, say, for inclusion in a narrative summary of an electronic patient record (Hallett et al., 2006) or for ontology developers or users who want to know the precise ontological representation of some part of the ontology. So far, our aim has been to generate descriptions that fall into the latter category. For this purpose we retain the full expressions of the pseudo-English labels found in the official SNOMED Descriptions document[6], representing them within our generation process as "quotes" (Mellish et al., 2006) and signalling them through the use of italics. The texts still need to be grammatical, however, and achieving this can be challenging. In what follows we give a few examples of why this is so.

It is a convention of ontology design to treat each class as singular; we follow this convention, introducing each class with the indefinite article. So, for example, the SNOMED labels

<Intrathoracic cardiovascular structure>

and

< Structure of thoracic viscus>

can be expressed straightforwardly as "a *structure of thoracic viscus*" and "an *intratrathoracic cardiovascular structure*". However, matters are not so simple. For example,

<Heart structure>

will require the definite article ("the *heart structure*") and while

<Structure of thoracic viscus>

will attract an indefinite article at its front, it would read much better if it also had a definite article within it, giving "a *structure of* the *thoracic viscus*". A similar story holds for

<Abdomen and pelvis>

which properly should be "the *abdomen and pelvis*" or "the *abdomen and* the *pelvis*". Achieving this level of grammaticality will rely on knowledge that, for example, the human body contains only one heart and abdomen. Interestingly, this information is not captured within the SNOMED ontology, and so external resources will be required. Additionally, introducing articles within the labels (as in "*abdomen and* the *pelvis*", above) will require some level of natural language interpretation of the labels themselves.

The same applies to number. While we currently follow the SNOMED convention of describing entities in the singular, there are occasions where the plural is called for. For example:

<Abdominal vascular structure>
    SubClassOf <Abdomial structure>
        SubClassOf <Lower body part structure>

<Abdominal cavity structure>
    SubClassOf <Abdominal structure>
        SubClassOf <Lower body part structure>

would be better expressed as "Lower body part structures include all abdominal structures", instead of as currently "A lower body part structure includes an abdominal structure".

Another issue to consider is the roles of properties in SNOMED. This problem can be characterised by the following example:

<Hypertension secondary to kidney transplant>
    EquivalentTo (<Hypertension associated with transplantation>
    and (<After> some <Transplant of kidney>))>

which is currently verbalised as

> *Hypertension secondary to kidney transplant* is defined as *hypertension associated with transplantation* that has an *after* in a *transplant of kidney*.

In SNOMED, the property *after* is used to give an after-effect (i.e., "*Hypertension associated with transplantation*" is an after-effect of a kidney transplant), and for a non-SNOMED expert, this meaning is not at all clear in the generated text. This applies to many properties in SNOMED. Consider for example, the properties "*finding site*" and "*clinical course*" as in:

> "*Chronic heart disease* is defined as a *chronic disease of cardiovascular system* that is a *heart disease*, and has a *clinical course* in a *chronic*."

and

> "*Abdominal organ finding* is a *general finding of abdomen* that has a *finding site* in a *structure of abdominal viscus*."

---

6 http://www.nlm.nih.gov/research/umls/licensedcontent/snomedctarchive.html

The extent to which issues such as these are treated within the generation process will, as we mentioned before, be a matter of how fluent the text needs to be for a given purpose.

6 Conclusion

We have described a method for generating coherent and fairly fluent natural language descriptions of ontologies, and have shown how the method can be applied successfully to SNOMED CT, a medical terminology whose use is widely mandated. Through the application of Rhetorical Structure Theory, the ontological content is organised into a discourse schema that allows us to generate appropriate discourse markers, pronouns, punctuation and layout, thereby making it more easily accessible to those who are not fully familiar with the ontology language in use. In its current form, the system is aimed at readers who care how the SNOMED is constructed – for example, those wishing to know the precise meaning of a given class. We believe there is no single solution to satisfying a wider range of user interests, and thus of text types. While we continue to work towards improving the output of our system, evaluating the output with non-ontology specialists, and testing our method with other ontologies and ontology languages, achieving fully fluent natural language is beyond the scope of our system. We are not at this point overly concerned by this limitation, as the need for clarity and transparency of ontologies is, we believe, more pressing than the need for fully fluent natural language descriptions.

Acknowledgments

This work has been undertaken as part of the Semantic Web Authoring Tool (SWAT) project (see www.swatproject.org), supported by the UK Engineering and Physical Sciences Research Council (EPSRC) grant EP/G032459/1 to the University of Manchester, the University of Sussex, and the Open University.

References

Franz Baader, Ian Horrocks and Ulrike Sattler. 2005. Description logics as ontology languages for the semantic web, Lecture Notes in Artificial Intelligence, 2605: 228-248.

Charles B. Callaway. 2003. Integrating discourse markers into a pipelined natural language generation architecture, 41st Annual Meeting on Association for Computational Linguistics, 1: 264-271.

Catalina Hallett, Richard Power and Donia Scott. 2006. Summarisation and Visualisation of e-Health Data Repositories. UK E-Science All-Hands Meeting, pages 18-21.

Matthew Horridge, Nicholas Drummond, John Goodwin, et al. 2006. The Manchester OWL syntax. 2006 OWL: Experiences and Directions (OWLED'06).

Holger Knublauch, Ray W. Fergerson, Natalya Fridman Noy, et al. 2004. The Protégé OWL plugin: an open development environment for Semantic Web applications. International Semantic Web Conference, pages 229-243.

William C. Mann and Sandra A. Thompson. 1987. Rhetorical Structure Theory: a theory of text organization, USC/Information Sciences Institute Technical Report Number RS-87-190 Marina del Rey, CA.

William C. Mann and Sandra A. Thompson. 1988. Rhetorical Structure Theory: toward a functional theory of text organisation, Text, 8(3): 243-281.

Chris Mellish, Donia Scott, Lynne Cahill Daniel Paiva, et al. 2006. A reference architecture for natural language generation systems, Natural Language Engineering, 12(1): 1-34.

Megan Moser and Johanna D. Moore. 1996. Toward a synthesis of two accounts of discourse structure, Computational Linguistics, 22(3): 409-420.

Hyun Namgoong and Hong-Gee Kim. 2007. Ontology-based controlled natural language editor using CFG with lexical dependency. 6th international, the Semantic Web, and 2nd Asian Conference on Asian Semantic Web Conference, pages 353-366. Springer Verlag Berlin, Heidelberg.

Richard Power, Robert Stevens, Donia Scott, et al. 2009. Editing OWL through generated CNL. 2009 Workshop on Controlled Natural Language (CNL'09), Marettimo, Italy.

Kent A. Spackman and Keith E. Campbell. 1998. Compositional concept representation using SNOMED: Towards further convergence of clinical terminologies, Journal of the American Medical Informatics Association: 740-744.

Caroline Sporleder and Alex Lascarides. 2008. Using automatically labelled examples to classify rhetorical relations: an assessment, Natural Language Engineering, 14(3): 369-416.

Self-training and co-training in biomedical word sense disambiguation

Antonio Jimeno-Yepes
National Library of Medicine
8600 Rockville Pike
Bethesda, 20894, MD, USA
antonio.jimeno@gmail.com

Alan R. Aronson
National Library of Medicine
8600 Rockville Pike
Bethesda, 20894, MD, USA
alan@nlm.nih.gov

Abstract

Word sense disambiguation (WSD) is an intermediate task within information retrieval and information extraction, attempting to select the proper sense of ambiguous words. Due to the scarcity of training data, semi-supervised learning, which profits from seed annotated examples and a large set of unlabeled data, are worth researching. We present preliminary results of two semi-supervised learning algorithms on biomedical word sense disambiguation. Both methods add relevant unlabeled examples to the training set, and optimal parameters are similar for each ambiguous word.

1 Introduction

Word sense disambiguation (WSD) is an intermediate task within information retrieval and information extraction, attempting to select the proper sense of ambiguous words. Supervised learning achieves better performance compared to other WSD approaches (Jimeno-Yepes et al., 2011). Manual annotation requires a large level of human effort whereas there is a large quantity of unlabeled data. Our work follows (Mihalcea, 2004) but is applied to the biomedical domain; it relies on two semi-supervised learning algorithms.

We have performed experiments of semi-supervised learning for word sense disambiguation in the biomedical domain. In the following section, we present the evaluated algorithms. Then, we present preliminary results for self-training and co-training, which show a modest improvement with a common set-up of the algorithms for the evaluated ambiguous words.

2 Methods

For self-training we use the definition by (Clark et al., 2003): "a tagger that is retrained on its own labeled cache on each round". The classifier is trained on the available training data which is then used to label the unlabeled examples from which the ones with enough prediction confidence are selected and added to the training set. The process is repeated for a number of predefined iterations. Co-training (Blum and Mitchell, 1998) uses several classifiers trained on independent views of the same instances. These classifiers are then used to label the unlabeled set, and from this newly annotated data set the annotations with higher prediction probability are selected. These newly labeled examples are added to the training set and the process is repeated for a number of iterations. Both bootstrapping algorithms produce an enlarged training data set.

Co-training requires two independent views on the same data set. As first view, we use the context around the ambiguous word. As second view, we use the MEDLINE MeSH indexing available from PubMed which is obtained by human assignment of MeSH heading based on their full-text articles.

Methods are evaluated with the accuracy measure on the MSH WSD set built automatically using MeSH indexing from MEDLINE (Jimeno-Yepes et al., 2011) [1] in which senses are denoted by UMLS concept identifiers. To avoid any bias derived from

[1] Available from: http://wsd.nlm.nih.gov/collaboration.shtml

Proceedings of the 2011 Workshop on Biomedical Natural Language Processing, ACL-HLT 2011, pages 182–183,
Portland, Oregon, USA, June 23-24, 2011.

the indexing of the UMLS concept related to the ambiguous word, the concept has been removed from the MeSH indexing of the recovered citations.

10-fold cross validation using Naïve Bayes (NB) has been used to compare both views which achieve similar accuracy (0.9386 context text, 0.9317 MeSH indexing) while the combined view achieves even better accuracy (0.9491).

In both algorithms a set of parameters is used: the number of iterations (1-10), the size of the pool of unlabeled examples (100, 500, 1000) and the growth rate or number of unlabeled examples which are selected to be added to the training set (1, 10, 20, 50, 100).

3 Results and discussion

Results shown in Table 1 have been obtained from 21 ambiguous words which achieved lower performance in a preliminary cross-validation study. Each ambiguous word has around 2 candidate senses with 100 examples for each sense. We have split the examples for each ambiguous word into $2/3$ for training and $1/3$ for test.

The baseline is NB trained and tested using this split. Semi-supervised algorithms use this split, but the training data is enlarged with selected unlabeled examples. Self-training and the baseline use the combined views while co-training relies on two NB classifiers, each trained on one view of the training data. Even though we are willing to evaluate other classifiers, NB was selected for this exploratory work since it is fast and space efficient. Unlabeled examples are MEDLINE citations which contain the ambiguous word and MeSH heading terms. Any mention of MeSH heading related to the ambiguous word has been removed. Optimal parameters were selected, and average accuracy is shown in Table 1.

Method	Accuracy
Baseline	0.8594
Self-training	0.8763 (1.93%)
Co-training	0.8759 (1.88%)

Table 1: Accuracy for the baseline, self-training and co-training

Both semi-supervised algorithms show a modest improvement on the baseline which is a bit higher for self-training. Best results are achieved with a small number of iterations (< 5), a small growth rate (1-10) and a pool of unlabeled data over 100 instances. Noise affects the performance with a larger number of iterations, which after an initial increase, shows a steep decrease in accuracy. Small growth rate ensures a smoothed increase in accuracy. A larger growth rate adds more noise after each iteration. A larger pool of unlabeled data offers a larger set of candidate unlabeled examples to choose from at a higher computational cost.

4 Conclusions and Future work

Preliminary results show a modest improvement on the baseline classifier. This means that the semi-supervised algorithms have identified relevant disambiguated instances to be added to the training set.

We plan to evaluate the performance of these algorithms on all the ambiguous words available in the MSH WSD set. In addition, since the results have shown that performance decreases rapidly after few iterations, we would like to further explore smoothing techniques applied to bootstrapping algorithms and the effect on classifiers other than NB.

Acknowledgments

This work was supported by the Intramural Research Program of the NIH, National Library of Medicine, administered by ORISE.

References

A. Blum and T. Mitchell. 1998. Combining labeled and unlabeled data with co-training. In *Proceedings of the eleventh annual conference on Computational learning theory*, pages 92–100. ACM.

S. Clark, J.R. Curran, and M. Osborne. 2003. Bootstrapping POS taggers using unlabelled data. In *Proceedings of the seventh conference on Natural language learning at HLT-NAACL 2003-Volume 4*, pages 49–55. Association for Computational Linguistics.

A. Jimeno-Yepes, B.T. McInnes, and A.R. Aronson. 2011. Exploiting MeSH indexing in MEDLINE to generate a data set for word sense disambiguation(accepted). *BMC bioinformatics.*

R. Mihalcea. 2004. Co-training and self-training for word sense disambiguation. In *Proceedings of the Conference on Computational Natural Language Learning (CoNLL-2004).*

Medstract – The Next Generation

Marc Verhagen
Computer Science Department
Brandeis University, Waltham, USA
marc@cs.brandeis.edu

James Pustejovsky
Computer Science Department
Brandeis University, Waltham, USA
jamesp@cs.brandeis.edu

Abstract

We present MedstractPlus, a resource for mining relations from the Medline bibliographic database. It was built on the remains of Medstract, a previously created resource that included a bio-relation server and an acronym database. MedstractPlus uses simple and scalable natural language processing modules to structure text and is designed with reusability and extendibility in mind.

1 Introduction

In the late 1990s, the Medstract project (Pustejovsky et al., 2002) set out to use common Natural Language Processing techniques and employ them to access relational information in Medline abstracts. Medstract used a set of pipelined Python scripts where all scripts operated on in-memory objects. The output of this pipeline was a set of relations, indexed by the PubMed identifier of the abstract in which they appeared. A Perl script proposed potential acronyms using a set of regular expressions on named entities in Medline abstracts. Both relations and acronyms were fed into an Oracle database, where access to these datasources was enabled by a set of Perl CGI scripts. The code, however, was not made public and was not maintained in any serious fashion after 2004. Developers of the system dispersed over the world and the Medstract server fatally crashed in 2007.

Here, we describe the resurrection of Medstract. One goal was that code should be open source and that installation should not depend on idiosyncracies of the developer's machine, which was a problem with the inherited code base. Reusability and extendability are ensured by following the principles embodied in the Linguistic Annotation Format (LAF) (Ide and Romary, 2006). In LAF, source data are untouched, annotations are grouped in layers that can refer to each other and to the source, and each layer is required to be mappable to a graph-like pivot format. For MedstractPlus, each component is set up to be independent from other layers, although of course each layer may need access to certain types of information in order to create non-trivial output. This allows us to swap in alternative modules, making it easier to experiment with different versions of the tagger and chunker for example. We now proceed to describe the system in section 2 and finish with the current status and future work in section 3.

2 System Design and Implementation

The general design of MedstractPlus is presented in Figure 1. The Lemmatizer creates what LAF calls the base-segmentation, a first layer of tokenized text that is the input to processing modules associated with other layers. The Lemmatizer incorporates a Python version of the Brill Tagger, extended with entries from the UMLS Thesaurus.

The Semantic Tagger is a group of components using (i) regular expressions for finding simple types like URLs, (ii) dictionary lookup in the UMLS type and concept lists as well as other typed word lists, (iii) off-the-shelf components like the Abner gene tagger (http://pages.cs.wisc.edu/ bsettles/abner/) and (iv) a statistical disambiguation model for genes trained on the GENIA corpus.

Proceedings of the 2011 Workshop on Biomedical Natural Language Processing, ACL-HLT 2011, pages 184–185,
Portland, Oregon, USA, June 23-24, 2011. ©2011 Association for Computational Linguistics

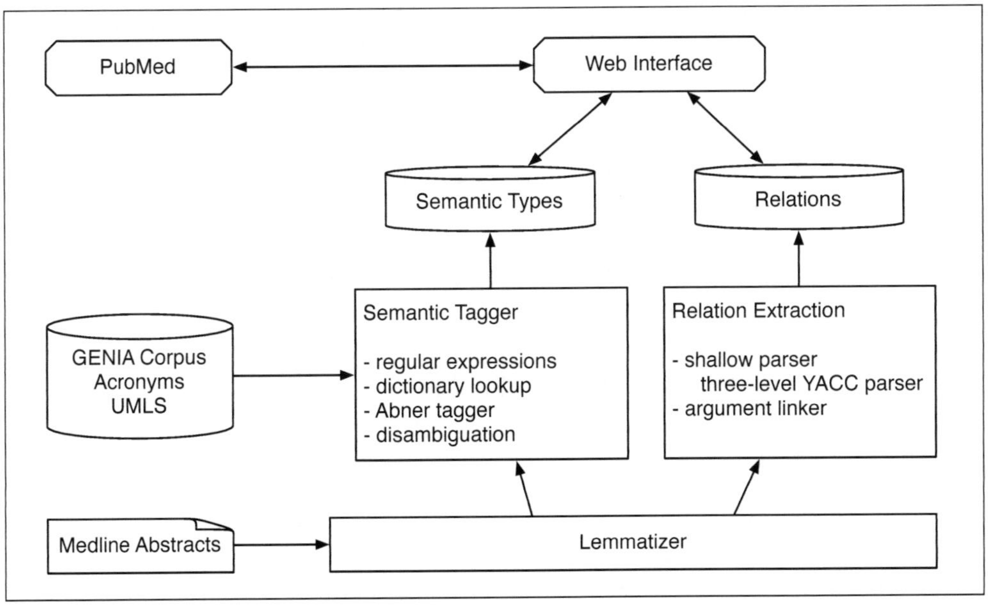

Figure 1: Overview of the MedstractPlus Architecture

The Relation Extraction component now contains a three-level 59-rule YACC parser that, starting with simple low-level chunking of noun and verb groups, proceeds to add more complex noun phrases and subordinated phrases. The argument linker produces binary relations, using a finite-state machine that runs on the data created by the shallow parser.

An advantage of this data-driven approach is that processing can be split up. A complete run of MedstractPlus on all Medline abstracts would take approximately 30 days on a entry-level desktop. But some relatively stable components like the Lemmatizer and the shallow parser (the latter being the most time-consuming component) can be run just once and subsequent runs can be restricted to those components that were changed.

The Web Interface gives access to the types and relations in a fairly standard way. In its current prototype form, it allows a user to type in a gene and then view all relations that the gene participates in. Alternatively, a pair of genes can be given.

3 Current Status and Future Work

The basic architecture depicted in Figure 1 is in place, but some components like the type disambiguator are in embryonic form. The web interface and the source code are or will be available at http://medstractplus.org.

Extensive additions to the basic typing and relation extraction component groups are in progress and the Relation Extraction component can be extended with specialized rule sets for specific relations like *inhibit* or *phosphorylate*. The interaction with the PubMed server is now limited to providing links. But the plan is that the MedstractPlus server will also query PubMed for relation pairs in case its own database provides little information. This approach can be extended to other relation servers like Chilibot (http://www.chilibot.net/), thereby moving towards a system than presents merged relations from the MedstractPlus database as well as relations from other servers.

Acknowledgments

This work was supported by the National Institutes of Health, under grant number 5R01NS057484-04.

References

Nancy Ide and Laurent Romary. 2006. Representing linguistic corpora and their annotations. In *Proceedings of the Fifth Language Resources and Evaluation Conference (LREC)*, Genoa, Italy.

James Pustejovsky, José Castaño, Roser Saurí, Anna Rumshisky, Jason Zhang, and Wei Luo. 2002. Medstract: Creating large-scale information servers for biomedical libraries. In *Proceedings of ACL'02*.

ThaiHerbMiner: A Thai Herbal Medicine Mining and Visualizing Tool

Choochart Haruechaiyasak† Jaruwat Pailai‡ Wasna Viratyosin∗ Rachada Kongkachandra‡

†Human Language Technology Laboratory (HLT),
National Electronics and Computer Technology Center (NECTEC), Thailand 12120

‡Department of Computer Science, Faculty of Science and Technology
Thammasat University, Thailand 12121

∗BIOTEC Central Research Unit,
National Center for Genetic Engineering and Biotechnology, Thailand 12120

Abstract

Thai Traditional Medicine (TTM) has a long history in Thailand and is nowadays considered an effective alternative approach to the modern medicine. One of the main knowledge in Thai traditional medicine is the use of various types of herbs to form medicines. Our main goal is to bridge the gap between the traditional knowledge and the modern biomedical knowledge. Using text mining and visualization techniques, some implicit relations from one source could be used to verify and enhance the knowledge discovery in another source. In this paper, we present our ongoing work, *ThaiHerbMiner*, a Thai herbal medicine mining and visualizing tool. ThaiHerbMiner applies text mining to extract some salient relations from a collection of PubMed articles related to Thai herbs. The extracted relations can be browsed and viewed using information visualization. Our proposed tool can also recommend a list of herbs which have similar medical properties.

1 Introduction

In 1993, the Royal Thai Government instituted the National Institute of Thai Traditional Medicine, under the supervision of the Ministry of Public Health. The goal of the institute is to systematize and standardize the body of Thai Traditional Medicine (TTM) knowledge. The main task is to gather, revise, verify, classify, and explain the TTM knowledge. There are many ongoing project collaboration to digitize the TTM knowledge, many of which are documented on palm leaves. The digitized contents contain information on Thai medical herbal formulations with the healing properties. A medical herbal formulation could contain more than one herb and combined with others for better effect.

Apart from the traditional knowledge, today biomedical research has advanced into the genetic level. Many researchers have performed in-depth studies of herbs' medical properties on disease treatment. The main goal of our research is to combine the knowledge from traditional and modern biomedical research. Using knowledge from one source could support the knowledge discovery in another source. To assist the researchers in Thai herbal medicine, we propose *ThaiHerbMiner*, a text mining and visualizing platform. ThaiHerbMiner's main task is to extract and visualize relations among herbs, properties and other entities. Our work is similar to the current ongoing research in mining Traditional Chinese Medicine (TCM) which has gained increasing attention in recent years (He et al., 2011; Lukman et al., 2007).

2 Design and implementation

Text mining has become a widely applied technique for analyzing biomedical texts (Cohen and Hersh, 2005). The proposed *ThaiHerbMiner* is designed with the standard text mining process. We started by collecting PubMed articles by using herb names as keywords. Currently, we have obtained approximately 18,000 articles related to Thai herbs such as garlic, curcuma and ginger.

Figure 1 shows the text mining process of extracting relations from given input texts. The process includes sentence segmentation, tokenization, POS

Proceedings of the 2011 Workshop on Biomedical Natural Language Processing, ACL-HLT 2011, pages 186–187,
Portland, Oregon, USA, June 23-24, 2011.

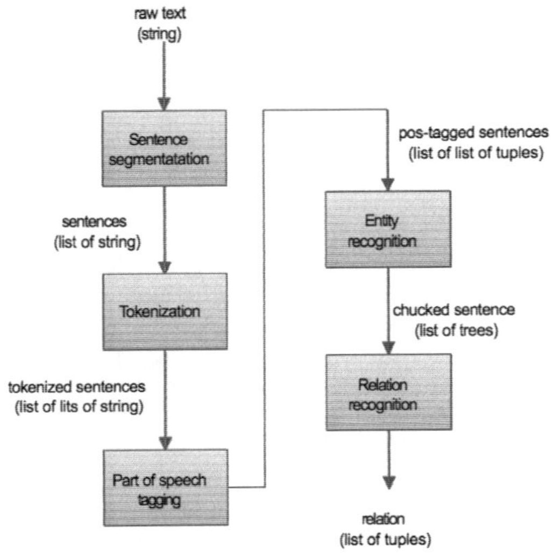

Table 1: The text mining process for extracting relations from input texts.

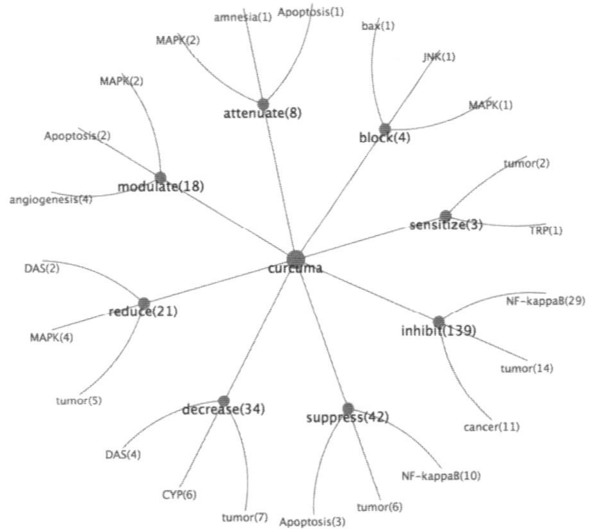

Table 2: An example of relations between curcuma and other relevant entities.

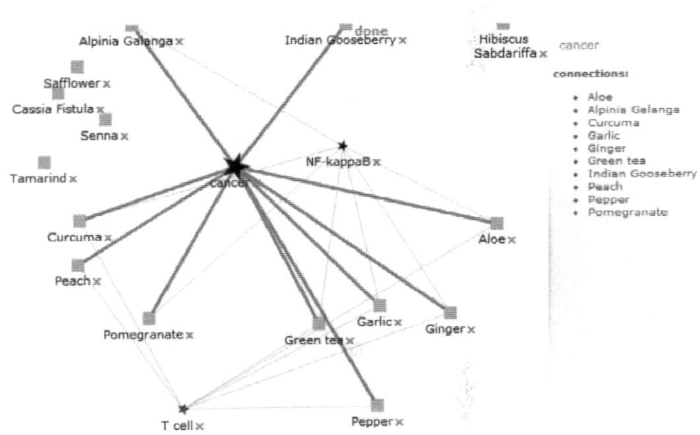

Table 3: An example of relations among different herbs sharing the same entities.

tagging and entity & relation recognition. We used *OpenNLP*[1] to perform all text processing tasks. For relation recognition based on syntactic structure, we focus on a group of causal verbs such as *activate*, *induce*, *inhibit*, *prevent*, *regulate* and *suppress*. Then the information visualization based on JavaScript [2] is applied to represent the extracted relations.

Figure 2 shows an example of a hyperbolic tree visualizing relations between curcuma and other entities. For example, *curcuma* has the property of *inhibit* with *NF-kappaB*, *tumor* and *cancer*. Figure 3 shows an example of a force-directed graph visualizing similar herbs sharing two entities, *cancer* and *NF-kappaB*. The visualizing result is useful to researchers for finding herbs which share similar medical properties.

3 Conclusion and future work

The results of literature mining can be potentially useful in revealing implicit relations underlying the knowledge in herbal medicine. In particular, the results can be used in screening the research in Thai herbal medicine to form a novel hypothesis. Our next step is to perform comparative analysis on the knowledge from Thai traditional medicine and the knowledge extracted from the modern research publications.

[1]The OpenNLP Homepage, *http://opennlp.sourceforge.net*

[2]The JavaScript InfoVis Toolkit, *http://thejit.org*

References

Cohen, Aaron M. and William R. Hersh. 2005. A survey of current work in biomedical text mining. *Briefings in Bioinformatics*, 6(1):57-71.

He, Ping, Ke Deng, Zhihai Liu, Delin Liu, Jun S Liu and Zhi Geng. 2011. Discovering herbal functional groups of traditional Chinese medicine. *Statistics in medicine*, March 17, 2011.

Lukman, Suryani, Yulan He, and Siu-Cheung Hui. 2007. Computational methods for Traditional Chinese Medicine: A survey. *Comput. Methods Prog. Biomed.*, 88(3): 283–294.

Author Index

Association for Computational Linguistics
209 N. Eighth Street
Stroudsburg, Pennsylvania 18360

Joint Workshop on Exploiting Synergies Between Information Retrieval and Machine Translation (ESIRMT) and Hybrid Approaches to Machine Translation (HyTra) 2012

Avignon, France
23-24 April 2012

ISBN: 978-1-62276-470-9